# Remote Sensing
# and
# Tropical Land Management

## Publications of the Commonwealth Geographical Bureau

*Problems of Regional Development in Eastern Africa.* East African Geographical Review (special issue) No. 15, June 1977. Published jointly by Uganda Geographical Association and Makerere University.

*Natural Resources in Tropical Countries.* Ooi, J. B. (ed.), Singapore University Press (1983).

*Remote Sensing and Tropical Land Management.* Eden, M. J. and Parry, J. T. (eds), John Wiley & Sons Ltd. (1986).

*The Impact of Tourism on Small Developing Countries.* Clarke, W. C. (ed.) (in preparation).

*Energy, Development and Environment in the Tropics.* Courtenay, P. P. and Abdul Samad Hadi (eds), Universiti Kebangsaan Press, Malaysia (in preparation).

The Commonwealth Foundation

# Remote Sensing and Tropical Land Management

*Edited by*

**M. J. Eden**

*Department of Geography,*
*Royal Holloway and Bedford New College,*
*University of London, UK*

*and*

**J. T. Parry**

*Department of Geography*
*McGill University,*
*Montréal, Canada*

Published by John Wiley & Sons with a grant from the Commonwealth Foundation and the Canadian International Development Agency

***Library of Congress Cataloging-in-Publication Data:***
Main entry under title:

Remote sensing and tropical land management.

Based on papers presented at a workshop of the
Commonwealth Geographical Bureau hosted by
McGill University, Montreal, Aug. 20–27, 1983.
Includes index.
1. Land use—Remote sensing—Congresses.
2. Land use—Tropics—Congresses.
I. Eden, M. J. (Michael John) II. Parry, J. T. (John T.)
III. Commonwealth Geographical Bureau.
HD108.8.R45 1986 333.73′13′028 85-17932

ISBN 0 471 90889 4

***British Library Cataloguing in Publication Data:***

Remote sensing and tropical land management.
1. Land use—Tropics—Remote sensing.
I. Eden, M. J. II. Parry, J. T.
333.73′028 HD108.8

ISBN 0 471 90889 4

Phototypeset by Dobbie Typesetting Service, Plymouth, Devon.
Printed and bound in Great Britain.

# *List of Contributors*

**Peter O. Adeniyi** *Department of Geography, University of Lagos, Lagos, Nigeria.*

**G. Ross Cochrane** *Remote Sensing Research Unit, Department of Geography, University of Auckland, Private Bag, Auckland, New Zealand.*

**J. Roosevelt K. Daniel** *Department of Geography, University of Guyana, Turkeyen Campus, Box 101110, Georgetown, Guyana.*

**Michael J. Eden** *Department of Geography, Royal Holloway and Bedford New College, University of London, Egham Hill, Egham, Surrey TW20 0EX, U.K.*

**Ayodele Fagbami** *Department of Agronomy, University of Ibadan, Ibadan, Nigeria.*

**Akin Fapohunda** *Department of Agronomy, University of Ibadan, Ibadan, Nigeria.*

**Peter A. Furley** *Department of Geography, University of Edinburgh, Drummond Street, Edinburgh EH8 9XP, Scotland.*

**Peter Laut** *Division of Water and Land Resources, C.S.I.R.O., P.O. Box 1666, Canberra City, ACT 2601, Australia.*

**Edward M. Makhanya** *Department of Geography, National University of Lesotho, P.O. Roma 180, Lesotho.*

**Jean-Paul Malingreau** *Code 623 NASA GSFC, Greenbelt, MD 20771, U.S.A.*

**Hezekiel M. Mushala** *Department of Geography, University of Dar es Salaam, P.O. Box 35049, Dar es Salaam, Tanzania.*

**John T. Parry** *Department of Geography, McGill University, 805 Sherbrooke Street West, Montreal, P.Q., H3A 2K6, Canada.*

**Stephen D. Prince** *School of Biological Sciences, Queen Mary College, University of London, Mile End Road, London E1 4NS, U.K.*

**Susan M. Ringrose** *Department of Environmental Science, University of Botswana, Private Bag 0022, Gaborone, Botswana.*

**Ashbindu Singh** *Indian Forest Service, Forest Department, Government of Manipur, Sanjanthong, Imphal 795001, Manipur, India.*

**Mark G. Williams** *School of Geography, University of New South Wales, P.O. Box 1, Kensington 2033, NSW, Australia.*

# Contents

# *Preface*

This volume is the second major publication from the Commonwealth Geographical Bureau, following *Natural Resources in Tropical Countries* (1983) edited by Professor Ooi Jin Bee. The volume is based on papers presented at a Workshop of the Bureau, hosted by McGill University, Montréal from 20 to 27 August 1983. The Workshop was concerned with the Application of Remote Sensing Techniques to Tropical Land Management, and involved 14 participants from ten Commonwealth countries. Seven of the participants were from developing countries, with a strong African representation. Other participants from Australia, New Zealand, Canada, and the United Kingdom provided case studies and experience from a wide range of tropical environments.

During the Workshop, material was presented covering the range of remote sensing techniques from conventional aerial photography to computer-processed satellite imagery, illustrating both the potential and constraints of advanced technology in relation to land management. When considering the possibility of publishing the Workshop papers, the editors were aware that, both systematically and regionally, significant gaps existed in the range of available material. Because of the value of individual case studies and the useful juxtaposition of developing and developed world viewpoints, it was felt that publication of the material was justified, but that it was essential to extend the coverage of the volume. Several additional contributors were approached and, with their assistance, the systematic framework of the volume has been expanded and some extra-Commonwealth material incorporated. In addition, introductory and concluding sections for the book have been prepared, which aim to summarize the status of and prospects for remote sensing in the context of tropical land management. The editors do not claim that all the original gaps have been filled, but a more comprehensive and balanced coverage has been achieved. As was originally the case, the primary focus remains on *renewable* resources and their management.

It is necessary to record the Bureau's appreciation and thanks to the Commonwealth Foundation for their support of the Workshop. Thanks are also due to the Canadian International Development Agency, to the International Development Research Centre, Ottawa, and to McGill University, Montréal who supported the venture. During the Workshop, visits were made to the Canadian

Centre for Remote Sensing and Kenting Earth Sciences Ltd, both in Ottawa; thanks are due to Ms Jean Heffernan (CCRS) and to Mr John McCartney (Kenting Earth Sciences Ltd) for their assistance and hospitable reception.

Many individuals have assisted in the organization of the Workshop and in the preparation of the book. Thanks are particularly due to Professor Ooi Jin Bee and Dr Garth Cant, then Director and Secretary, respectively, of the Commonwealth Geographical Bureau, for their sustained help and encouragement. Professor Brian Bird and Professor Theo Hills of McGill University gave their support to the Workshop and, in the latter case, organized an excellent field excursion. Dr Tony Allum of Inco Canada Ltd gave a stimulating evening talk on the use of remote sensing for geology and mineral exploration. Margo and Jim Critchley of the Gault Estate, McGill University deserve special thanks for handling the domestic arrangements during the seven days of the Workshop at Mont St Hilaire. The typing assistance of Mrs Rosemary Dawe and the cartographic assistance of Mr Ron Halfhide have been most helpfully provided.

The Bureau is deeply grateful to the Commonwealth Foundation and to the Canadian International Development Agency for helping to meet the costs of publishing the book.

J. T. Parry
*Montréal*

M. J. Eden
*London*

# PART 1

# Introduction

Remote Sensing and Tropical Land Management
Edited by M. J. Eden and J. T. Parry


# CHAPTER 1

# *The Management of Renewable Resources in the Tropics: the Use of Remote Sensing*

Michael J. Eden
*Department of Geography, Royal Holloway and Bedford New College, University of London, U.K.*

The tropical regions cover a multitude of natural environments, ranging from swamp and rain forest to savanna and grassland. In many areas, there is a long history of permanent land exploitation and relatively dense populations have existed from pre-modern times. Wetland habitats in parts of southeast Asia have long been used in this way, and contemporary management practices are necessarily focused on increasing the productivity and carrying capacity of land that is already occupied. Elsewhere in the tropics, relatively vacant land exists and in these areas low population densities and impermanent settlement and exploitation are more characteristic. This is the case across much of the Amazon region of South America, where only localized modern colonization has occurred. Between these two extremes there exists a range of other exploitation patterns, characterized by more or less intensive levels of technological achievement.

In parts of the tropics, land exploitation is sustaining or improving the socio-economic status of local populations, but in many areas low productivity is more characteristic, and there is concern about the damaging impact of contemporary resource use and the shortcomings of existing management. Many factors affect the success of tropical land management, ranging from the environmental to the socio-political, and few easy solutions exist to the numerous problems that arise. A significant feature of the last decade, however, is the continuing application of new technologies, which, if used appropriately, can provide a basis for more effective land management. The present volume is concerned with one particular technology, namely remote sensing, which is increasingly capable of providing precise and up-to-date information on land resources and land degradation in the tropics, thereby contributing to the process of land management. The technology is as relevant in settled and intensively exploited lands as it is in frontier regions where pioneer colonization is characteristic. The objective of the present volume is to illustrate the manner in which remote sensing is currently being employed in these areas, particularly in the context of renewable resources, and to consider the longer-term prospects of this technology for land management in the tropics.

Remote sensing may be broadly defined as the collection of information about an object without being in physical contact with the object. Aircraft and satellites are the common platforms from which remote sensing observations are made. The term *remote sensing* is restricted to methods that employ electromagnetic energy as the means of detecting and measuring target characteristics (Sabins, 1978). Remote sensing systems can provide information about the location, availability, and changing conditions of natural resources in specific areas, information that is essential for establishing appropriate priorities and effective plans for land management (Adrien and Baumgardner, 1977). Remote sensing techniques have been increasingly used in the field of resource management in recent decades. Even before the Second World War, black and white aerial photographs were commonly employed in tropical regions for land and resource surveys, and increasing use has latterly been made of colour and colour infrared photography. During the 1960s, airborne imaging radar techniques became available for large-area surveys (Jensen *et al.*, 1977), while the 1970s have seen the emergence of multispectral scanning systems as a major tool for surveying and monitoring changes in land resources. Landsat imagery is now generally available throughout the tropics, and, although there was some initial tendency to overrate the data (Baltaxe, 1980), satellite systems offer significant advances over previous remote sensing techniques, both in respect of the synoptic view that is provided and the monitoring capability.

## REMOTE SENSING APPLICATIONS IN THE TROPICS

Remote sensing makes diverse contributions to the management of renewable resources in the tropics. Major applications of the technology, which are reflected in the organization of the present volume, include:

1. the classification and evaluation of land resources;
2. the mapping of land use;
3. the monitoring of vegetation cover and crops.

Land classification and evaluation surveys have been completed for large areas of the tropics. Major projects, using aerial photographs, were undertaken by the Commonwealth Scientific and Industrial Research Organization (CSIRO) in Australia and New Guinea (Christian, 1958; Christian and Stewart, 1968; Stewart, 1968). These integrated land system surveys provided a model that has been followed elsewhere in the tropics by the United Kingdom Ministry of Overseas Development and other agencies (Cochrane, 1967; Bawden and Carroll, 1968; Ollier *et al.*, 1969).

Aerial photographs have been used in many parts of the tropics as a basis for the survey and mapping of soils. Projects of this kind have been undertaken by the Food and Agriculture Organization in eastern Colombia (FAO, 1965) and Guyana (FAO, 1966), and similar surveys have been made by the UK Ministry of Overseas Development in the Solomon Islands (Wall *et al.*, 1979). In many cases soil and land system surveys have been accompanied by the preparation of land capability maps. In addition, aerial photographs have been used extensively for vegetation mapping purposes (Paijmans, 1975), and for the preparation of forest inventories (Heinsdijk, 1958; Johnson and Chaffey, 1973).

Since the release of side-looking airborne radar (SLAR) systems for non-military purposes in the mid-1960s, such systems have provided a useful alternative to aerial photography for land classification and evaluation surveys, especially in equatorial and monsoon regions with persistent cloud cover (Jensen *et al.*, 1977). The largest survey of this kind, covering some 4 million $km^2$, was undertaken in the Brazilian Amazon during the 1970s. The *Radam* project provided information on geology, morphology, soils, vegetation, and land use potential, and offered the first overview of natural resources within this vast region (Projeto Radam, 1973; Hammond, 1977). SLAR imagery has been employed for similar investigations in adjacent areas of Panama, Venezuela, and Colombia (MacDonald, 1969; Aeroservice, 1972; Proradam, 1979), and for other reconnaissance surveys in Nigeria and Indonesia (Trevett, 1978; Parry and Trevett, 1979).

During the 1970s Landsat imagery has been used for land survey purposes in the tropics. Early studies included the inventory of land, water, and vegetation resources in savanna areas of the Sudan (Adrien and Baumgardner, 1977). More recently, terrain and vegetation units have been surveyed in southwest Kenya (Odenyo, 1979), and a forest inventory has been undertaken in the Philippines (Umali, 1980). In addition, more general land cover surveys have yielded data on vegetation types in Bolivia (Brockman, 1978) and Indonesia (Kalensky *et al.*, 1978).

The mapping of land use, using remote sensing techniques, often provides information of value in environmental planning and land management. Land systems analysis of the kind described above frequently incorporate land use data, and no precise methodological distinction exists between the two categories of survey. Recent land resource studies in Nigeria contain information on both land systems and land use (Hill, 1978), while land cover surveys commonly provide data on both vegetation types and existing land use (Brockman, 1978; Kalensky *et al.*, 1978).

Specific surveys of land use have been undertaken in various parts of the tropics. During the 1960s, aerial photographs provided the basis for land use surveys in peninsular Malaysia (Nor, 1978), and similar surveys have been made in Sri Lanka (Sridas, 1966), Chile (Saa Vidal, 1967) and India (Tomar and Maslekar, 1974). Landsat imagery has also been used for this purpose. Feasibility studies have been undertaken in parts of Thailand and Laos (Omakupt, 1978; Wacharakitti and Morain, 1978; Omakupt and Vunpiyarat, 1980), and the mapping of sugar cane and forestry plantations has been attempted in southeast Brazil (Shimabukuro *et al.*, 1978; Mendonça, 1981). In many parts of the tropics, the relatively poor spatial resolution of the Landsat multispectral scanner has limited the precision of such surveys, particularly in areas of small-scale subsistence farming, but sensor capabilities in this respect are likely to improve considerably with the new generation of satellites, beginning with SPOT in 1986 (Schultink, 1983; CNES, undated).

The monitoring of land resources represents a third major application of remote sensing in tropical regions, providing additional information of relevance for environmental planning and land management. The monitoring capability of earth-orbiting satellites has significant potential for the tropics, in respect of both long-term land cover monitoring and shorter-term crop and hazard monitoring. In the past, sequential aerial photography has been used for examining vegetational and agricultural changes in tropical areas (Lanly, 1969; Eden, 1974), and will have continuing value for this purpose. The introduction of satellite systems has revolutionized this procedure by providing repetitive coverage at shorter intervals and consistent image scale and quality. As with aerial photography, the utility of Landsat is diminished in cloudy regions, but even in the humid zones repetitive coverage is often available.

One important development in this respect has been the creation of a long-term monitoring programme for the moist forest regions of the tropics (FAO, 1975). A general baseline survey has been completed, using aerial photographs, SLAR, and Landsat imagery. This has been published as part of UNEP's Global Environment Monitoring System (FAO, 1981). Continuing forest surveys are being undertaken with Landsat in various countries, including Brazil, which contains more than 30% of the surviving tropical moist forests (Grainger, 1983).

In drier zones, investigations have commenced into the use of satellite imagery for monitoring vegetational and other changes relating to desertification. Landsat studies have been undertaken for this purpose in parts of the African tropics (Bannert, 1980; Coiner, 1980) and in northeast Brazil (Lombardo and de Carvalho, 1980). Long-term monitoring of the extent and impact of grass fires has been initiated in rangeland areas of East Africa (Norton-Griffiths, undated). The use of Landsat imagery for shorter-term monitoring of crop and pasture performance has also been examined. In the wet season, cloud incidence continues to be a problem with current imaging techniques (Malingreau, 1978), but studies of seasonal changes in pasture production have been initiated in drier African grasslands, using both Landsat and NOAA-7 imagery (Croze, undated; Tucker *et al.*, 1983).

The present volume contains studies that involve the application of remote sensing for both survey and monitoring purposes. In some instances, the studies seek to extend the application of established techniques into new regions, while others investigate new applications for remotely sensed data and imagery. The studies cover work undertaken by scientists from both developed and developing countries. The relevance of remote sensing as an aid to tropical land management is demonstrated in all the chapters, but from individual studies it is apparent that various technical and cultural factors affect, and at times constrain, the extent to which the potential of remote sensing is achieved. This is particularly true of developing countries where access to remotely sensed data and processing facilities is limited.

## TRANSFER OF REMOTE SENSING TECHNOLOGY

In spite of the diverse and growing application of remote sensing in the tropics, local factors frequently influence the effectiveness of its use. This applies to all forms of remote sensing, but it is especially relevant in the case of the Landsat sensors which provide much information beyond the capacity of visual interpretation and for optimum returns require access to a computer for digital enhancement and display (Kalensky *et al.*, 1978; Green, 1983). In these circumstances, it is understandable that, in spite of the established use of conventional aerial photography in the tropics, the transfer of newer Landsat technology has at times been disappointingly slow or incomplete (Green, 1983). The situation is by no means peculiar to remote sensing, however, for similar delays have occurred when attempts have been made to transfer other kinds of mid-latitude technology to the tropics. An obvious example is provided by the advanced agricultural cropping systems that constitute the Green Revolution. In some areas, conspicuous progress has been made in diffusing these systems, but elsewhere difficulties have been encountered in achieving their initial adoption by farmers or in sustaining their longer-term operation or performance (Chakravati, 1973; Pearse, 1980). Similar delays have been apparent in relation to new technology developed for improving birth control. Diffusion of the technology has been rapid in some Third World countries, but slow in many others, reflecting both social and government attitudes (Boserup, 1981). In these cases, the introduced technologies themselves have not always been entirely appropriate, and local cultural conditions have adversely affected the rate and pattern of technological transfer. This is as much the case with remote sensing as it is with other technologies.

In the field of remote sensing, some countries, such as India and Brazil, have recently achieved significant technological advances, and they are developing major applications in the context of renewable resources. In Brazil, for example, SLAR imagery was adopted as the basic tool for reconnaissance land survey (Projeto Radam, 1973), while Landsat imagery is being employed for a variety of survey and monitoring purposes. In 1979 there were more than 900 Brazilian users of Landsat imagery, which is currently acquired through a national receiving station (Adrien and Bartolucci, 1978; Parada, 1980). Much of the early usage of Landsat in the country depended on visual analysis (Hammond, 1977), but more sophisticated digital processing and enhancement techniques are now being adopted (Shimabukuro *et al.*, 1978; Filho *et al.*, 1980; Santos and de Carvalho, 1980; Mendonça, 1981). Brazil is also planning the construction of two dedicated environmental satellites for launching in the 1990s (NOAA, 1984). Comparable progress has been achieved in India with new remote sensing techniques. A national Landsat receiving station is operational, and plans exist for launching, in 1986, the Indian remote sensing satellite (IRS); this is to be the first in a series that will provide environmental data over an 8–10 year period (Rao, 1978; Bhavsar, 1983; NOAA, 1984).

In other countries there has been less progress in the adoption of newer remote sensing techniques. In West Africa some use has been made of both SLAR and Landsat imagery. The latter was readily adopted, for example, in UNEP's pilot project for forest cover monitoring in Benin, Cameroon, and Togo (FAO, 1980). However, little progress was made in developing common procedures for continuous monitoring, and independent manual analysis of Landsat imagery was recommended in preference to automated analysis using computers (Grainger, 1983). Similar preference was shown for manual analysis of Landsat in forest inventory and monitoring studies in Costa Rica (Knaebel, 1980). In other countries, such as Malaysia, interest in Landsat itself has been slow to develop. Aerial photography, which was employed as the basis of the land use survey of peninsular Malaysia in 1966, was updated and used in preference to Landsat when revision of the survey was undertaken in the mid-1970s (Nor, 1978). Interest in Landsat is emerging in Malaysia (Nor, 1978; Roque, 1980) but, as in a number of tropical countries, significant reliance continues to be placed on conventional aerial photography. Progress towards computer-based processing of Landsat imagery is thus uneven, with many users finding it preferable, and at times more effective, to persist with traditional types of imagery and manual modes of analysis.

The reasons for differential rates of transfer of remote sensing technology are complex, and reflect both technical and cultural conditions in individual countries. More detailed consideration of this question occurs in the final section of the volume, but some preliminary factors may be identified. One immediate consideration is the fact that the technical potential to acquire, store, and analyse remotely sensed data has increased vastly in recent years. Beneficial as this may appear, it often presents difficulties in tropical countries, where existing information systems are commonly antiquated and trained technical personnel are in short supply. In the case of Landsat, ready access generally exists to the products themselves, either from the United States or from low-latitude receiving stations (Fig. 1), but full exploitation of these products is rarely achieved. Even

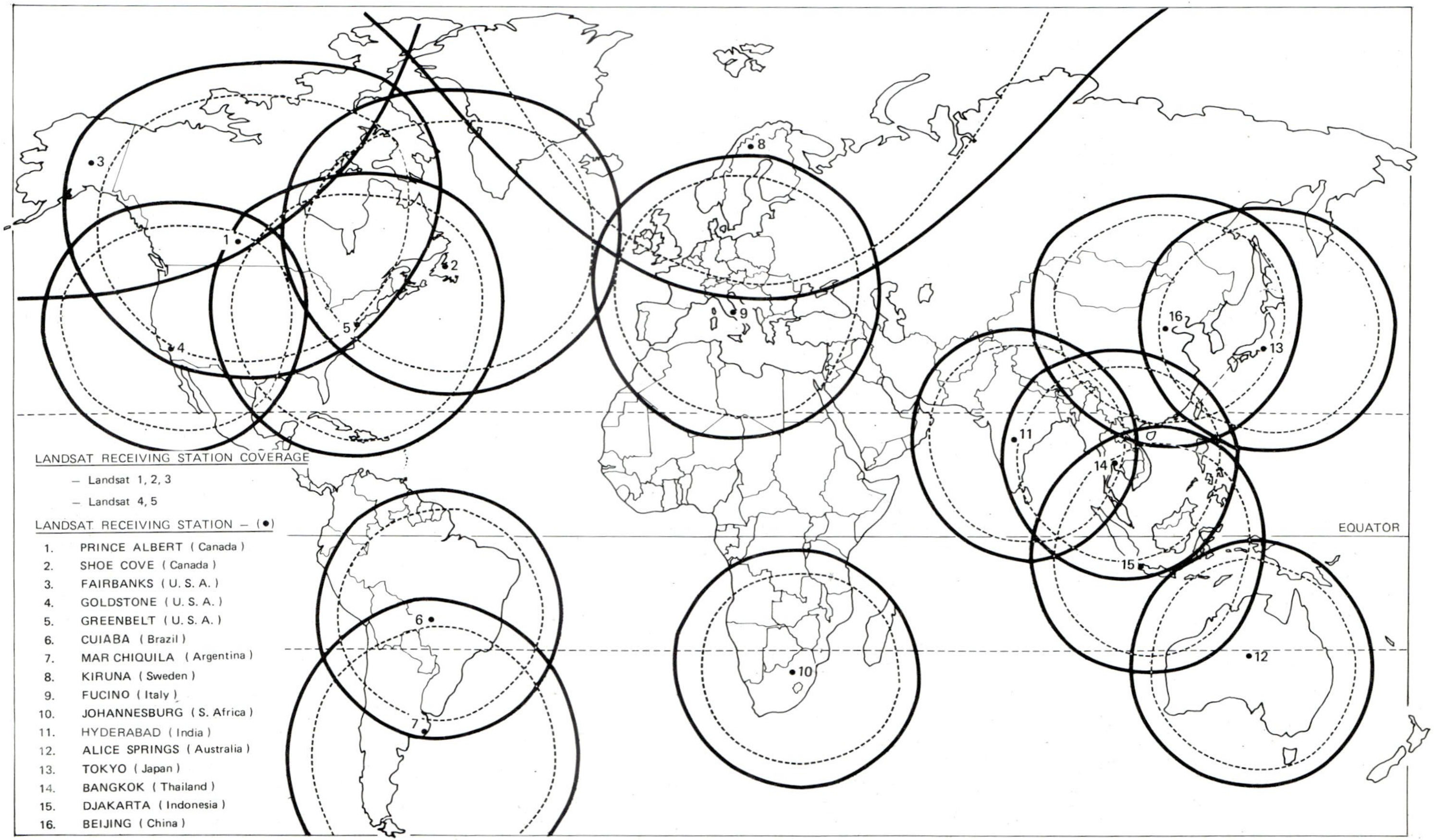


Fig. 1 Landsat receiving stations (after NASA, 1982).

if trained personnel with image-processing experience acquired in North America or Europe are available, the provision and maintenance of the requisite computing and processing equipment is not always sufficient to match the available human skills. As indicated, sophisticated results are being achieved in some countries, but elsewhere severe restraints exist and major improvements in general information systems are needed before optimal use can be made of multispectral, digital products. The current development of microcomputing systems for image processing is creating an appropriate technology for many resource-orientated investigations in the tropics (Green, 1983; Schultink, 1983), but there still remains the urgent need to upgrade ancillary information systems.

The acquisition of ground truth in support of remote sensing investigations presents additional problems in many tropical areas. Even in countries with relatively efficient ground and air transport networks, the natural hazards and other difficulties associated with undertaking fieldwork in remote or unpopulated areas can be considerable (Hammond, 1977; Odenyo, 1979; Sadowski and Danjoy, 1980). In places, fundamental problems are also encountered relating to the availability of trained field staff or to the provision of vehicles, fuel or field equipment. Even though sophisticated imagery analysis may be achieved in the laboratory, it can only be validated to the extent that the requisite field checking is undertaken.

The technical and logistical problems associated with remote sensing programmes, both in the field and the laboratory, can often be reduced by devolving responsibility for the tasks to overseas agencies or consultancy groups. This approach, which has often been adopted in the past, allows effective use of the experience and specialist skills of such groups, but an overdependence on foreign expertise will inhibit the transfer of the relevant skills and technology. In many cases, a compromise is sought whereby a local training element is incorporated into an operational programme. An example of this is provided by the Nigerian vegetation mapping project, based on SLAR imagery, that was undertaken during the late 1970s. The imagery itself was acquired by Motorola (MARS) Inc. of the United States, while the British company, Hunting Technical Services Ltd, was contracted to interpret the available imagery and also to train Nigerian staff in its use (Trevett, 1978). In such cases, a balance is preserved between exploiting the available foreign expertise and developing national skills for the future.

As well as resolving the basic technical problems associated with the use of newer remote sensing technology in tropical areas, wider diffusion of information and expertise is required at the national level, in order to back up any elite group of foreign-trained specialists and to facilitate the application of the technology itself. While some countries have made progress in this direction, many encounter difficulties, particularly in identifying and informing the array of potential users. As the range of imagery has diversified and its information content increased in both quantity and complexity, the task of diffusing the information in a comprehensible and relevant form has become correspondingly more difficult. An instructive example of formal attempts at diffusion is provided in Kenya, where a Regional Remote Sensing Facility was established in 1976, with financial aid from the United States. The Facility is located in Nairobi, and provides technical training and assistance for countries in eastern and southern Africa. Regular training courses are organized for specialist groups, and a range of user services are provided. The Facility benefits from being located within an established survey and mapping agency, where the newer remote sensing technology can be linked to the traditional approach of topographic survey. Useful as the Facility may be, its regional impact is at times constrained by local socio-economic conditions, which, in the view of Conitz *et al.* (1981) may make the transfer of remote sensing technology in the area a longer-term process than most people recognize. This problem is by no means confined to tropical countries but, as is evident in Kenya and elsewhere, the diffusion of remote sensing depends upon more than the simple provision of technical equipment and service facilities.

In discussing the general problem of information transfer, Cochrane (1980) indicates that there is an onus on the users themselves to become familiar with the new remote sensing technology. While this is realistic in many parts of the developed world, the process is much less readily achieved in countries where the background level of technical knowledge is poor, and where there is less experience in resolving problems on the basis of technical criteria. The situation is further aggravated by the diversity of potential users of remote sensing and the frequent lack of assigned operational responsibilities (Tuyahov *et al.*, 1982). As previously indicated, some tropical countries have been receptive to newer remote sensing technology, but in many instances the transfer has been slow and major opportunities for the improvement of the data base for land management have been neglected or lost.

An additional factor that has at times affected the availability and use of remote sensing products has been a concern for national security. This has been the case in the past with aerial photographs and other aerial imagery, and in places continues to be so. National governments clearly have the right to impose restrictions on access to such material acquired over their territories but, in the case of satellite systems, authority of this kind is greatly diminished. Local control may still be exercised but Landsat products are today generally available for all tropical regions and, within the limits imposed by their resolution, can be freely used for the acquisition of environmental data. The improved resolution and spectral sensitivity of the Landsat thematic mapper and the SPOT system will further increase the supply of alternative data compared with that obtainable from aerial photographs.

Increased acquisition of remotely sensed data will not of itself guarantee any improvement in the quality of land management. Indeed, the acquisition of new imagery may in places encourage or accelerate the over-exploitation of newly sensed resources. However, in the long term, remote sensing should make the mismanagement of resources more difficult, because of the ability of modern systems to alert the resource planner to many kinds of environmental stress or damage. Under optimum conditions remote sensing can be expected to make a major contribution to environmental planning and land management, and hence to the socio-economic well-being of tropical populations. It must be remembered, however, that as with any development-orientated technology, the benefit is as much dependent on local conditions as on the technology itself. For this reason it is essential to provide remote sensing systems that are appropriate, not only in the sense that they are technically efficient, but also capable of being adapted to local usage. Since the technology and the associated information flux are continuing to develop apace, the opportunities offered by remote sensing to politicians, planners, and land managers, although frequently daunting, should be seen as a challenge with demonstrable benefits for the planning and rational utilization of tropical resources.

## REFERENCES

Adrien, P-M. and Bartolucci, L. A., 1978. Remote sensing technology and new development trends in Latin America. *Proc. 12th int. Symp. Remote Sensing Environ.*, **1**, 187–201.

Adrien, P-M. and Baumgardner, M. F., 1977. Landsat, computers, and development projects. *Science*, **198**, 466–470.

Aeroservice, 1972. *Levantamiento de Radar de Region Sur de Venezuela*. Caracas: Comision para el Desarrollo del Sur de Venezuela.

Baltaxe, R., 1980. *The Application of Landsat Data to Tropical Forest Surveys*. Rome: Food and Agriculture Organization.

Bannert, D., 1980. An operational system to monitor desertification using Landsat data. *Proc. 14th int. Symp. Remote Sensing Environ.*, **2**, 783–785.

Bawden, M. G. and Carroll, D. M., 1968. *The Land Resources of Lesotho. Land Resource Study 3*. Tolworth: Land Resources Division, Directorate of Overseas Surveys.

Bhavsar, P. D., 1983. The importance of remote sensing from space to the Indian subcontinent. *Int. J. Remote Sensing*, **4**, 529–536.

Boserup, E., 1981. *Population and Technology*. Oxford: Basil Blackwell.

Brockman, C. E., 1978. *Mapa de Cobertura y Uso Actual de la Tierra, Bolivia; Memoria Explicativa*. La Paz: Geobol.

Chakravati, A. K., 1973. Green Revolution in India. *Ann. Ass. Am. Geogr.*, **63**, 319–330.

Christian, C. S., 1958. The concept of land units and land systems. *Proc. 9th Pacif. Sci. Congr.*, **20**, 74–81.

Christian, C. S. and Stewart, G. A., 1968. Methodology of integrated surveys. In *Aerial Surveys and Integrated Studies. Proceedings of the Toulouse Conference*. Paris: UNESCO, pp. 233–280.

CNES, undated. *SPOT. Système de Télédétection par Satellite*. Toulouse: Centre National d'Etudes Spatiales.

Cochrane, G. R., 1980. The contribution of remote sensing to resource evaluation. In Anderson, A. G. (Ed.), *The Land Our Future: Essays on Land Use and Conservation in New Zealand*. Auckland: Longman Paul and New Zealand Geographical Society, pp. 43–73.

Cochrane, T. T., 1967. *A Land Systems Map of Central Tropical Bolivia*. La Paz: Misión Británica en Agricultura Tropical, Ministerio de Agricultura.

Coiner, J. C., 1980. Using Landsat to monitor changes in vegetation cover induced by desertification processes. *Proc. 14th int. Symp. Remote Sensing Environ.*, **3**, 1341–1351.

Conitz, M., Falconer, A., Hart, T. and Anderson, R., 1981. Remote sensing training and technical assistance at the regional level in East Africa. *Proc. 15th int. Symp. Remote Sensing Environ.*, **3**, 1509–1517.

Croze, H., undated. *Ecological Monitoring of Rangeland*. Nairobi: Regional Remote Sensing Facility.

Eden, M. J., 1974. The origin and status of savanna and grassland in southern Papua. *Trans. Inst. Brit. Geogr.*, **63**, 97–110.

FAO, 1965. *Soil Survey of the Llanos Orientales, Colombia*, vol. 1, *General Report*. Rome: Food and Agriculture Organization.

FAO, 1966. *Report on the Soil Survey Project of British Guiana*, vol. 3, *Reconnaissance Soil Survey*. Rome: Food and Agriculture Organization.

FAO, 1975. *Report on the Formulation of a Tropical Forest Cover Monitoring Project*. Rome: Food and Agriculture Organization.

FAO, 1980. *Global Environment Monitoring System. Pilot Project on Tropical Forest Cover Monitoring. Benin — Cameroon — Togo. Project Implementation: Methodology, Results and Conclusions*. Rome: Food and Agriculture Organization.

FAO, 1981. *Tropical Forest Resources Assessment Project*, vols 1–4. Rome: Food and Agriculture Organization.

Filho, P. H., Santos, A. P. dos, Novo, E. L. M. de M., Shimabukuro, Y. E., Duarte, V., Medeiros, J. S. de, Alves, E. C. M. and de Santana, C. C., 1980. The use of Landsat data for evaluation and characterization of deforested pastureland and reforested areas in Brazil. *Proc. 14th int. Symp. Remote Sensing Environ.*, **3**, 1723–1729.

Grainger, A., 1983. Improving the monitoring of deforestation in the humid tropics. In Sutton, S. L., Whitmore, T. C. and Chadwick, A. C. (Eds), *Tropical Rain Forest: Ecology and Management*. Oxford: Blackwell Scientific Publications, pp. 387–395.

Green, K. M., 1983. Using Landsat to monitor tropical forest ecosystems: realistic expectations of digital processing technology. In Sutton, S. L., Whitmore, T. C. and Chadwick A. C. (Eds), *Tropical Rain Forest: Ecology and Management*. Oxford: Blackwell Scientific Publications, pp. 397–409.

Hammond, A. L., 1977. Remote sensing. *Science*, **196**, 511–516.

Heinsdijk, D., 1958. *Report to the Government of Brazil on a Forest Inventory in the Amazon Valley. Pt. 2 Region between Rio Xingú and Rio Tocantins*. Rome: Food and Agriculture Organization.

Hill, I. D. (Ed.), 1978. *Land Resources of Central Nigeria. Agricultural Development Possibilities*, vol. 1B, *The Bauchi Plains. Land Resource Study 29*. Surbiton: Land Resources Development Centre, Ministry of Overseas Development.

Jensen, H., Graham, L. C., Porcello, L. J. and Leith, E. N., 1977. Side-looking airborne radar. *Scient. Am.*, **237** (4), 84–95.

Johnson, M. S. and Chaffey, D. R., 1973. *A Forest Inventory of Part of the Mountain Pine Ridge, Belize. Land Resource Study 13*. Surbiton: Land Resources Division, Overseas Development Administration.

Kalensky, Z. D., Darmoyuwono, K., Potts, T. F. and Michino, T., 1978. Thematic map of Lombok Island from Landsat computer compatible tapes. *Proc. 12th int. Symp. Remote Sensing Environ.*, **2**, 1349–1365.

Knaebel, S., 1980. Remote sensing for development. *Proc. 14th int. Symp. Remote Sensing Environ.*, **1**, 9–13.
Lanly, J. P., 1969. Regression de la fôret dense en Côte d'Ivoire. *Bois Fôrets Trop.*, **127**, 45–59.
Lombardo, M. A. and de Carvalho, V. C., 1980. Preliminary analysis of the potential of Landsat imagery to study desertification. *Proc. 14th int. Symp. Remote Sensing Environ.*, **2**, 645–652.
MacDonald, H. C., 1969. Geologic evaluation of radar imagery from Darien province, Panama. *Mod. Geol.*, **1**, 1–63.
Malingreau, J. P., 1978. Remote sensing inputs into a wetland rice regional production model (Java). *Proc. 12th int. Symp. Remote Sensing Environ.*, **3**, 1617–1642.
Mendonça, F. J., 1981. Use of Landsat data for automatic classification and area estimation of sugar-cane plantation in São Paulo state, Brazil. *Int. J. Remote Sensing*, **2**, 361–368.
NASA, 1982. *Landsat Data Users Notes*, **21**, 1–12.
NOAA, 1984. 13th LGSOWG meeting held in Washington, D. C. *Landsat Data Users Notes*, **30**, 6–7.
Nor, S. M., 1978. Remote sensing activities in Malaysia. *Proc. 12th int. Symp. Remote Sensing Environ.*, **1**, 133–138.
Norton-Griffiths, M., undated. *Grass Fires in Serengeti National Park Tanzania*. Nairobi: Regional Remote Sensing Facility.
Odenyo, V., 1979. *Application of Landsat Data to the Study of Land and Range Resources in the Narok Area, Kenya. Research Report 79–4*. Ottawa: Canada Centre for Remote Sensing.
Ollier, C. D., Lawrence, C. J., Beckett, P. H. T. and Webster, R., 1969. *Terrain Classification and Data Storage: Land Systems of Uganda*. MEXE Rept. 959. Oxford: Soil Science Laboratory, Department of Agriculture.
Omakupt, M., 1978. Land use inventory of north Thailand using Landsat imagery. *Proc. 12th int. Symp. Remote Sensing Environ.*, **3**, 2297–2306.
Omakupt, M. and Vunpiyarat, V., 1980. Remote sensing and land use inventory. *Proc. 14th int. Symp. Remote Sensing Environ.*, **2**, 897–904.
Paijmans, K., 1975. Explanatory notes to the vegetation map of Papua New Guinea. *Land Res. Ser. C.S.I.R.O. Aust.*, **35**, 1–25.
Parada, N. J., 1980. Collection, processing and distribution of remote sensing data for the Brazilian receiving station. *Proc. 14th int. Symp. Remote Sensing Environ.*, **1**, 55–64.
Parry, D. E. and Trevett, J. W., 1979. Mapping Nigeria's vegetation from radar. *Geogrl J.*, **145**, 265–281.
Pearse, A., 1980. *Seeds of Plenty, Seeds of Want; Social and Economic Implications of the Green Revolution*. Oxford: Clarendon.
Projeto Radam, 1973. *Levantamento de Recursos Naturais*, vol. 1. *Parte das Folhas SC 23 Rio São Francisco e SC 24 Aracaju*. Rio de Janeiro: Departmento Nacional da Produção Mineral, Ministério das Minas e Energia.
Proradam, 1979. *La Amazonia Colombiana y sus Recursos*, vols 1–5. Bogota: Proyecto Radargrametrico del Amazonas.
Rao, K. R., 1978. Remote sensing activities in India: a country report. *Proc. 12th int. Symp. Remote Sensing Environ.*, **1**, 43–54.
Roque, C. R., 1980. ASEAN remote sensing program. *Proc. 14th int. Symp. Remote Sensing Environ.*, **1**, 65–70.
Saa Vidal, R., 1967. Land-use mapping in Chile. *Photogrammetria*, **22**, 153–159.
Sabins, F. F., 1978. *Remote Sensing. Principles and Interpretation*. San Francisco: W. H. Freeman.

Sadowski, F. G. and Danjoy, W. A., 1980. Some observations on the utility of remote sensors for humid tropical forests. *Proc. 14th int. Symp. Remote Sensing Environ.*, **3**, 1799–1802.

Santos, J. R. dos and de Carvalho, V. C., 1980. Characterization of cerrado vegetation using automatically classified Landsat MSS data. *Proc. 14th int. Symp. Remote Sensing Environ.*, **2**, 1089–1099.

Schultink, G., 1983. Integrated remote sensing and information management procedures for agricultural production potential assessment and resource policy design in developing countries. *Can. J. Remote Sensing*, **9**, 4–18.

Shimabukuro, Y. E., Hernandez, P. F., Koffler, N. F. and Chen, S. C., 1978. Automatic classification of reforested *Pinus* spp. and *Eucalyptus* spp. in Mogi-Guaçu, S. P., Brazil using Landsat data. *Proc. 12th int. Symp. Remote Sensing Environ.*, **3**, 2091–2100.

Sridas, S., 1966. Interpretation and mapping of rural land use from air photographs in Ceylon. *Photogrammetria*, **21**, 77–82.

Stewart, G. A. (Ed.), 1968. *Land Evaluation*. Melbourne: Macmillan.

Tomar, M. S. and Maslekar, A. R., 1974. *Aerial Photographs in Land Use and Forest Surveys*. Dehra Dun: Jugal Kishore.

Trevett, J. W., 1978. Vegetation mapping of Nigeria from radar. *Proc. 12th int. Symp. Remote Sensing Environ.*, **3**, 1919–1935.

Tucker, C. J., Vanpraet, C., Boerwinkel, E. and Gaston, A., 1983. Satellite remote sensing of total dry matter production in the Senegalese Sahel. *Remote Sensing Environ.*, **13**, 461–474.

Tuyahov, A. J., Weinstein, R., Weber, J., Vitale, J., Bracken, D. and Rowely, R., 1982. Technology transfer of space applications by the National Aeronautics and Space Administration. In Johannsen, C. J. and Sanders, J. L., *Remote Sensing for Resource Management*. Ankeny: Soil Conservation Society of America, pp. 379–588.

Umali, R. M., 1980. Application of satellite remote sensing to natural resources management in the Philippines. *Proc. 14th int. Symp. Remote Sensing Environ.*, **1**, 307–313.

Wacharakitti, S. and Morain, S. A., 1978. Procedures for land-use analysis in developing countries: examples from southeast Asia. *Proc. 12th int. Symp. Remote Sensing Environ.*, **1**, 587–595.

Wall, J. R. D., Hansell, J. R. F., Catt, J. A., Ormerod, E. C., Varley, J. A. and Webb, I. S., 1979. *The Soils of the Solomon Islands*, vol. 1, *Technical Bulletin 4*. Surbiton: Land Resources Development Centre, Ministry of Overseas Development.

Remote Sensing and Tropical Land Management
Edited by M. J. Eden and J. T. Parry


# CHAPTER 2

# *The Technology of Remote Sensing: System Components, Data Processing, and Prospects*

G. Ross Cochrane
*Department of Geography, University of Auckland, New Zealand*

This chapter seeks to provide a general introduction to the technology of remote sensing of the environment, and to comment on likely future developments. The theory, techniques, and instruments of remote sensing are comprehensively documented in volume 1 of the recent *Manual of Remote Sensing* (Colwell, 1983), and an enormous range of applications is presented in the companion volume 2. Excellent general texts on remote sensing are provided by Sabins (1978) and Lillesand and Kiefer (1979). Swain and Davis (1978) focus on quantitative procedures in remote sensing, while Green (1983) and Levialdi (1984) augment and complement these works. The increasing emphasis on merging large, spatial data sets, acquired by remote sensing with geographic information systems (GIS), is well covered by Marble and Peuquet (1983).

The increasing use of remote sensing in environmental analysis is evident in the resource applications compiled by Johannsen and Sanders (1982), the geological studies in Siegal and Gillespie (1980), and the terrain analyses in Townshend (1981). The *Proceedings of the International Symposia on Remote Sensing of the Environment* by the Environmental Research Institute of Michigan (ERIM) and the *Proceedings of the William T. Pecora Memorial Symposia* provide a wealth of information on the increasingly diverse applications of remote sensing. Other specialized information is available in remote sensing periodicals which include *Remote Sensing of the Environment, Photogrammetric Engineering and Remote Sensing,* the *International Journal of Remote Sensing*, and *Photogrammetria*. In addition, a large number of technical papers, reports, and special publications emanate from the United States National Aeronautics and Space Administration (NASA), the United States National Oceanographic and Atmospheric Administration (NOAA), the

European Space Agency (ESA), the French Centre National d'Etudes Spatiales (CNES), the French Groupement pour le Développement de la Télédétection Aérospatiales (GDTA), the Canadian Centre for Remote Sensing (CCRS), and the Japanese National Space Development Agency (NASDA). Many other technical and professional groups, particularly in aeronautics, electrical engineering, optics, cartography, surveying, photogrammetry, and image processing, provide specialist publications.

Remote sensing describes the range of techniques used to record information about an object from a distance. The sensor or sensing instrument is not in physical contact with the object and normally measures some property of its reflected, emitted or transmitted electromagnetic energy. Radiant energy moves with the constant velocity of light ($3 \times 10^8$ m $s^{-1}$), in a harmonic wave pattern. It ranges from very high-frequency gamma and cosmic rays through X-rays, ultraviolet radiation, visible, infrared and microwave radiation to low-frequency radio waves. In a narrow sense, remote sensing is measurement that makes use of visible, infrared, and microwave radiation in the acquisition of environmental information, which is stored in the form of images or digital tapes.

The purpose of remote sensing is not only to acquire data but also to process them into usable form. This involves the display and subsequent analysis of the data which are frequently in the form of images. For this reason, image analysis techniques, particularly the recognition of patterns, play an important part in remote sensing. Analysis can be either manual, machine-assisted or totally automated.

In a broader sense, remote sensing comprises the collection of data, their display, their analysis or interpretation, and their subsequent use for purposes of inventory, survey, monitoring, planning, and management. Remote sensing is especially useful where speed, repetitive observation and a broad synoptic

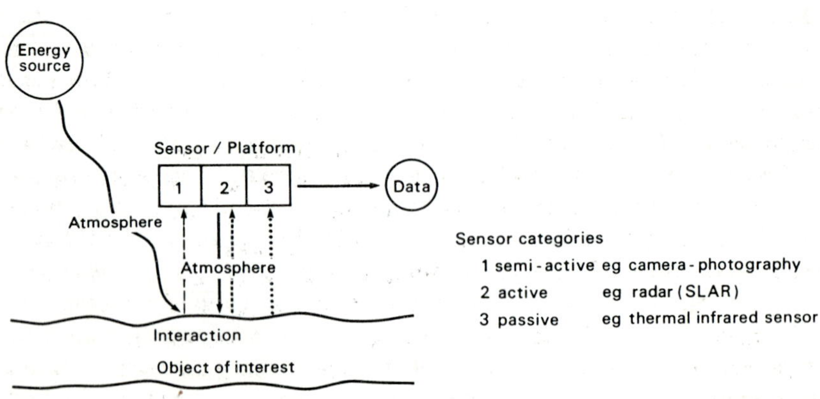


Fig 1. Major components of remote sensing systems.

view are required; and it is providing an important new dimension in the analysis of the global environment and the monitoring of environmental problems.

## COMPONENTS OF REMOTE SENSING SYSTEMS

Regardless of the wide diversity in remote sensing instrumentation and the even more varied display systems and associated data products, there are four major components of a remote sensing system: the energy source, the platform/sensor, the object of interest, and the data which are generated by energy interactions at the surface of the object or as a result of the passive emission of energy stored in the object (Fig. 1). A fifth factor, the atmosphere, can modify the energy flux between the energy source and the object and also between the object and the sensor.

### Energy

The major source of energy for remote sensing is the sun. All objects are bathed in electromagnetic energy or radiation (EMR) covering a broad continuum of wavelengths. On reaching the earth's surface, the energy is selectively transmitted, reflected or scattered, absorbed, and emitted. Variation in the EMR from different objects provides information which is directly related to the properties of those objects. Table 1 shows the more common remote sensors and the wavelengths in which they operate.

### Atmosphere

The atmosphere has a significant effect on the intensity and spectral composition of the energy recorded by a sensor. Because the atmosphere is not completely transparent the outgoing spectral radiance of objects at the earth's surface is modified before it reaches the sensor. The atmospheric gases, carbon dioxide, ozone and water vapour, cause reflection, scattering, and attenuation of energy, so that only a few 'windows' are available through which part or all the pertinent radiation is transmitted. One important atmospheric window exists in the visible spectral region (0.4–0.7 $\mu$m). There are three windows in the infrared region (3.0–4.5 $\mu$m, 8.5–14.0 $\mu$m, and 17.5–19.0 $\mu$m). The window at 8.5–14.0 $\mu$m is particularly important because this happens to coincide with the wavelength of energy radiated by a black body at 270–300° K, which corresponds to the approximate range of ambient terrain temperatures that are encountered. Fig. 2 provides a schematic diagram of atmospheric absorption for the wavelength interval 0–22 $\mu$m. In general, remote sensing with a major interest in terrestrial phenomena avoids those wavebands in which atmospheric absorption is strongly marked and matches the sensor sensitivity with an appropriate window.

Table 1 The electromagnetic spectrum.

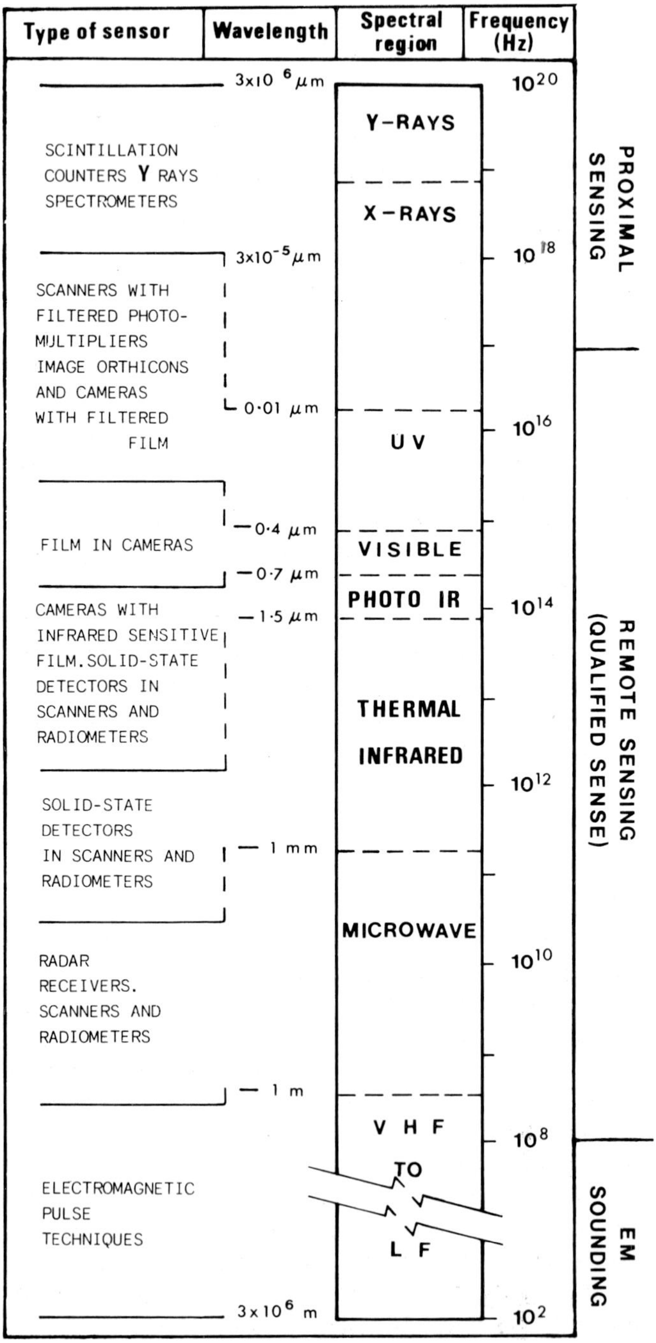

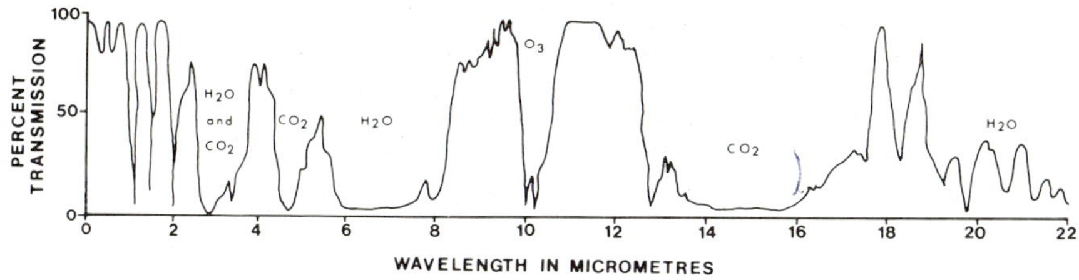


Fig. 2 Schematic diagram of atmospheric transmission. 'Windows' for remote sensing occur in the bands of high transmission.

## Objects

Many features of objects can be measured or recorded by remote sensing. Multispectral scanning allows the recording of a wider range of data than is possible from single-band sensors. By choosing appropriate sensors, the range of information that is obtained can be extended to include object size, shape, colour, surface characteristics, chemical composition, thermal properties, mineral properties, and moisture status. Using long wavelength airborne or spacecraft radars, it has been demonstrated that it is possible to penetrate dry surface materials and obtain details of subsurface features.

A great variety of details about plant cover can be acquired by remote sensing. This may relate to external geometry, including plant lifeform, size, shape and leaf attitude, foliage type, and trunk-branch-foliage percentages. Information about the internal metabolism of plants can provide useful insights into plant health, vigour, and dormancy, and some features not apparent at visible wavelengths can be recorded with colour infrared film or multispectral scanning systems.

## Platforms and sensors

Platforms for remote sensing can be fixed or moving, operate from low or orbital altitudes, and be either manned or unmanned; they may be temporary, semipermanent or virtually permanent. The elevation or altitude of a platform has a direct effect on the scale of the data and the size of the area that is imaged. The latter can vary from a few square metres with a scanner mounted on a vehicle, such as a cherry picker, through tens of square kilometres with an aerial camera, thousands of square kilometres with a Landsat MSS to a continental scale coverage with a GOES meteorological satellite.

The sensor is an instrument that is physically distant from the object of interest. It measures electromagnetic radiation from the object within a specific wavelength band and the field of view of the sensor is designed so as to provide appropriate coverage from aerial or orbital platforms. One of the oldest and simplest, but still very widely used, sensors is the camera recording on panchromatic film in the visible waveband (0.4–0.7 μm). Long experience with aerial photography has made interpreters familiar with the typical tones, patterns, and surface textures of objects in respect of this narrow waveband. Other sensors have been developed that record the response of objects across a much greater spectral range. The sensors most widely used by environmental scientists include: cameras with quartz lenses for use with ultraviolet film, cameras for use in the visible and photo infrared spectral regions, multiband cameras, optical mechanical scanners and radiometers for the thermal infrared and passive microwave regions, and active microwave sensors such as side-looking airborne radar (SLAR) and synthetic aperture radar (SAR) which are not dependent on daylight and possess an all-weather capability.

### *Cameras*

Cameras were first carried by balloons and pigeons more than a century ago. Initially, their major use was for military reconnaissance. Although military uses are still important, aerial photography now has many and diverse applications. The aerial camera records an instantaneous image of a complete scene and is particularly useful for the mapping of both natural and cultural features.

Cameras mounted in aircraft are used to obtain both oblique and vertical aerial photographs. The former are generally used in reconnaissance studies, while the latter provide the photogrammetric base for many kinds of resource surveys. It is difficult to achieve optimum conditions for the acquisition of high quality aerial photographs, particularly in respect of cloud and other atmospheric factors. However, cameras and film materials have improved significantly over the last two decades, and aerial photographs are still widely used for a range of resource applications (Paine, 1981). Four types of aerial photography will be considered: panchromatic (black and white); colour; infrared, both false colour and black and white.

*Black and white aerial photography*. Panchromatic aerial photography records details of the landscape in a range of grey tones between black and white. The grey scale differences within an aerial photograph provide spectral records of specific objects, and so recognition of their tonal variation is critical in any interpretation. Inevitably, a great amount of environmental variability is coalesced into just a few categories of grey values. Texture and pattern provide additional details which aid the interpreter in discriminating between various surface conditions.

*Colour aerial photography*. Colour film has significant advantages over panchromatic film. There is, firstly, a greatly increased range of information on colour film. In practice, it is not difficult to recognize many thousand variations in colour, which represent real differences in ground conditions. Such variation is useful for identifying local soil differences or contrasts in crop type, stage or vigour. Secondly, more accurate analysis is possible with colour aerial photography. The various components of colour—hue, value, and chroma—can be objectively measured for comparative purposes, and standardized colour systems are available for reference and classification purposes. Thirdly, colour aerial photography records features in their 'natural colour', as seen by the human eye. In practical terms, this facilitates their interpretation, because it provides an indication of quality or condition, and it is aesthetically more pleasing for display purposes.

*Infrared aerial photography*. Black and white and colour infrared film are sensitive to both the visible wavelengths (0.4–0.7 $\mu$m) and infrared wavelengths as far as 0.9 $\mu$m. Energy-object responses in the longer wavelengths are markedly different from those in the visible range. This is particularly so in respect of vegetation, which has a very high reflectance in the former, and water, which has minimal or no reflectance. The particular advantage of colour infrared film is that it is responsive to changes in infrared reflectance which signal disease or stress in plants and crops.

### *Single band and multispectral scanning systems*

Scanning systems are operated in either aircraft or spacecraft. The forward motion of the platform permits the scanning of successive strips of terrain orthogonal to the flight or orbit path. In contrast to cameras, which record a full scene instantaneously with each exposure, scanning systems use thermal or photon detectors and optical mechanical scanning devices to record incoming radiation in a raster mode.

The spectral sensitivity of photon detectors covers a broader range than that of photographic film, and so it is possible to construct multispectral scanners (MSS) that operate at a range of wavelengths. From 4- to 18-band MSS systems are widely used in aircraft and spacecraft, providing a variety of environmental information.

The multispectral scanner (MSS) on Landsats 1, 2, and 3 operates from an orbital altitude of 913 km, recording in four different wavelength bands, 0.5–0.6 $\mu$m, 0.6–0.7 $\mu$m, 0.7–0.8 $\mu$m, and 0.8–1.1 $\mu$m, over a swath width of 185 km. The MSS data are recorded in electronic form and relayed by radio to earth, where images are reconstituted at ground receiving stations. The ground resolution of the system is 80 m. A second sensor on Landsats 4 and 5 is the more complex thematic mapper (TM), which has greatly improved spatial and spectral resolution (Table 2). Recording in seven different wavebands, the TM has a ground resolution of 30 m for six of its bands. The system is designed to measure specific surface characteristics in individual bands, but it also provides more general information.

Table 2 Generalized spectral reflection curve of healthy vegetation, the factors controlling leaf reflectance, and the sensor bands adopted for the MSS in Landsats 1, 2, and 3 and the TM in Landsats 4 and 5.

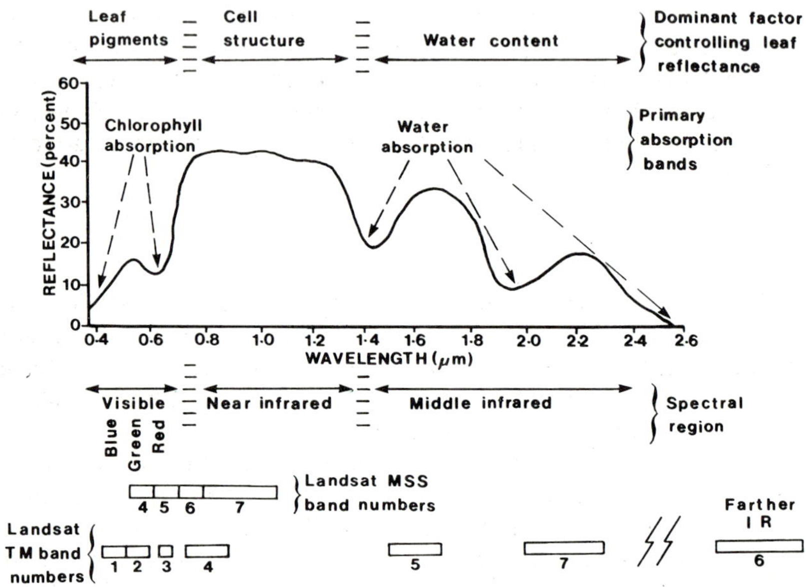


| TM Band | Spectral Range μm | Ground Resolution metres | Principal applications |
|---|---|---|---|
| 1 | 0.45–0.52 | 30 | Coastal water mapping, soil/vegetation differentiation, deciduous/coniferous differentiation |
| 2 | 0.52–0.60 | 30 | Green reflectance by healthy vegetation |
| 3 | 0.63–0.69 | 30 | Chlorophyll absorption for plant species differentiation |
| 4 | 0.76–0.90 | 30 | Biomass surveys, water body delineation |
| 5 | 1.55–1.75 | 30 | Vegetation moisture measurement, crop differentiation, snow/cloud differentiation. |
| 6 | 10.4–12.5 | 120 | Plant heat stress measurement, soil moisture, thermal mapping |
| 7 | 2.08–2.35 | 30 | Hydrothermal mapping, lithological studies. |

The 3–5 μm and 8–24 μm spectral bands are important 'atmospheric windows' for monitoring the thermal characteristics of surface materials from both aircraft and spacecraft. In the case of geological studies, thermal flux is generally recorded in pre-dawn and early afternoon missions. The Heat Capacity Mapping Mission (HCMM) of NASA's AEM satellites was an 18-month experiment designed to obtain data on thermal inertia for geological and vegetation cover mapping (Kahle *et al.*, 1981). Meteorological satellites of the TIROS/NOAA, Nimbus and SMS series (GOES, Meteosat), and the Japanese Himawari also acquire much useful data from the thermal channels. Simultaneous images in the visible bands are also obtained to aid in the interpretation of the thermal data.

### *Side-looking airborne radar (SLAR)*

SLAR imagery is dependent upon successive, short pulses of microwave-frequency energy which are directed in a narrow beam orthogonal to the aircraft or satellite path. Successive strips of ground to the side of the platform are 'illuminated' and the specular return or backscatter is used to make up a continuous swath of microwave imagery.

SLAR systems usually operate at wavelengths between 0.86 cm and 3.3 cm (K- and X-bands) and more recently, on Seasat and the Shuttle flights, at 23 cm (L-band). Radar pulses are emitted at millisecond intervals for distant targets and microseconds for closer targets. The backscattered signals are received by the same antenna which switches from transmitter to receiver. The length of the antenna, measured in units of the operative wavelength, is directly related to image resolution—the longer the antenna the higher the resolution. There are obvious physical limits to the antenna length possible on aircraft and satellites, and so synthetic aperture systems (SAR) have been developed which electronically combine the phase histories of terrain features while they lie within the radar beam. Ground resolution of 2–3 m in both range and azimuth is achieved. Radar backscatter is amplified and recorded either digitally or as a data film which is processed in a ground-based correlator to produce an image film (Ulaby *et al.*, 1981).

Radar systems are more complex than either aerial photography or multispectral scanning systems. Radar system parameters that strongly influence the image detail include the depression angle of the antenna, the wavelength of the pulsed energy, and the polarization of transmitted and received signals. Object parameters that can be examined using SLAR imagery include the dielectric properties, surface roughness relative to the incident wavelength, surface slope relative to the incidence angle of the radar beam, the physical resonance of the object, and subsurface characteristics in the case of longer wavelength systems (S- and L-bands). Although these variables make the interpretation of radar data more complex, they allow information to be acquired about certain object parameters that is not available from shorter wavelength sensors. The information relates, for example, to surface roughness, material composition, and moisture content.

### Data

The critical function of remote sensing is to provide the user with information that has a good signal/noise ratio, so that an object is clearly recognizable against its background. Information in remote sensing is generally provided either as a continuous tone image, i.e. a photograph or photo-product, or as digital data. In the case of aerial photographs, the camera records reflected radiation directly onto film. The ground reflectance varies greatly, depending upon surface characteristics. Values for agricultural crops and forests are relatively low, so that black and white photography is not ideal for differentiating subclasses. Greater reliance must be placed on other characteristics, namely, texture, size, shape, pattern, and shadow. The altitude at which aerial photographs are taken is also critical because the object/background contrast is significantly reduced by haze effects.

In the case of digital data, whether derived from scanning or radar systems, electrical analogue signals are usually sampled and then recorded as digital output on tape. Digital image processing has evolved rapidly over the past two decades, triggered by the development of satellite sensors capable of very high data collection rates and by parallel growth in computing technology (Andrews, 1970). Digital image processing combines computing and image scanning techniques for the purpose of image enhancement, correction of distortion, pattern recognition, and object measurement (Gonzalez and Wintz, 1977). This technology avoids many of the inherent problems associated with visual analysis of images and objects.

The basic components of an image processing system, digital or otherwise, are image acquisition, image processing, and image display. The principles of

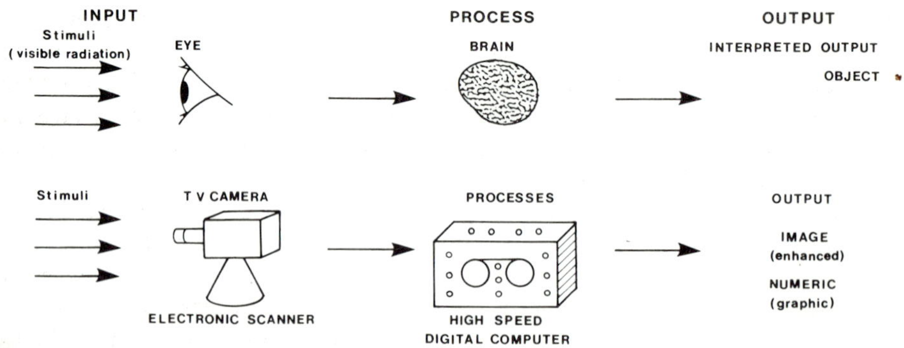


Fig. 3 Principles of image processing.

digital image processing are comparable to the object/human eye/brain responses and the object/camera/film process (Fig. 3). However, the important difference is the capability of digital image processing to offer numeric and/or image output, with the possibility of a variety of enhancement techniques, as required. The emergence of digital image processing, as a popular and sophisticated methodology in remote sensing, arises from several factors:

1. The development of faster and cheaper computers, larger data storage devices, and improved display technology systems (Moik, 1980).
2. Digital systems have a wider dynamic range than either the human eye or photographic film. Digital imaging systems commonly use 512 or 1024 discrete shades of grey, and portray a much greater range of colour tones than can be resolved by the human eye or reproduced on film.
3. A vast amount of digital information can be shown in a compact and readily interpretable form on a visual display unit. This represents a very effective mode of display for both trained and untrained observers.
4. A wide range of manipulative processing can be employed that is not possible with non-digital imagery, i.e. data transformations, enhancements, and compressions.
5. Photographic products can themselves be digitized and subsequently processed.

## IMAGE DATA PROCESSING

Digital image processing has become an important research tool in many scientific disciplines. Extractive processing is based on the proposition that the information required by the analyst may be characterized in terms of the properties of perceived patterns. Information extraction involves the detection and recognition of patterns. Precise algorithms for directing automatic processing in pattern recognition are not well developed and so it is not surprising that most information extraction tasks require human interaction. The human eye/brain system has extraordinary power in pattern recognition, although it is not always capable of extracting all the information from an image. Radiometric degradation, geometric distortion, and noise introduced during transmission, recording or display can severely reduce the usefulness of the data. The purpose of image processing is to remove these distortions and thus aid the information extraction process.

Many interactive image analysis systems are available, with varying facility for image display and interpretation. The general objective of the systems is to improve the usefulness of the data and the quality of the extracted information. The general structure of such systems is illustrated in Fig. 4.

A number of standard procedures are often applied in the course of digital image processing. They can be considered in four groups:

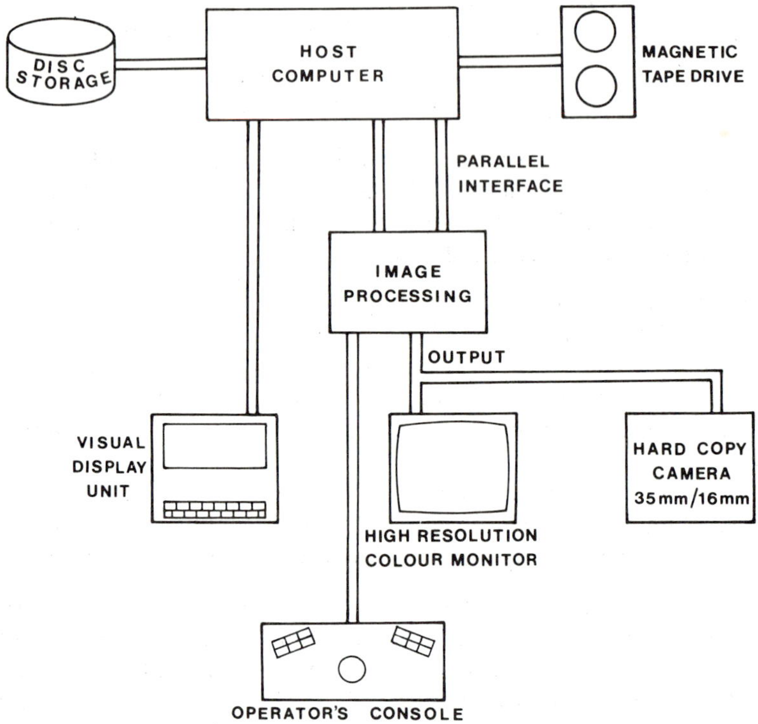


Fig. 4 Block diagram of an interactive image analysis system.

1. Image restoration procedures. These are concerned with the corrections for distortion, degradation, and noise.
2. Image registration procedures. In many applications it is necessary to compare images of the same scene taken from different sensors at the same time or from one or several sensors at different times. In all these cases it is essential that the component images are registered either mutually or in terms of a superordinate grid or projection.
3. Image enhancement procedures. These involve the improvement of an image in the context of a particular application. There are many well-known enhancement procedures including contrast stretch, band to band ratioing, spatial filtering in both the frequency and spatial domain, and data compression techniques.
4. Image classification procedures. In the same way that it is possible to classify a geographical area according to a particular theme so an image can be classified using the spectral responses collected by a remote sensing device. The most popular methods of image classification are density slicing and

supervised classification using the minimum distance to means classifier, the parallelepiped classifier or the maximum likelihood classifier.

## REMOTE SENSING APPLICATIONS

In all remote sensing systems the five basic components of energy, the atmosphere, the platform/sensor, the object, and the data are interlinked. Lillesand and Kiefer (1979) have suggested that in an optimal system the linkages can be defined in relatively simple terms:

1. a uniform energy source;
2. a non-interfering atmosphere;
3. a series of unique energy/matter interactions at the earth's surface, giving each object a distinct and definable spectral signature;
4. a super-sensor that would be simple, reliable, accurate, and economical to operate;
5. a real-time data handling system;
6. multiple data users with a profound knowledge and ability to make wise decisions about management of the environment and its resources.

As the authors indicate, such a system does not exist. Real remote sensing systems rarely achieve the optimum at any stage. Atmospheric factors, for example, do not constitute a problem with radar sensors, but they significantly modify the spectral signature of objects in both photographic and scanning systems.

Initially, remote sensing systems were relatively simple and focused on a single theme. Black and white photography was obtained for a particular resource or mapping purpose and was processed and interpreted accordingly. Thermal scanners were used to measure differences in the radiation temperature of objects. Radar applications were generally single-purpose, and early satellite projects were designed for the simple testing of hypotheses or for data collection.

In recent years, remote sensing systems and applications have become increasingly complex and diverse, and the subject is now multidisciplinary in character. No longer can an expert in one field cope with all the ramifications of the technology, the image processing, and the applications. However, in any application of remote sensing to resource evaluation, there is a broad general pattern of procedural steps or linkages, as shown in Fig. 5. The specific details depend on the equipment available, the skills of the operators, the technical and financial resources available, and the requirements of the user. Not all those wishing to use remote sensing for resource applications may have access to sophisticated digital interactive processing and display systems, especially in the developing world. Nevertheless, the basic procedures remain the same.

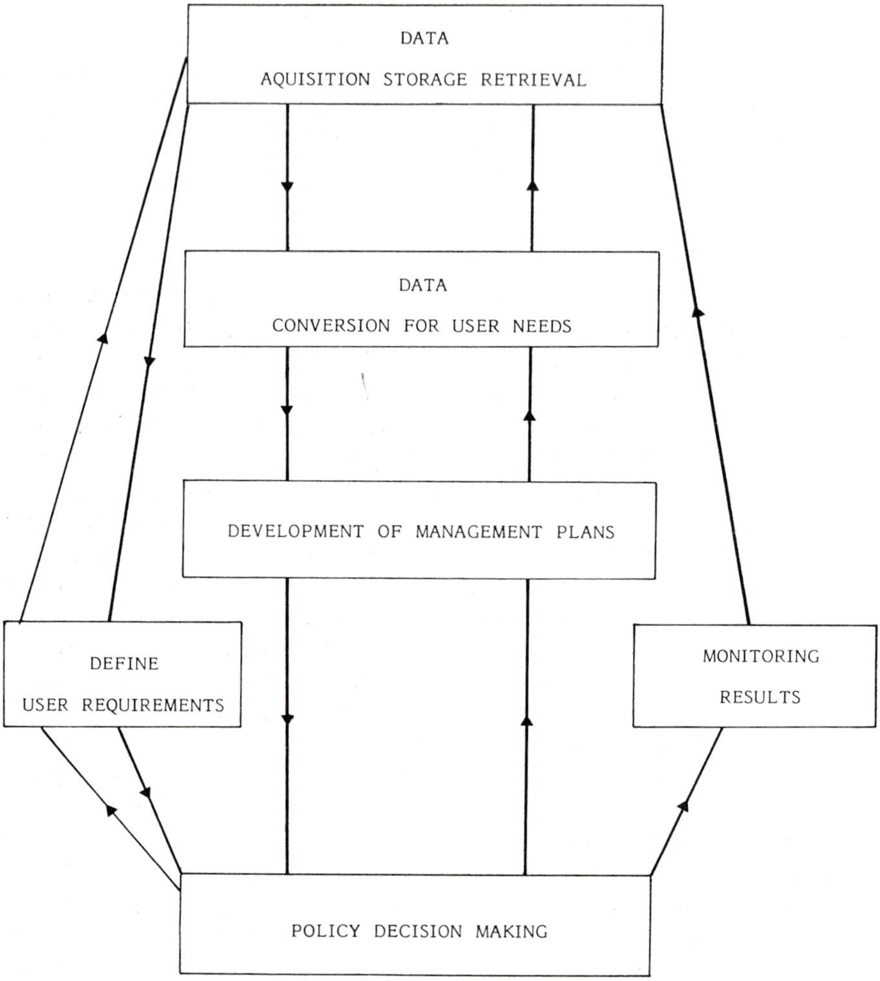


Fig. 5 Linkages in a sensor-user model.

As remote sensing became better understood during the 1960s and 1970s, some degree of integration in its application became apparent. Thus, a widely adopted three-tier model for resource applications was based on the use of (1) satellite data for broad synoptic coverage, (2) selected aerial photography for sample studies, and (3) ground survey within the sample areas. While this approach has some continuing value, as indicated by later papers in this volume, improvements in the spatial resolution of satellite data, coupled with better spectral and radiometric resolution and refinements in image processing, have reduced the need for the second tier. For many resource applications, therefore, a two-tier model, based on satellite data and field checking of selected areas, is both practical and time- and cost-efficient. Nevertheless, the complementary roles of different sensing systems, whether based on visible, infrared or radar techniques, are clearly recognized and firmly established as a major feature of satellite remote sensing of the 1980s.

## RETROSPECT AND PROSPECT

In terms of length of operation and volume of data recorded, the operational, polar-orbiting meteorological satellites from TIROS in 1960 to the current NOAA-9 (TIROS-N series) and the SMS-GOES geostationary satellites are the most important. However, in terms of wide acceptance and diversity of applications, the Landsat satellite series has had the most dramatic impact. It has revolutionized the collection, transmission, processing, and use of remotely sensed data for the inventory, monitoring, planning, and management of earth resources in little over a decade. The emphasis upon Landsat MSS applications in this volume is indicative of its importance. The improved spatial resolution and wider range of spectral bands of the Landsat thematic mapper (TM) have further increased this potential for earth resource applications.

Improved spectral resolution, as an aid to more detailed survey, will also become available with the French earth resources satellite, SPOT (*Système Probatoire pour Observation de la Terre*), which is scheduled for launching in 1986. The improved spatial resolution to 20 m and 10 m in the SPOT system should prove valuable for detailed agricultural mapping and urban analysis. The Landsat MSS system has been unable either to resolve the complex agricultural patterns of Europe or many tropical areas, or to record detail within urban areas. SPOT simulation studies by the Centre National d'Etudes Spatiales (CNES) and other government agencies over the past three years have demonstrated the potential of SPOT data for detailed agricultural, forestry, coastal, and urban studies as well as for other resource applications. The high resolution will also be important for many cartographic applications. Additionally, SPOT will be ground controlled to view either vertically (nadir-viewing) or at an angle (off-nadir viewing). In the latter mode, the sensors are directed in an across-track direction in order to allow rapid access to any point lying within the arc $\pm 27°$ from the satellite ground track, thereby providing a very flexible multiple-look capability. This innovation also allows the acquisition of stereoscopic image pairs from different satellite passes. A fundamental feature of the SPOT IMAGE approach to second generation satellite remote sensing is concern for the user and for the encouragement of an expanding range of applications (Fig. 6).

While both the Landsat TM and SPOT systems involve improved spatial resolution, earth resource applications for the NOAA Advanced Very High

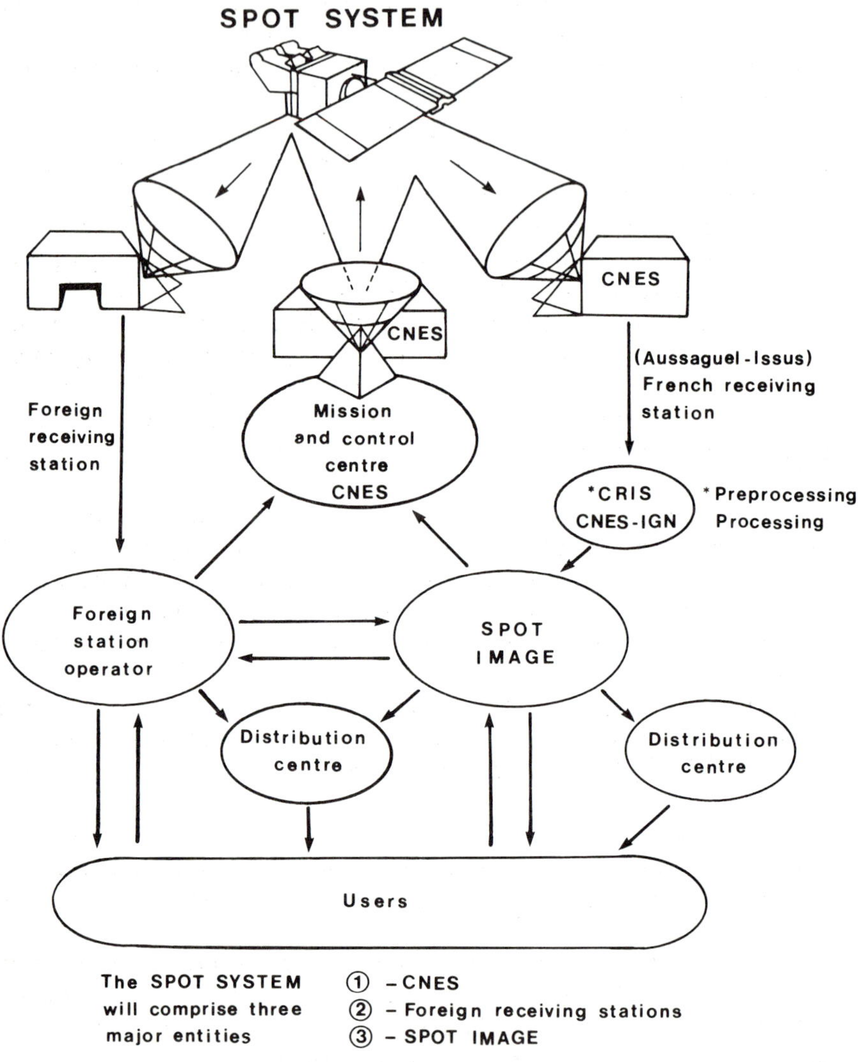


Fig. 6 Sensor-user links in the SPOT system.

Resolution Radiometers (AVHRR) have also developed apace during the 1980s. Established basically as meteorological satellites, the NOAA systems, with their AVHRR 5-channel multispectral scanners, have a low ground resolution (1.1 km), but offer the advantage of frequent repetitive cover over almost all of the earth's surface and a broad swath width of 3000 km. Data are available four times daily from the current NOAA satellites. Important applications of the NOAA AVHRR are in biomass studies and surface albedo/temperature surveys.

Most orbital remote sensing is based on semipermanent platforms. In contrast, the NASA Shuttle provides a valuable short-term platform with a wide range of uses. Earlier research in the 1970s concentrated mainly on the development of imaging techniques in the visible and infrared portions of the spectrum, but with the Shuttle (first launched in April 1981) it was possible to consider other sensors. These have been selected by NASA's Office of Space and Terrestrial Applications (OSTA) and include three potentially useful systems: a synthetic aperture radar, a pair of television cameras, and an optical imager. There has been particular interest in the L-band (23 cm) imaging radar experiments, SIR-A in 1981 and SIR-B in 1984, because of the radar's ability to penetrate both cloud and surface materials (Holmes, 1983). A wide range of experiments in geography, geology, hydrology, oceanography, agronomy, and vegetation have been incorporated in the Shuttle radar experiments, and many unique contributions have been made. In addition, the Shuttle provides a flexible, experimental platform for testing instruments and data collection procedures.

Plans for the late 1980s and 1990s show a continuation of the innovative, technical developments that have characterized remote sensing during the 1970s and early 1980s. The European Space Agency (ESA) plans to continue Spacelab developments and to launch ERS-1 in 1987, which will be the first of a series with microwave instruments. France has already approved SPOT follow-on satellites.

Canada began development of Radarsat in 1978. This is a cooperative venture with the ESA. The satellite, which will be launched in 1989, will use a C-band SAR system to analyse sea ice movements. Prompt availability of the data is of fundamental importance for the safety of off-shore drilling rigs and a three-hour period between data collection and delivery is promised. This represents a major breakthrough in data processing which will have an effect on all subsequent systems concerned with the monitoring of environmental hazards and disasters.

Japan has an ambitious programme of space development, involving a marine observation satellite (MOS 1) to be launched in 1986 and an earth resources satellite (ERS 1) to be launched in 1990. The latter will carry a stereoscopic linear array camera and a synthetic aperture radar.

The United States has plans for a wide range of remote sensing developments. These include using the Shuttle to test a scanning radar altimeter for digital topographic mapping, and a 5–10 channel thermal infrared multispectral scanner, operating in the 8–13 $\mu$m wavelength range, for geological purposes. A dedicated satellite, Geosat, has been proposed for geological survey, and another, Mapsat, is being considered for cartographic purposes and for the provision of stereoscopic coverage. Satellites for resource monitoring at current Landsat ground resolutions, but at geostationary orbital altitudes, have also been suggested.

Information about space developments in the USSR is not readily available, but the nation that led space exploration with Sputnik in 1957 and has subsequently paralleled U.S. achievements is likely to continue in space applications.

Of the countries in the tropical world, only India has a well-developed satellite sensing programme. A joint reconnaissance and earth resources satellite, Bhaskara 1, was launched in June 1979. This carries low-resolution RBV cameras similar to Landsats 1 and 2. Bhaskara 2 was launched in November 1981, and improved resolution sensors are planned for land use mapping and disaster monitoring especially flooding.

In the mid-1980s, many nations are developing new remote sensors to provide better spectral and spatial resolutions for a wide range of resource applications. A worldwide user-community is poised to receive such data for the better management of the world's resources.

## REFERENCES

Andrews, H., 1970. *Computer Techniques in Image Processing*. London: Academic Press.

Colwell, R. N. (Ed.), 1983. *Manual of Remote Sensing*, vols 1 and 2. Falls Church: American Society of Photogrammetry.

Gonzalez, R. and Wintz, P. 1977. *Digital Image Processing*. Reading, Massachusetts: Addison-Wesley.

Green, W. B., 1983. *Digital Image Processing. A Systems Approach*. New York: Van Nostrand.

Holmes, A. L., 1983. Shuttle imaging radar—information and data availability. *Photogramm. Engng Remote Sensing*, **49**, 65–76.

Johannsen, C. J. and Sanders, J. L. (Eds), 1982. *Remote Sensing for Resource Management*. Ankeny: Soil Conservation Society of America.

Kahle, A. B., Schieldge, J. P., Abrams, M. A., Alley, R. E. and Levine, C. J., 1981. *Geological Applications of Thermal Inertia Imaging using HCMM Data*. Publ. 81–55, Jet Propulsion Laboratory, Pasadena.

Levialdi, S. (Ed.), 1984. *Digital Image Analysis*. London: Pitman.

Lillesand, T. M. and Kiefer, R. W. 1979. *Remote Sensing and Image Interpretation*. New York: John Wiley.

Marble, B. F. and Peuquet, D. J., 1983. Geographic information systems and remote sensing. In Colwell, R. N. (Ed.), *Manual of Remote Sensing*, vol. 1. Falls Church: American Society of Photogrammetry, pp. 923–958.

Moik, J. G., 1980. *Digital Processing of Remotely Sensed Images*. NASA SP-431, Washington.

Paine D. P., 1981. *Aerial Photography and Image Interpretation for Resource Management*. New York: John Wiley.

Sabins, F. F., 1978. *Remote Sensing Principles and Interpretation*. San Francisco: Freeman.

Siegal, B. S. and Gillespie, A. R. (Eds), 1980. *Remote Sensing in Geology*. New York: John Wiley.

Swain, P. H. and Davis, S. M., 1978. *Remote Sensing: the Quantitative Approach*. New York: McGraw Hill.

Townshend, J. R. G. (Ed.), 1981. *Terrain Analysis and Remote Sensing*. London: George Allen and Unwin.

Ulaby, F. T., Moore, R. K. and Fung, A. K., 1981. *Microwave Remote Sensing, Active and Passive*, vol. 1, *Fundamentals and Radiometry*. Reading, Massachusetts: Addison-Wesley.

# PART II

# Land Classification and Evaluation

Remote Sensing and Tropical Land Management
Edited by M. J. Eden and J. T. Parry


# CHAPTER 3

# *Aerial Photography and Land Development Projects in Guyana*

ROOSEVELT K. DANIEL
*Department of Geography, University of Guyana*

Aerial photography was the forerunner of the present array of remote sensing devices. Although aerial photography was developed and refined for military purposes in the course of two world wars, its benefits in providing the planimetric detail for topographic maps and the visual data needed for resource inventory and analysis were realized more than half a century ago. The contribution of the aerial camera and the aircraft in the task of gathering data about resources and land use is by no means exhausted, and aerial photography still has a major role to play. Despite the rapid growth of more advanced techniques in remote sensing, conventional aerial photography is still one of the most economic and versatile operational sensing systems. In this paper, the record of aerial photography in Guyana is examined and some of the limitations of spacecraft remote sensing, specifically Landsat, are reviewed in the context of the environmental conditions and economic problems faced by developing countries in the equatorial zone.

## GUYANA: THE REGIONAL SETTING

Guyana is located in the northern part of South America between 0° 41′N and 8° 33′N and 56° 32′W and 61° 22′W. It occupies an area of 213 000 $km^2$ and experiences a humid tropical climate. The mean annual temperature at Georgetown, in the coastal zone, is 27 °C, and only slight spatial variation in this value is evident across the country. The rainfall is distributed throughout the year in most areas, with two marked wet seasons. The maximum rainfall is recorded in the Pakaraima mountains, which receive over 4400 mm annually. Rainfall totals decrease to 1400 mm in the Rupununi savannas of the southwest where rains are seasonal. The annual rainfall at Georgetown is 2400 mm.

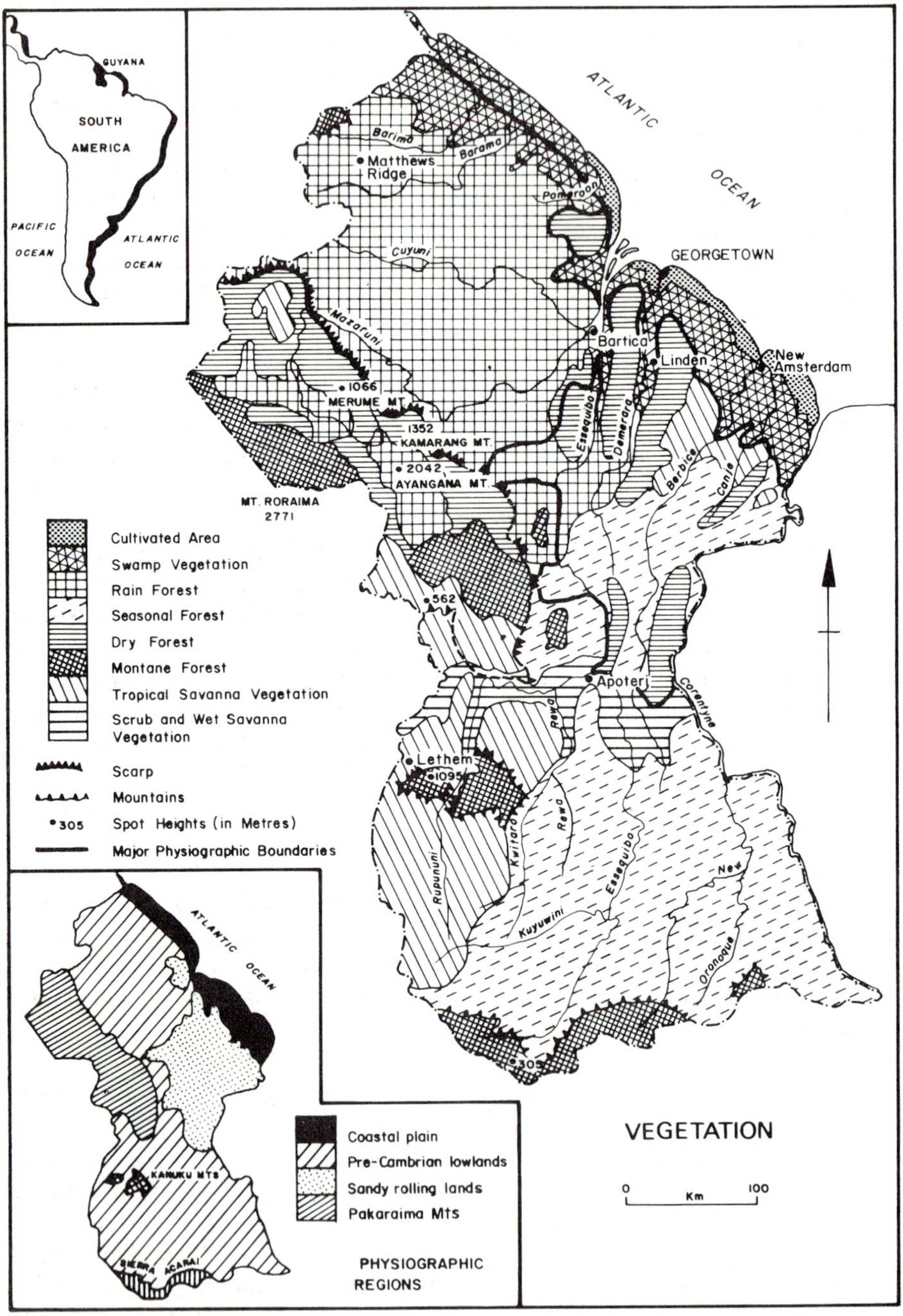


Fig. 1 Vegetation and physiography of Guyana.

Humid conditions and associated cloud formation commonly result in restricted hours of sunshine in Guyana. In the coastal zone, mean bright sunshine averages 7 hours per day, while in mountainous areas of the interior 5 hours per day are typical. The highest mean values occur in the Rupununi savannas, where 8 hours per day are recorded.

Guyana may be divided into four physiographic regions: the coastal plain, the sandy rolling lands, the Pakaraima mountains, and the Pre-Cambrian lowlands (Fig. 1).

The coastal plain is flat and contains fertile, alluvial soils. It extends from Punta Playa in the west to the Corentyne river in the east, a distance of 435 km. The width of the plain is 77 km in the west and 26 km in the east. Although this forms only a narrow strip along the coast, the area is an important agricultural zone. The land is about a metre below sea level at high tide and extensive drainage, irrigation, flood control, and sea defence facilities have been provided along much of the coast. The natural vegetation of the zone includes areas of both herbaceous and forested wetland, and the coastline itself is fringed by mangrove vegetation.

The rolling topography of the White Sands comprises an area to the south of the coastal plain. The sandy soils are porous and infertile, with limited agricultural potential. The natural vegetation ranges from evergreen forest in wetter areas to xeromorphic scrub and savanna.

The Pakaraima mountains, occupying the mid-western section of the country, present a series of plateaux and tablelands bounded by steep scarps and gorges. Horizontally bedded sandstones are present, with intruded dolerite sills. The fertile soils developed over the latter provide favourable conditions for the growth of rain forest, but on well-drained sandy soils a sparser forest cover prevails. On higher plateaux, stunted tree growth and grassland appear.

The Pre-Cambrian lowlands, underlain by crystalline basement rocks, form the largest physiographic unit in Guyana. The land surface is generally 90–120 m above sea level, with occasional peaks rising to 300 m in the north and to 1000 m or more in the south. Most of the area is covered by rain forest, which is part of the main Amazonian *hylaea*; in the southwest, the Rupununi savanna forms an extension of the larger Rio Branco savanna of Brazil.

## AERIAL PHOTOGRAPHY IN GUYANA

Guyana was one of the few developing countries that made use of aerial photography from a relatively early date. The chronology of photographic surveys in Guyana is given in Table 1. Three broad directions of survey can be recognized. In the earlier period, before the country gained independence in 1966, aerial photographs were taken primarily for military purposes. Subsequently, emphasis was placed on topographic mapping, with the major requirement being the small to medium scale photographic coverage of the whole country. In addition to

Table 1. Aerial photography of Guyana.

| Year | Photo Agency or Organization | Scale | Areas covered | Type of survey |
|---|---|---|---|---|
| 1940/41 | Keystone Aviation Co. (USA) | 1:60 000 | Lower reaches of rivers | Air photo (panchromatic) |
| 1950 | Air Survey Co. Ltd (England) | 1:30 000 | Coastal area | Air photo (panchromatic) |
| 1954 | Spartan Air Services (Canada) | 1:30 000 | Coastal area and parts of Northwest District | Air photo (panchromatic) |
| 1956 | Hunting Aero Surveys Ltd (England) | 1:30 000 | Coastal area and parts of Northwest District | Air photo (panchromatic) |
| 1959 | Fairey Surveys Ltd (England) | 1:30 000 | Coastal area and parts of Northwest District | Air photo (panchromatic) |
| 1959/60 | RAF | 1:60 000 | Entire country | Air photo (panchromatic) |
| 1962/65 | Geophysical Survey Co. (USA) | 1:50 000 | Parts of Northwest District | Aeromagnetic |
| 1966/69 | Geophysical Survey Co. (USA) | 1:50 000 | Parts of Northwest District and parts of southern Guyana | Aeromagnetic |
| 1967/70 | Terra Surveys (Canada) | 1:40 000 to 1:60 000 | Most parts of the country | Air photo (panchromatic and infrared) |
| 1971/77 | Geological Survey of Canada—Geophysics division (Canada) | 1:50 000 | Northwest coast and eastern parts of the country | Aeromagnetic |
| 1979/82 | BKS Surveys Ltd (England) | 1:20 000 | Coastal belt | Air photo (panchromatic and infrared) |

this, aerial photography at various scales was obtained for specific resource evaluation and development projects.

In 1940–41, the Keystone Aviation Company of America obtained aerial photographs of the lower reaches of all the rivers at a scale of 1:60 000 solely as a war exercise. Atkinson airfield near Georgetown, now a civilian airport, was established at this time. The Royal Air Force completed a coverage of the entire country at the same scale in 1959–60. These photographs were subsequently made available for civilian purposes (Kissoon, 1975), as for example in the FAO soil survey of Guyana (FAO, 1966). In the early period, some aerial photographs of particular parts of the country were obtained under contract with private companies and several areas were rephotographed on different occasions.

In the post-independence period, a new aerial photographic coverage of the country was acquired by Terra Surveys Limited of Canada. The company covered about 70% of the country at a scale ranging from 1:40 000 to 1:60 000. The contract also included the production of topographic maps at a scale of 1:50 000. These maps, which gradually became available during the 1970s, are reasonably accurate and have been used as base maps during the initial phases of many development projects. In addition, they have been widely used for the evaluation of natural resources.

The first application of aerial photography and other remote sensing techniques to resource evaluation was in respect of mineral exploitation. The ultimate objective of the programme was to assess the mineral potential of the country and to produce a detailed geological map. The first phase of the programme was carried out before independence. The Geophysical Survey Company of America was contracted to carry out airborne geomagnetic surveys of selected areas and to prepare aeromagnetic maps at a scale of 1:50 000. The second phase of the programme was undertaken in the early 1970s with the participation of the Geological Survey of Canada and Terra Surveys Limited of Canada. With the production of a further set of aeromagnetic maps at a scale of 1:50 000, more than two-thirds of the country had been covered by 1972.

Aeromagnetic maps and aerial photographs were extensively used in the production of geological maps in the difficult Amazonian rain forest areas and the Rupununi savannas to the south of latitude 4° N. The geological mapping programme was completed by the United Kingdom Directorate of Overseas Surveys in collaboration with the Institute of Geological Sciences of London in 1971.

A programme to assess the forest resources of Guyana was first carried out between 1966 and 1970 by the UNDP and the FAO in a joint effort with the Guyanese government (Welch *et al.*, 1972). Lack of base maps forced the surveyors to rely heavily on aerial photographs. In this study, the commercially productive forest area was divided into zones, and each zone was studied in detail. The basic aim of this project was to assess the frequency of occurrence of the rare hardwood greenheart (*Ocotea rodiaei*), and to examine the distribution of *dalli* (*Virola surinamensis*), a swamp tree used for a variety of commercial purposes. In recent years, medium-scale aerial photographs (1:20 000) were obtained to re-examine the forest resources of the areas that have been commercially harvested.

The Ministry of Housing and Planning has been using aerial photographs to compile and update large-scale plans of the major towns since the 1940s. The photographic coverage from different dates is an important historical resource in the study of urban change and development.

The coastal zone of Guyana has been photographed several times in the past. The most recent coverage of the coastal belt was concluded in 1982, following a delay of about three years. The project, aided by the United Kingdom

Directorate of Overseas Surveys, was contracted to B.K.S. Surveys Limited of Britain, who were to obtain panchromatic and infrared aerial photographs at a scale of 1:20 000. These have been used to evaluate the effectiveness of the irrigation, drainage, and sea defence projects that have been carried out in the coastal belt in recent years and to update the existing maps of the area.

## THE APPLICATION OF AERIAL PHOTOGRAPHY TO DEVELOPMENT PROJECTS

In Guyana, development projects that have had a major requirement for aerial photography can be considered in their geographic context—those of the coastal areas and those of the interior. The coastal projects have been related either directly or indirectly to agricultural developments on the coastal plain. The planning and execution of land development schemes, under which large areas have been brought into cultivation through drainage and irrigation, have relied upon aerial photography and photogrammetrically prepared topographic maps, at least during the initial phases. Other projects, such as the preparation of large-scale maps (1:10 000) of the coastal zone, the semi-detailed soil survey of the coastal and near-coastal areas, and the sea defence project helped the land development schemes indirectly. Special photographic missions flown over the coastal zone in the past and in recent times have also helped to set up and improve communication networks and town planning.

In the interior of the country considerable attention has been given to the possibility of generating hydroelectricity in sufficient quantity to meet the national demand. The feasibility study of the northwest district, carried out by the UNDP in 1975, relied upon existing aerial photographs and topographic maps for a general survey and upon specially flown aerial missions and specially compiled maps for a detailed survey of potential hydroelectric sites.

Other projects planned for the interior were of a more general nature and covered extensive areas. One such project was the survey of the mineral potential of the country, in which aeromagnetic surveying and aerial photographic interpretation techniques were combined. Similar methods were used in geological mapping and the compilation of the revised geology map of Guyana. The reconnaissance soil survey map and the land capability map, prepared during the FAO soil survey project of Guyana, were also largely based on aerial photographic interpretation.

### Aerial photography and topographic mapping

The production of topographic maps is of vital importance to the economic development of Guyana because in most development projects, at least in their initial phases, topographic maps are the most useful type of base map that is available. The most suitable maps for development projects are at medium scales

(1:50 000–1:100 000). In Guyana, all the maps in the 1:50 000 topographic series have been produced from aerial photographs and the coverage of the country is now complete.

Topographic and other maps are produced by the Department of Lands and Surveys, which is part of the Ministry of Agriculture. The Department is housed in a modern building, and is organized in three sections: photogrammetric, cartographic, and printing. The Department is relatively well equipped with modern stereoplotters and is capable of producing accurate topographic maps at a variety of scales.

Since there is a dearth of qualified personnel, the Department gives in-service training in photogrammetry and cartography. Department staff trained in the Netherlands at the International Training Centre at Enschede are responsible for the supervision of the trainees. Specialized additional training has been provided by the United Nations. Under this programme, experienced personnel from the United Nations have been involved in training young Guyanese photogrammetrists and cartographers. The training period for each batch of students lasts about six months, with a sequence of lecture courses, examinations, and practical experience. The successful trainees are admitted to either the cartography or photogrammetric sections of the Department.

**Aerial photography and soil survey**

One of the earliest projects in which aerial photographs were extensively used was the soil survey project carried out by the FAO in the early 1960s. Before the project began, only 20% of the country had soil maps. The project concentrated on the coastal and near-coastal areas because of their agricultural potential. However, a general survey of soils, at the reconnaissance level, was also carried out for the country as a whole (FAO, 1966).

The major objective of the semi-detailed survey was to explore further expansion of agriculture and also the development and conservation of forests. In addition, the survey helped to determine the optimum size of holdings and the appropriate crops for particular soil types. The areas that had diminishing crop returns were also identified, as a basis for conducting further research on soil fertility. Since aerial photographs were used, it was possible to collect detailed information on drainage, irrigation and land use as well as on soils.

Aerial photographs at a scale of 1:30 000 were used to identify the soil mapping units, which were initially delineated on the photographs. After field checking, the boundaries were transferred to topographic maps at a scale of 1:50 000. The final, semi-detailed soil maps, produced at a scale of 1:60 000, covered a total area of 827 600 ha in the coastal and near-coastal zone.

A reconnaissance survey of the soil types of the whole country was carried out to investigate the potential for agricultural development of the unknown or little-known areas. The boundaries of the great soil associations were delineated on photomosaics at a scale of 1:125 000. The accuracy of the boundaries was then ascertained in the field by spot checks and random sampling, and the boundaries were transferred to maps at a scale of 1:500 000. The final map was published at a scale of 1:1 000 000 and covered almost the whole country, although a few strips where photographic coverage was inadequate or non-existent were necessarily omitted. A land capability map covering all of Guyana was also prepared using standard photointerpretation methods. Both the soil and land capability maps were used as base maps for projects involving agricultural and forestry development in the interior of the country for several years because of the absence of topographic maps at that time.

**Aerial photography and hydroelectric power survey**

The largest single survey for which aerial photographs have been used both directly and indirectly was the hydroelectric power survey of Guyana carried out between 1974 and 1976. The survey was undertaken by the Montreal Engineering Company Limited (1976) under the United Nations Development Programme with funding from the World Bank. The survey had two main purposes—a reconnaissance and inventory of the hydroelectric resources of the country, and pre-feasibility studies of a limited number of sites with a view to developing low-cost power for mining, for an aluminium smelter, and for the electrification of the coastal zone. To meet these requirements a total power installation of 2610 MW with an energy capability of 1600 MW at four sites was recommended. The capital cost of such an undertaking was estimated at $1700 million, based on mid-1974 price levels.

Unfortunately, the hydroelectric power project was never undertaken because of increasing costs and the difficulty of obtaining funds. Nevertheless, the survey methodology is discussed in this review because it provides a good example of the way in which aerial photography is used in a reconnaissance survey of a particular resource.

At the preliminary reconnaissance level, the aerial photographs taken by Terra Surveys Limited between 1967 and 1970, at a scale of 1:40 000 to 1:60 000, were used. For the coastal areas, other available medium-scale photography was also examined. The level of detail required from the aerial photographs and any existing maps varied according to the level of survey, as indicated in Table 2. At the pre-inventory level, medium to small scale data sources were studied. Drainage basins were outlined and preliminary estimates of river gradient and discharge were obtained. Possible power sites were identified for more detailed study at a later stage. Reconnaissance flights were carried out to confirm the accuracy of the interpretation and to obtain additional information. Preliminary estimates of the cost of the project and the power potential of each site were made at the pre-inventory stage.

Table 2 The application of aerial photographs and topographic maps at different stages in the hydroelectric power survey of Guyana.

| Level of survey | Source materials | Map scale | Type of information required | Engineering use |
|---|---|---|---|---|
| Pre-inventory | Topographic maps (1:50 000) Aerial photographs (1:40 000) | 1:500 000 | River profiles and drainage areas; regional topography | Drainage area calculations; preliminary site selection; estimates of costs and power output |
| Inventory (phase 1) | Topographic maps (1:50 000) Aerial photographs (1:40 000) | 1:50 000 | General layout of sites; dam site identification; full and low supply level head ponds; tail water elevation | Cost estimates for each site; power potential estimate for each site; cost estimate for roads and power transmission lines |
| Inventory (phase 2) | Topographic maps (1:50 000) | 1:1 000 000 | Layout of project and powerline distribution network | Powerline route selection |
| | On-site survey | 1:5000 to 1:63 000 | Project plans | Details of dam sites, roads, rock fills, spillways, power house, etc. |
| | On-site survey | 1:10 000 to 1:20 000 | Geological map—lithology and structure | Drilling, blasting, and foundation engineering |
| Pre-feasibility | Topographic maps (1:50 000) modified by ground surveys | 1:20 000 | General plan of dam sites | To estimate costs more accurately |
| | Horizontal and vertical control | 1:50 000 to 1:120 000 | Reservoir plan | Planning shoreline modifications |
| | Horizontal and vertical control points | 1:500 to 1:5000 | Main dam site plan | Location of tunnels and other underground facilities |

At the inventory level, the same sources of information were used, but a more exhaustive analysis was carried out. Each drainage basin was assessed and all sites with a power potential of more than 15 MW were identified. Route selection was undertaken for roads and power transmission lines from these sites and cost estimates were made. At the most promising sites additional information was obtained from field survey of topography and both surficial and bedrock geology. Contouring was done at 10 ft (3.05 m) intervals using the best available photography in conjunction with ground control.

At the pre-feasibility level, more detailed information about individual sites was required. Dam sites were surveyed and the extent of head ponds was determined. The existing 1:50 000 topographic maps were found to be inadequate for this purpose, because of their relatively large contour interval—50 ft (15.24 m)—and also because of the errors introduced by the dense vegetation. In areas of rain forest, the operator of a stereoplotting instrument engaged in contouring has no opportunity for adjustment of the floating mark to ground level because the vegetation is generally unbroken. In such circumstances, contouring was carried out at canopy level, and adjustments were based on the best estimate of tree height. The contour accuracy is probably of the order of ±25 ft (7.6 m), but in some areas the deviation could be considerably greater, and may exceed 50 ft (15.25 m). At the pre-feasibility level, vertical control for the sites under study was provided by altimeter survey using helicopters and fixed-wing aircraft, where these could land at locations close to the sites. Horizontal control was established using the Aerodist Trilateration Net system.

### Aerial photography and special projects in the coastal belt

The bulk of the population of Guyana (over 90%) lives on the narrow coastal plain. The major agricultural crops, sugar and rice, are produced in this zone. The area requires good drainage and sea defence to be able to maintain its agricultural productivity because it is flat, and many sectors lie up to one metre below sea level (Fig. 2).

Several agricultural and sea defence schemes have been undertaken in this area, and several more are under investigation. Because of its vital importance to the economy of the country, this area has been frequently photographed since colonial times. Accurate topographic maps are available for the whole coastal zone at a scale of 1:50 000. The map sheets have been revised to include recent developments in drainage and irrigation. At present, large-scale maps (1:10 000) are being compiled from recent aerial photographs and detailed plans at a scale of 1:5 000 have been prepared for special project areas.

The first major land development scheme, dating from 1957, was the Black Bush polder, where 12 545 ha of land were drained and brought into cultivation. This was closely followed by the Boerasirie extension project where 52 610 ha of land were made available for cultivation. Another project of importance was

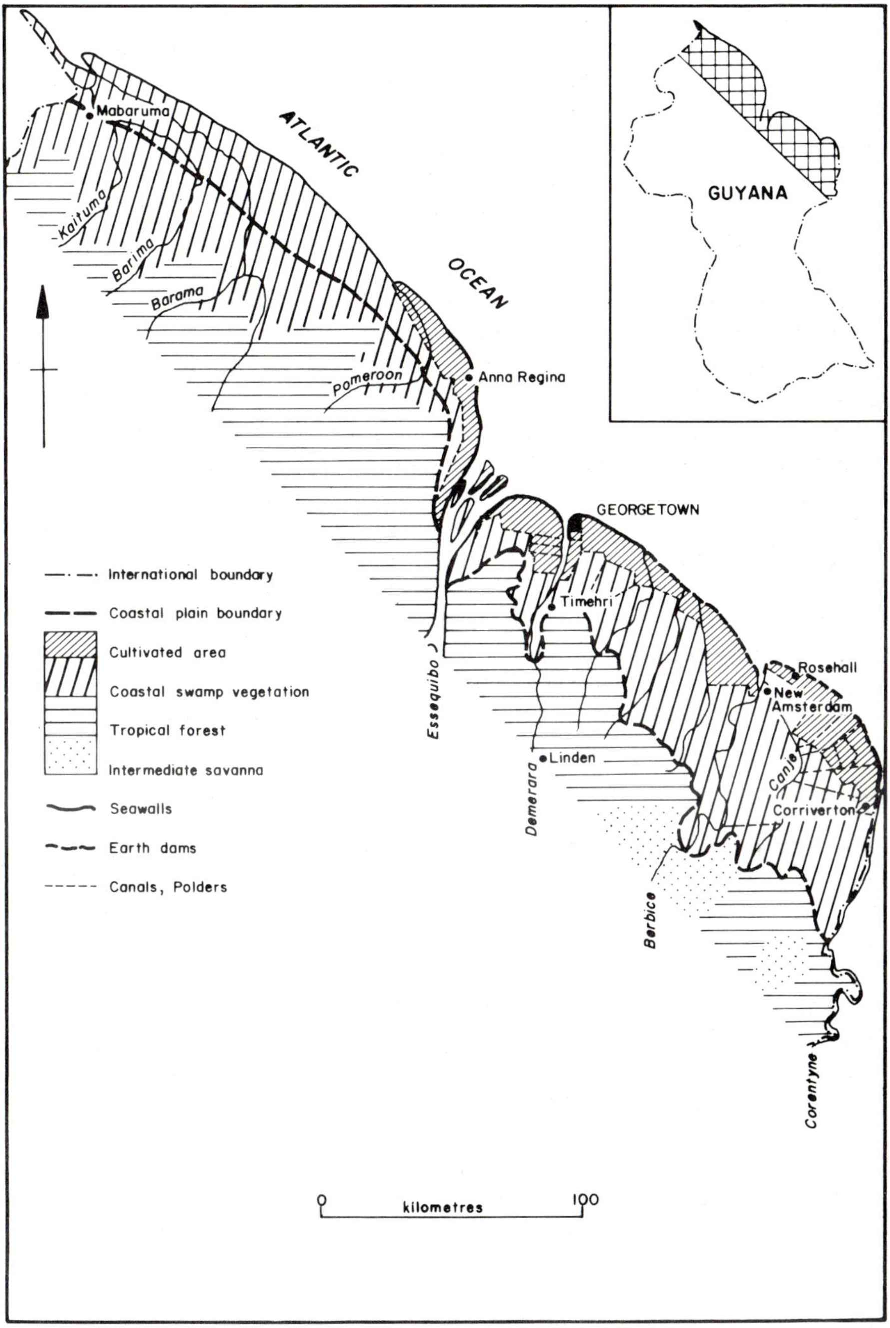


Fig. 2 Coastal plain of Guyana.

the Tapakuma project in which 14 570 ha of land were reclaimed for cultivation. Several other schemes are currently at various stages of development. Of these, the Mahaica Mahaicony Abary project and the Canje Reservoir project are the most advanced.

These development schemes require large scale topographic maps. In addition, specific maps are required showing soil, vegetation, land use, engineering works, and settlement detail. For planning engineering works, such as canals, drains, gates, and other structures, planimetrically accurate maps with a small contour interval are essential. Such maps are also required for planning the layout of building and access roads. The preparation of maps of this type requires large scale aerial photography and ground survey for horizontal and vertical control. These maps are often used as base maps for other specific surveys and for cadastral purposes, involving the issue of land titles and the delineation of administrative boundaries.

The sea defence projects are mainly based on surveys of the near-shore zone. Soundings off the coast of Guyana are carried out on a continuing basis and the movement of mud banks is monitored from year to year. The location of these banks in the near-shore zone influences the rate of erosion and deposition along the coast. Maps prepared from sequential aerial photography clearly show the changing details of the coastline. The dynamic nature of the coastal zone necessitates constant maintenance of sea defence works. Large scale maps prepared from aerial photographs are of immense importance for this purpose.

## LIMITATIONS ON THE USE OF AERIAL PHOTOGRAPHY IN GUYANA

Environmental conditions in the humid tropics impose some very severe restrictions on aerial photography and on the usefulness of photographs in providing information for major development projects. The main environmental problems are cloud cover and vegetation cover.

There is always the problem of obtaining cloud-free aerial photographs in the humid tropics because of climatic conditions. For photogrammetric purposes it is essential to obtain aerial photographs with the principal point unobscured by cloud and at least 80% of the image area cloud-free. It is very difficult to meet this requirement in tropical regions unless the aircraft remains below cloud level. The recent attempt to photograph the entire coastal strip of Guyana provides an excellent illustration of the cloud cover problem. The photographic mission, first planned for 1979, had to be postponed for three consecutive years because of persistent cloud cover. As a result, the revision of the large scale maps of the coastal zone had to be delayed.

Tropical vegetation can almost completely obscure ground conditions, and so the aerial photograph is of considerably less value as a data source for interpreting geology, landforms, soil conditions, and even topography in the rain forest than in other environments. In general, there is an inverse relationship between the density of the vegetation and the accuracy of maps prepared from aerial photographs. For example, a comparison of the accuracy of the contours on map sheets from different vegetation zones in Guyana shows that in the open savanna of the southern part of the country the maximum error in contour elevation was ±25 ft (7.6 m), whereas in the rain forest region the accuracy of the contours was reduced to ±50 ft (15.2 m) or more (Montreal Engineering Co. Ltd, 1976).

Very similar effects are found when other elements of topographic detail are considered. In an analysis of drainage patterns the author found that the drainage density measured on aerial photographs of the savanna area was accurate to within about 4% of the actual value, whereas in rain forest areas errors of as much as 33% were not uncommon (Daniel, 1981).

Careful interpretation of large scale aerial photographs of the rain forest can reveal some useful information about ground conditions, but considerable local knowledge is necessary, and the linkage between the surrogate features of the vegetation and the actual edaphic controls must be established by field reconnaissance. In Surinam, fairly detailed forest type maps have been prepared from medium scale (1:40 000) aerial photography, and it has been shown that certain vegetation features can be used as indicators of particular ground conditions (Zonneveld *et al.*, 1952; Heinsdijk, 1953a and b, 1955; Dillewijn, 1957). For example, rain forest dominated by *wallaba* (*Eperua* spp.) always occurs in Guyana on the unconsolidated parent materials of the White Sands series. These areas are generally characterized by gently rolling topography of low relief. At the reconnaissance level it would be possible to produce a map of physiographic units by careful interpretation of the forest vegetation. Areas with different relief and drainage characteristics can be delineated and certain aspects of the soil and lithology can be described, depending on the linkages that have been established between particular tree species and their site preferences. Analysis of this type demands a very high level of technical skill and experience in photographic interpretation, an expertise that is not readily available in many developing countries.

In addition to the problems posed by the tropical environment, there are those resulting from the economic situation in the developing world. In many poor countries, limited funds, shortage of trained personnel, and the problem of coordinating physical and economic planning have resulted in less than optimum aerial photographic coverage. National coverage has only infrequently been acquired in Guyana, and the areas covered by special film types, such as colour and colour infrared, are very limited. Part of the reason for the episodic and patchy nature of the coverage is the lack of an indigenous aerial photographic capability and the high cost of chartering aircraft and aerial cameras from private survey firms in Europe or North America. The delays encountered is calling tenders for an aerial survey and in receiving loans from institutions or governments for such a survey often seriously jeopardize the success of development projects.

Since aerial photographs of the required quality and scale are not always available at the initial stages of a project, much reliance is placed on topographic maps. For these maps to be useful, they have to be revised and updated at regular intervals. The Lands and Surveys Department responsible for this work is faced with many problems that are all too common in developing countries. Firstly, there are inadequate funds to purchase photographic, printing, and cartographic materials. Since most materials have to be imported, map production is impaired whenever foreign exchange required for the purchase of materials cannot be obtained. Secondly, the difficulty of retaining trained technicians in Guyana, particularly in the photogrammetric section, frustrates the department's ability to improve or even maintain its standards. Most of the recruits trained by the department seek employment abroad. As a result, new recruits have to be employed who lack advanced training and experience.

## THE PROSPECTS FOR OTHER TYPES OF REMOTE SENSING IN GUYANA

Within the last decade, developing countries have had the opportunity to assess the potential benefits of the more sophisticated types of remote sensing, including orbiting spacecraft such as Landsat. Faced with the dilemma of whether or not to become involved in the new technology, developing countries have been forced to take a hard look at their priorities. The nature, range, and level of detail of the resource information required for development varies from one country to another. Several factors set practical limits to the amount of resource information a country can acquire and use, including the size and geographical setting of the country, the level of resource information already available, the capacity of the infrastructure to assimilate the information and make effective use of it, the cost of acquiring the information, and the development objectives of the country.

Guyana is a small country with a land area slightly less than that of the United Kingdom. Over 84% of the area is tropical rain forest and, as noted above, there is a persistent cloud cover. The present generation of spacecraft remote sensing devices cannot penetrate cloud cover, and even when cloud-free imagery is obtained relatively little useful information can be derived from a small scale view of the rain forest canopy. Satellite images cannot be enlarged beyond a certain limit and so their use for detailed study of small areas is limited. Although the theoretical resolution of the Landsat multispectral scanner is 79 m, improving to 30 m with the thematic mapper on Landsat 4, the effective resolution that is achieved in routine operation is two to three times worse than the theoretical. The effective resolution is further reduced if the features in the area imaged have low reflectance values and limited contrast.

There is an inverse relationship between the usefulness of the small scale, synoptic view provided by Landsat and the level of resource information already available for the area. In spite of many problems, Guyana is relatively well-mapped for a tropical country, and resource surveys of various types have been completed in the last decades. Landsat imagery has less to offer than in countries where base maps are entirely lacking and the resource base is unexplored. Guyana has an established topographic survey with the technical ability to integrate Landsat imagery into its mapping programme at the planimetric level. At the present time, however, no convincing arguments can be sustained for a greater involvement in the Landsat programme. The benefits to be derived from an investment in the hardware, software, and personnel training required for a digital image processing and enhancement facility appear to be minimal. The problems of establishing and maintaining a facility of this type are well exemplified in the case of Brazil, a country with a much more elaborate infrastructure than Guyana. Brazil has good reason to invest in the Landsat programme. Its territory is nearly forty-two times larger than that of Guyana, and much of it is unmapped. Brazil was one of the first developing countries to have its own Landsat receiving station. Financial assistance and technical advice were available from the United States, and yet Brazil experienced all the problems and delays that beset developing countries when advanced technology is introduced (Hammond, 1977).

## CONCLUSION

An examination of recent literature on remote sensing indicates that conventional aerial photography is still widely used in many countries, particularly in the developing world. The main reasons for the continuing reliance on the aerial photograph as an essential data base for resource analysis lie in its proven worth, its versatility, and its low cost. In many development projects, at least during the reconnaissance stage, aerial photography is the only sensing system that offers a suitable image scale and sufficient resolution for the identification of ground control and the interpretation of surface detail (Table 3).

Although the operational use of non-photographic sensing techniques has been demonstrated in technologically advanced countries, conventional aerial photography is likely to remain the most important remote sensing procedure in developing countries for many years to come. The problems limiting the use of aerial photography, such as persistent cloud cover and the presence of dense vegetation, particularly the rain forest, are common to all sensing systems operating in the visible wavelengths. Side-looking airborne radar offers the best prospects for a major breakthrough in the analysis of resources in the equatorial region. The SIR A radar swaths obtained by the Space Shuttle have offered a glimpse of what the future may hold. For the present, many developing countries, such as Guyana, set greater store on solving the problems of financing

Table 3. The applicability of different types of remote sensing at various stages in engineering projects.

| Project phase | Required scale | Terrestrial photogrammetry | Aircraft | | | | Spacecraft | |
|---|---|---|---|---|---|---|---|---|
| | | | Photo | MSS | TIR | SLAR | Photo | MSS |
| Preliminary | 1:10 000 to 1:50 000 | × | ■ | × | × | ○ | ×–○ | ○–● |
| Reconnaissance | 1:1000 to 1:10 000 | ○ | ■ | × | × | × | × | × |
| Construction—main phase | 1:100 to 1:1000 | ● | ○–● | × | × | × | × | × |
| Construction—special projects | 1:100 to 1:1000 | × | × | × | × | × | × | × |
| Maintenance and monitoring | 1:100 to 1:1000 | × | × | × | × | × | × | × |

× inadequate, ○ limited, ○–● useful, ● very useful, ■ essential.
MSS: multispectral scanning system, SLAR: side-looking airborne radar, TIR: thermal infrared.

aerial photographic missions, training photogrammetrists and photointerpreters, and obtaining materials and supplies for photographic processing and map production.

## ACKNOWLEDGEMENT

The author wishes to thank Ms Claudette Foo, in the Department of Geography, University of Guyana, for cartographic assistance.

## REFERENCES

Daniel, J. R. K., 1981. Drainage density as an index of climatic geomorphology. *J. Hydrol.*, **50**, 147–154.

Dillewijn, F. J. van, 1957. *Sleutel voor de Interpretatie van Begroeiingsvormen uit Luchtfoto's 1:40,000 van het Noordelijk Deel van Suriname*. Paramaribo: Deenst's Landsbosbeheer Suriname.

FAO, 1966. *Report on the Soil Survey Project, British Guiana*, 5 vols. Rome: Food and Agriculture Organization.

Hammond, A. L., 1977. Remote sensing in Brazil. *Dev. Dig.*, **15**(3), 9–16.

Heinsdijk, D., 1953a. Begroeiing en luchfotographie in Suriname. *Publ. v.h. Centraal Bureau Luchtkaartering te Paramaribo*, **12**, 1–19.

Heinsdijk, D., 1953b. Bosbouwkundige fotointerpretatie. *Publ. v.h. Centraal Bureau Luchtkaartering te Paramaribo*, **13**, 1–19.

Heinsdijk, D., 1955. Forest type mapping with the help of aerial photographs in the tropics. *Trop. Woods*, **102**, 27–46.

Kissoon, O. M. L., 1975. Mapping in Guyana since 1940: a review of its relevance to national development. *Occ. Pap., Dep. Geog. Univ. Guyana*, **5**, 1–84.

Montreal Engineering Company Limited, 1976. *Hydroelectric Power Survey of Guyana*, Final Contract Report for the International Bank for Reconstruction and Development. Montreal: Montreal Engineering Company Limited.

Welch, I. A., Sampson, O. R. and Bell, G. S., 1972. Photo interpretation key to the vegetation types of Guyana. *For. Bull. Guyana*, N.S. **4**, 1–59.

Zonneveld, J. I. S., Cohen, A., Eijk, J. J. and Beltman, B. J., 1952. The use of aerial photographs in a tropical country (Surinam)—a symposium. *Photogramm. Engng*, **18**, 144–168.

Remote Sensing and Tropical Land Management
Edited by M. J. Eden and J. T. Parry


# CHAPTER 4

# *SLAR Imagery for Soil Mapping and Regional Planning in Western Nigeria*

AYODELE FAGBAMI and AKIN FAPOHUNDA
*Department of Agronomy, University of Ibadan, Nigeria*

The value of Landsat imagery for reconnaissance soil mapping in the cloud-free regions of West Africa has been demonstrated in several recent studies (Bleeker, 1978; Mainguet *et al.*, 1978; Fagbami, 1980). In the more equatorial regions, side-looking airborne radar (SLAR) provides a more effective tool because of its ability to penetrate cloud and haze (Fagbami, 1981a and b; Fagbami and Olaniyi, 1982). Preliminary studies, using SLAR imagery, were first undertaken in the eastern Niger delta region of Nigeria, and the results indicated the desirability of a similar exercise to the west of the Niger river. The work reported here is from the latter area, namely, Bendel State. It represents a further application of remote sensing to reconnaissance soil mapping in Nigeria, and is a contribution to the more general objective of producing a national soil map. The present findings are formulated in terms of landform and land type units, for each of which information on soil conditions is provided. Additional data are presented for individual units in respect of vegetation and land use. Finally, a broader evaluation is made of the land use potential of the region. The absence of any existing, comprehensive soil map of Bendel State makes the current project of particular importance, and of relevance for regional planning in the area.

## THE ENVIRONMENT

Bendel State lies between 5°00′ N and 7°00′ N, and covers an area of 38 293 km$^2$. It is bounded on the west by Ondo State and to the north by Kwara State; to the east lies the Niger river and to the south the Atlantic ocean.

## Climate

Bendel State has a hot, humid climate that displays the bi-modal rainfall pattern that is characteristic of the southern part of West Africa. The coastal area of the State has the highest rainfall in the country, with an annual total at Forcados of 3690 mm; the total decreases northward, falling to 1322 mm at Auchi. Rainfall seasonality also displays a south-north gradient; the south has two main dry months in the year (December and January), while the central part of the State has three (December to February), and the northern part four (November to February). The short, dry period in August, which is associated with the passage of the Inter-tropical Discontinuity, is more marked in the south than in the north. The rainfall maxima occur in July and September in the southern and central parts of the State, and in June and September in the north.

Table 1. Mean hourly sunshine per day, January to December at Warri, Bendel State, Nigeria.

| | Mean hourly sunshine per day |
|---|---|
| January | 4.2 |
| February | 4.8 |
| March | 4.4 |
| April | 5.2 |
| May | 5.4 |
| June | 3.9 |
| July | 2.4 |
| August | 2.0 |
| September | 2.7 |
| October | 4.0 |
| November | 5.6 |
| December | 5.4 |

The mean monthly temperature is relatively constant throughout the year, ranging from 25–28°C. Mean monthly sunshine hours at Warri show a relatively high incidence of overcast conditions (Table 1), although sunshine hours are probably somewhat higher in northern parts of the State. The months of the rainy season experience the greatest cloud cover.

## Geology

Sedimentary rocks underlie 90% of Bendel State. They range in age from Tertiary in the north to Holocene in the coastal mangrove swamps and the Niger floodplain. The Niger delta is composed of both fluviatile and marine sediments

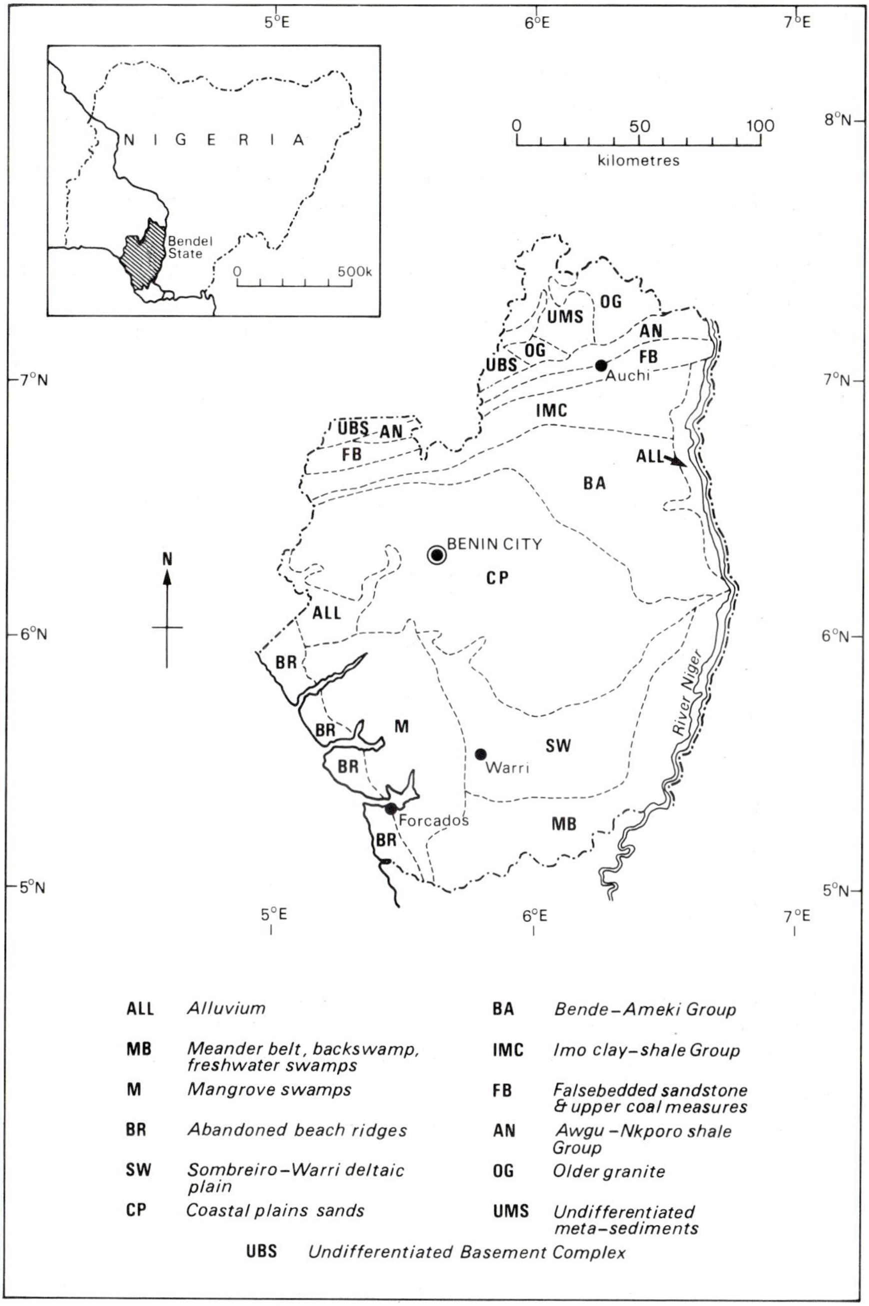


Fig. 1 Geology of Bendel State, Nigeria.

which have accumulated since the Upper Cretaceous period (Anderson, 1967). The Pre-Cambrian crystalline Basement Complex occupies the northern margin of the State (Fig. 1).

### Vegetation

Bendel State has a narrow, northern belt of derived savanna. To the south lies an area of rain forest, although in many places the latter has been degraded to secondary forest as a result of shifting cultivation. The southern coastal area is under tidal influence and distinctive mangrove communities have developed. The mangrove zone is broadest in the delta, where the tidal influences bring saline waters farthest inland.

## SLAR IMAGERY AND INTERPRETATION PROCEDURES

SLAR coverage of Nigeria was acquired in 1977 by Motorola Aerial Remote Sensing Inc. of the United States. The AN/APS-94D SLAR is an X-band (3.2 cm wavelength) real aperture system with HH polarization. Dual antennas simultaneously acquire images on both sides of the flight path. The swath width is 25 km at a range delay of 20 km from the flight path. Resolution varies with range. In the near range the system can resolve objects with a 30–35 m spacing, while in the far range the resolution is reduced to more than 100 m. The capability of the system to acquire imagery from both sides of the flight path resulted in the double coverage of each ground scene at a scale of 1:250 000.

The imagery was acquired in the first instance to facilitate the production of a comprehensive forest inventory map of Nigeria (FDF 1978), but is here employed for the purpose of reconnaissance soil mapping. Eight mosaic sheets covering Bendel State have been examined and interpreted, using a Fluorescent-illuminated Magnifier (Forest Supplies Inc.). A detailed interpretation of the drainage characteristics and the land units of the area was carried out. The analysis of drainage pattern was supplemented by data extracted from topographic maps. Differences in drainage density and texture (Fig. 2) provide information on geological and morphogenetic conditions and by extension this information can be used in the examination of soil conditions. The analysis of land units was based on the physiographic approach outlined by Buringh (1960), Goosen (1967), and Vink (1968), which requires an integrated assessment of drainage pattern, relief, and vegetation. The evaluation of relief on the radar imagery is a critical step in the identification of major landform units, especially in areas of moderate to high relief in the central and northern parts of the State. Vegetation characteristics, evident in the grey tone and texture of the imagery, were employed for subdividing landform units into their component land types. The resulting pattern is shown in Fig. 3.

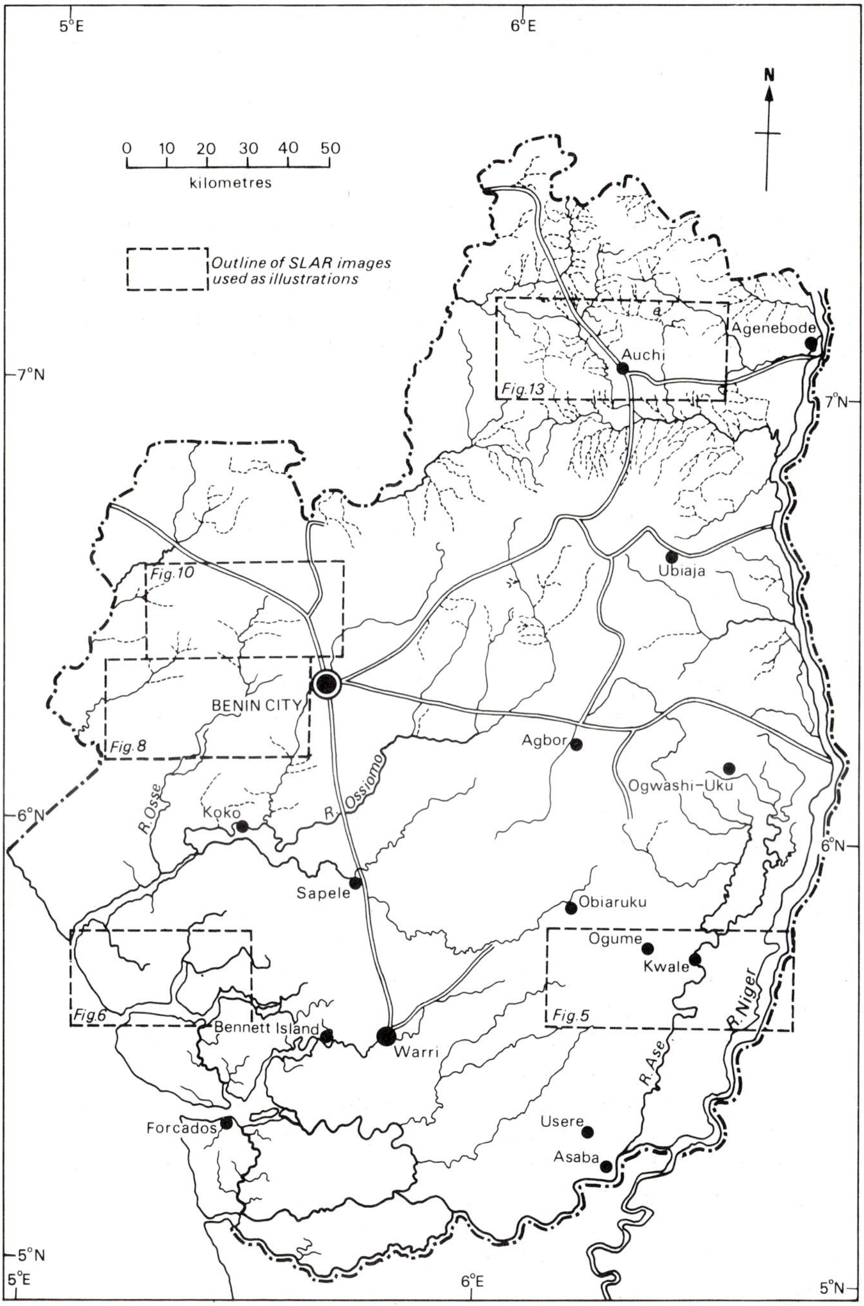


Fig. 2 Drainage pattern of Bendel State and location of sample SLAR images.

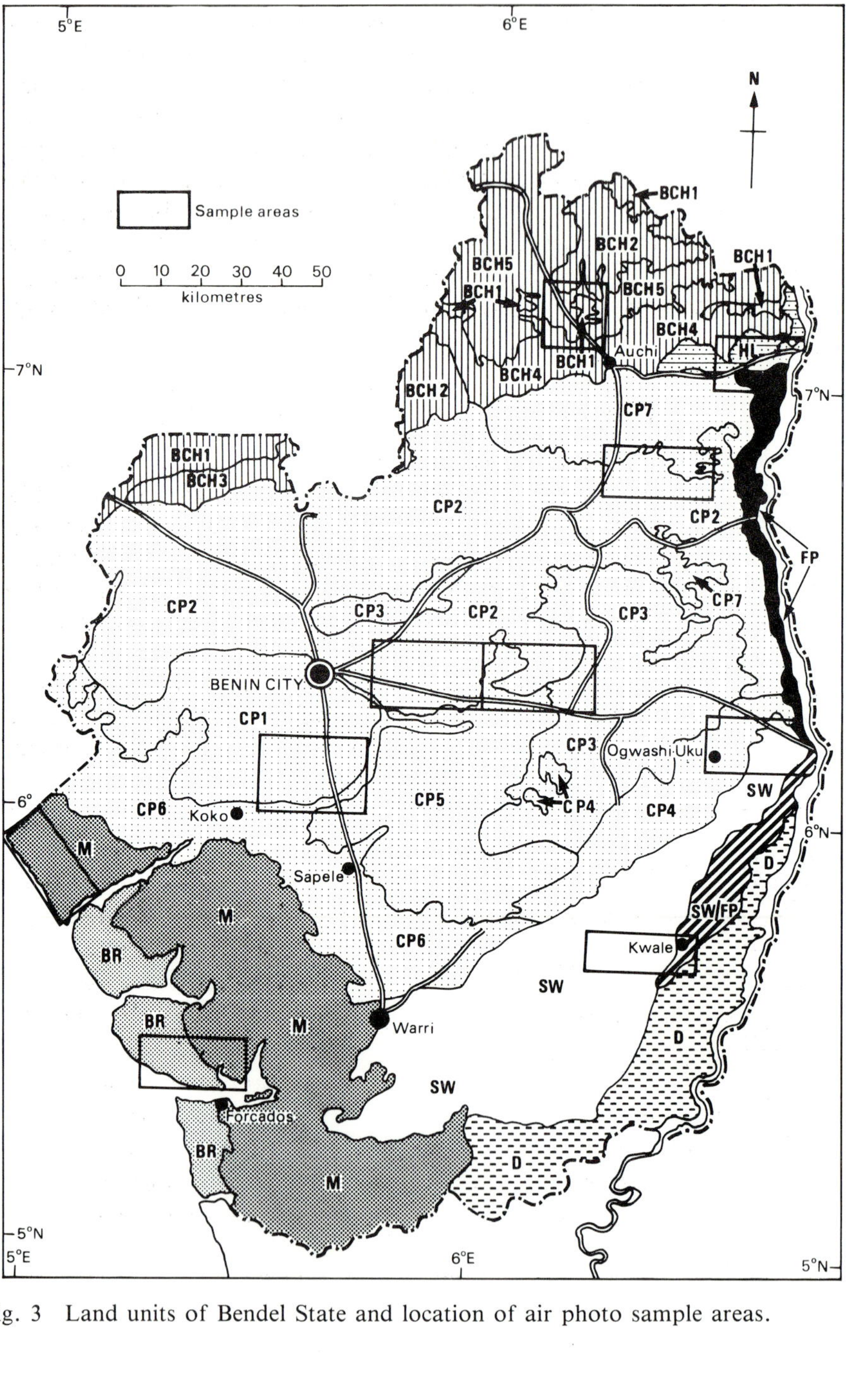


Fig. 3 Land units of Bendel State and location of air photo sample areas.

Having mapped the land units from the SLAR imagery at a scale of 1:250 000, aerial photographs were employed for checking the results at a local level as a basis for more detailed investigation of specific areas. For this exercise, blocks of aerial photographs, comprising 10 to 18 stereo pairs, were independently interpreted for ten sample areas (Fig. 3). The photographs were at scales of 1:25 000 or 1:40 000. Although it was not practical to use aerial photographs for mapping the whole area of Bendel State, their local use was of value in checking the validity of the initial SLAR interpretation. In particular, an increased understanding of the relationships between landforms, soils, and vegetation was gained from the photographs, which, as a function of scale, also provided a more favourable basis than SLAR imagery for field checking.

## FIELD CHECKING AND VERIFICATION OF SLAR IMAGERY INTERPRETATION

The collection of ground truth included data on morphology, soils, vegetation, and land use. The initial task of fieldwork involved the local checking of land unit maps along selected road or footpath transects within, and to some extent outside, the sample areas. In wetland areas, field checking was accomplished by over-flying inaccessible mangroves in a helicopter or by boat trips through the creek and lagoon network. Partially accessible areas at the coast were traversed by motor cycle or by trekking.

Within each land unit identified in the sample areas soil conditions were investigated. Soils were first examined at auger holes or mini-pits sited at 250 m intervals along survey transects; this provided an indication of the component

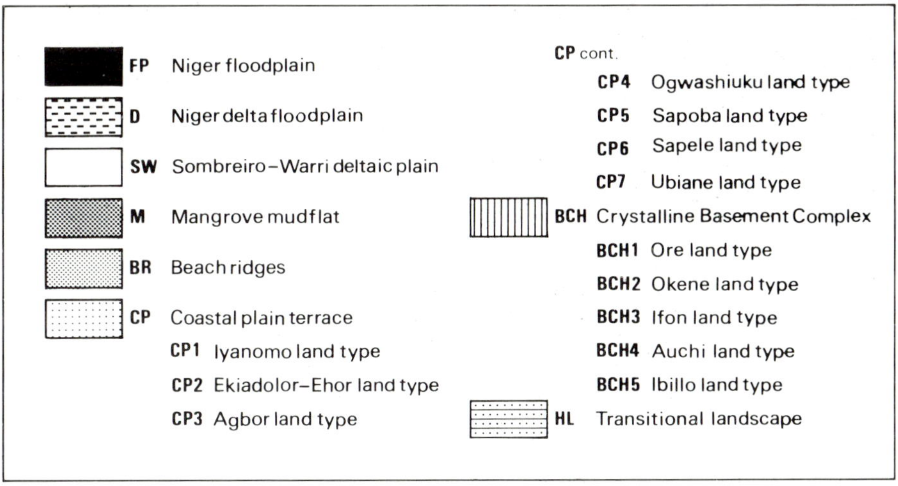


Key to Fig. 3

soil types within each land unit. Thereafter, modal soil profiles were dug and described, according to the FAO guidelines for soil profile description (FAO, 1974). General terrain conditions at each site were also recorded, including rockiness, stoniness, slope, vegetation and land use. Altogether some 600 auger profiles and mini-pits were examined across Bendel State, and 95 soil pits were excavated.

Soil samples were collected and analysed for particle size, pH, exchangeable cations, base saturation, organic matter and available phosphorus.

## RESULTS AND DISCUSSION

### Description of land units

The drainage systems and land units mapped from the SLAR imagery are shown in Figs. 2 and 3. Seven major landform units have been recognized, namely (1) the Niger floodplain, (2) the Niger delta floodplain, (3) the Sombreiro-Warri deltaic plain, (4) the mangrove mudflat, (5) the beach ridges, (6) the coastal plain terrace, and (7) the crystalline Basement Complex. Within the landform units, subdivisions known as land types have also been designated. These units are identified in the following pages and their soil characteristics are described. Additional information on the vegetation and land use is provided.

#### *Niger floodplain (FP)*

The upper limit of the Niger floodplain is encountered at Lokoja, which is situated to the north of the study area. The floodplain extends over a distance of some 300 km to a point a few kilometres south of Kwale. Along its course the Niger river flows across both Basement Complex and sedimentary rocks, and this is reflected in the character of the valley. To the north, the valley is narrow and rocky, while in the sedimentary zone below Agenebode the valley widens considerably. In the latter area, the floodplain is characterized by alluvial deposits which are uniformly sandy in texture and partly covered by dense rain forest. The soils are mainly Entisols, with variable drainage that reflects their proximity to the river.

To the south of Agenebode, yams are cultivated on a large scale and are the most important crop in the area. Cassava, maize, swamp rice, and sweet potatoes are also cultivated. The floodplain soils are moist throughout most of the year and, in order to avoid the annual Niger floods in August and September, crops are planted earlier than in surrounding areas. As a result yams and maize appear in local markets from late May.

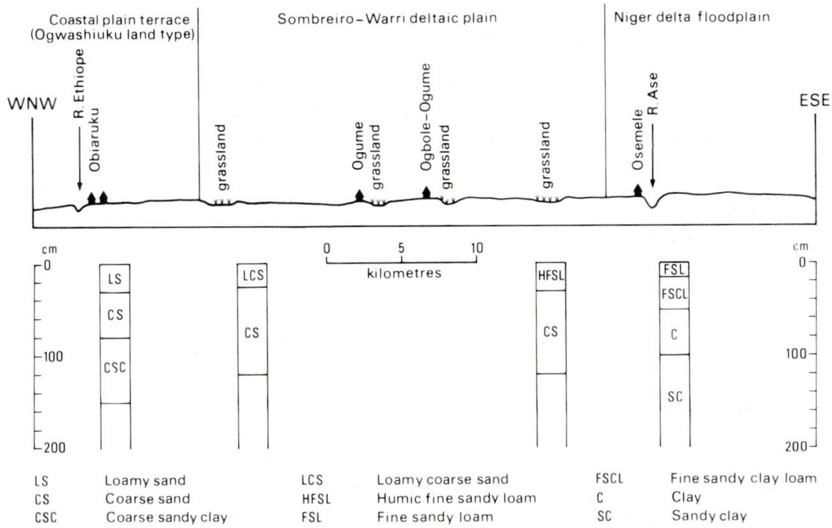


Fig. 4 Cross section of land types in southeastern Bendel State showing the relation between topography and soil texture.

### *Niger delta floodplain (D)*

To the east of the Ase river, there is a dense network of Niger river distributaries that meander across the delta floodplain. A cross section through this land unit and the adjacent Sombreiro-Warri deltaic plain is shown in Fig. 4.

The unit is covered by dense tropical forest, with *Terminalia* sp., *Albizia* sp., and *Elaeis guineensis* as the main species. *E. guineensis*, the oil palm, commonly reaches 30 m or more in height in the forest, making harvesting of its fruit very difficult. *Eupatorium odoratum* and the grass *Andropogon* sp. often establish themselves in fallow areas along the rivers.

Topsoils in the delta floodplain are commonly dark brown to black in colour. They consist of humic loamy sand to fine sandy loam, with weak to moderate subangular blocky structure. Subsoil horizons are mottled grey in colour and the texture is sandy on levee ridges and sandy clay to clay on levee slopes. Basin soils are very clayey. The coarser-textured delta floodplain soils are Ultisols, while the finer-textured soils of the levee slopes and basin areas are Inceptisols with an aquic moisture regime.

The Niger delta floodplain is intensively cultivated for maize and root crops (mainly cassava and yams) for which the soils are generally suitable. Rubber is the only significant tree crop in the area, although fewer trees are planted close to the Ase and Niger rivers because of the high water table.

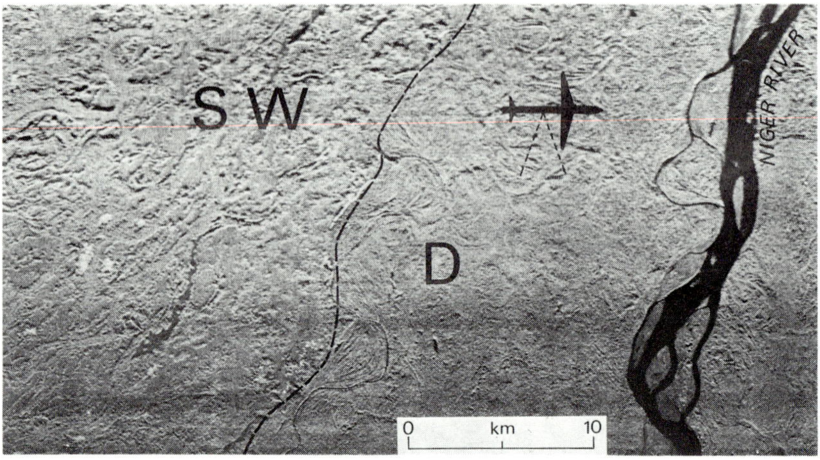


Fig. 5 Land units in southeastern Bendel State. D: Niger delta floodplain, SW: Sombreiro-Warri deltaic plain. AN/APS-94D SLAR image, February 1977. Approx. scale 1:250 000

*Sombreiro-Warri deltaic plain (SW)*

Drained by the Ase river, the Sombreiro-Warri deltaic plain commences about 10 km north of Asaba; it extends in a south-westward direction to the vicinity of Warri and Usere.

In the west, the plain is nearly flat, with slopes of $<2°$. The slopes are long, and have a very low density of flowing streams. However, a dense network of fossil stream channels is evident on the radar imagery (Fig. 5), creating a distinctive mesorelief pattern in the landscape. To the east, the land slopes down towards the Ase river and, at the boundary with the Niger delta floodplain, slopes are approximately 2–4°. In general, this area remains above the Niger floods.

The soils of the Sombreiro-Warri plain are dark brown loamy sands at the surface, with yellowish-red to red subsoils of sandy clay loam to sandy clay texture. The soils are moderately well drained to well drained. Strong leaching has occurred and is responsible for the low base saturation of the soils, which are mostly classified as Ultisols.

The Sombreiro-Warri plain is mainly covered by oil palm forest (*Elaeis guineensis*). Much of the primary forest has been cut down, and there has been extensive secondary forest regrowth. Grass-dominated areas occur along some of the fossil stream channels. In places, the plain is intensively cultivated for cassava, yams and maize, for which the soils are quite suitable. Rubber is the main tree crop in the area, although few trees are currently being planted near the Ase river on account of the rising water table.

*Mangrove mudflat (M)*

The mangrove mudflat covers 11% of the area of Bendel State and is a maze of anastomosing drainage channels. Occasional lakes are present, which may be 200–300 m wide and up to 2.5 km in length. In general, the area displays a simple depression-ridge-depression topography, created by the numerous distributaries of the Niger, Benin, and other rivers that flow from the hinterland.

An area examined on Bennett island, near Warri is illustrative of the unit. At the margin of the creeks, the levee is covered by very tall mangrove, beyond which the ground gradually slopes into a basin area that is under water for most of the year and contains short mangrove. The tall mangrove community contains relatively few species, principally red mangrove (*Rhizophora racemosa*), white mangrove (*Avicennia africana*), and salt water fern (*Acrostichum aureum*). *Rhizophora racemosa* occurs in the most saline areas near the sea, while *A. africana* grows in the less saline transitional zone. In individual sites, red mangrove (*R. racemosa*) may be succeeded by salt water fern (*A. aureum*) or white mangrove (*A. africana*). In the short or stunted mangrove of the basin areas, *Rhizophora mangle* and *R. harrisonii* are dominant, although the canopy is more open. *Acrostichum aureum* occasionally appears here among the mangrove species. The short mangrove gives way to a tall mangrove community as another levee is reached, and so the alternation goes on across the mangrove zone, as indicated in Fig. 6. Farther inland, less saline mudflat areas grade into rain forest, with a conspicuous abundance of *Raphia* sp. and *Elaeis guineensis*. The following species were also encountered in forest areas: *Chrysobalanus ellipticus*, *Dalbergia ecastaphyllum*, *Ixora* sp., *Nitragyna ciliata*, and *Paullinia pinnata*.

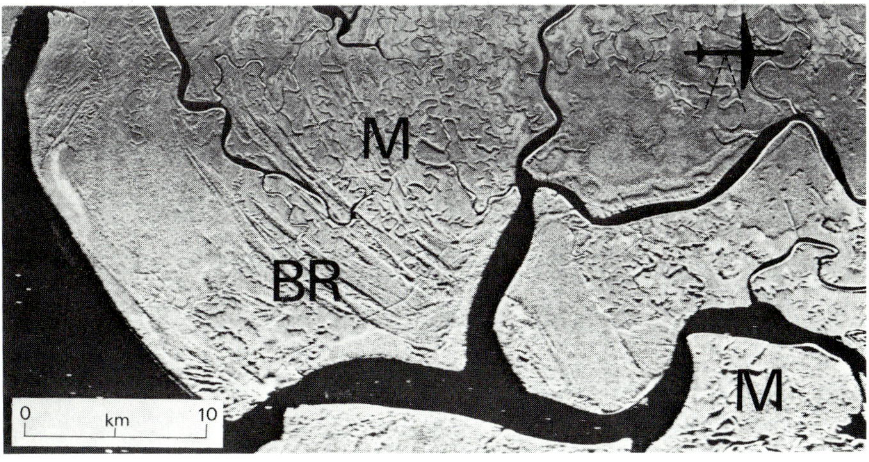


Fig. 6 Land units in southwestern Bendel State. BR: Beach ridges, M: Mangrove mudflat. AN/APS-94D SLAR image, February 1977. Approx. scale 1:250 000

The soils under the tall mangrove consist of saturated organic material, black to dark grey in colour, containing silt and clay bands. Where tall mangrove gives way to short, stunted mangrove, this soft mud is replaced by peaty clay soils, locally called *chikoko*.

### *Beach ridges (BR)*

In contrast to the mangrove mudflat, which is the product of fine sedimentation by rivers flowing from the hinterland, the beach ridges are indicative of an actively prograding shoreline. The beach ridge zone displays a distinctive mesorelief of ridges and troughs running parallel to the coastline. The beach ridge zone is clearly distinguished from the mangrove mudflats on the SLAR imagery by the subparallel banding generated by the individual ridges (Fig. 6). In general, it is somewhat darker in tone than the mangrove zone which is characterized by strong returns from the sinuous channels because of the scattering effect produced by the lattice of interlaced stilt-roots. The forest is floristically rich and contains tall trees with dense crowns. *Elaeis guineensis* and raffia palms (*Raphia* sp.) are present, with many shrubs and climbers.

Two main soil types occur within the land unit. Free to imperfectly drained areas are generally characterized by Entisols consisting of fine sands overlying coarse sandy material. Evidence of podzolization is also sometimes apparent. Soils of the intervening troughs are poorly drained, and may develop iron pans at depth (Fig. 7).

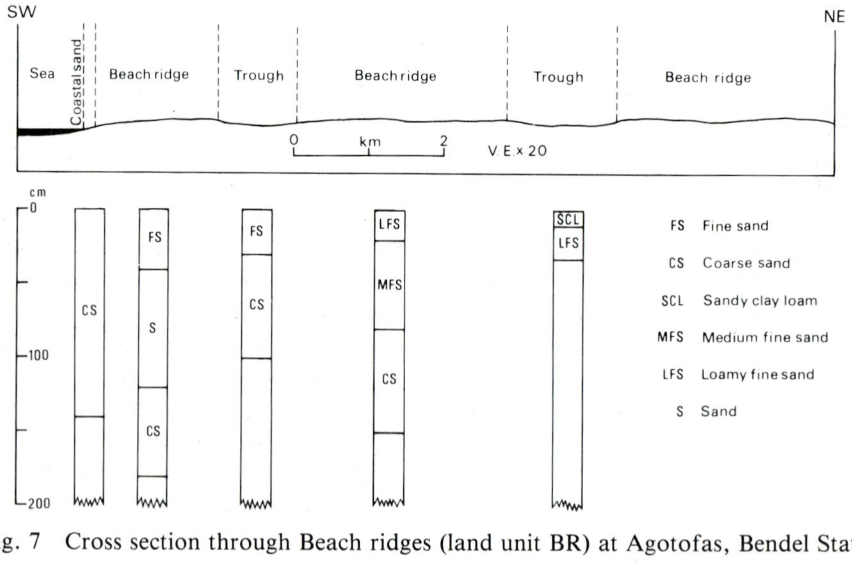


Fig. 7 Cross section through Beach ridges (land unit BR) at Agotofas, Bendel State.

In places, there are no distinct beach ridges separating the mangrove mudflat from the sea. In these cases, the mangrove zone may be fronted by a lower sandy area covered by rain forest containing *Terminalia catapa*, *Elaeis guineensis*, and *Mangifera indica* or by a saline sandy marsh containing *Paspalum vaginatum*, *Acrostichum aureum*, and *Avicennia africana*. The soils are generally coarser in texture than those on the beach ridges, and resemble those of the estuarine levee sites further inland.

The cultivation of cassava, maize, sugar cane, and green vegetables is undertaken on the beach ridges. A few coconut plantations have been established in the vicinity of Bonny and Brass, and there is further potential in the area for this crop and also for groundnuts. Less well-drained sites within the zone are unsuited to crop production, unless expensive drainage works can be undertaken.

### *Coastal plain terrace (CP)*

The term coastal plain is applied to areas underlain by sediments of the Ogoni sands (Anderson, 1967) or the Benin Formation (Kogbe, 1976). These sediments of Miocene to Pleistocene age were deposited by the Niger river and its distributaries. The coastal plain terrace, which slopes gently to the south, exhibits greater dissection in its northern part. The effect of post-depositional tectonic activity and erosion has produced a diversified landscape in which seven land types are distinguished. Three major land types, namely, the Iyanomo, the Ekiadolor-Ehor, and the Agbor, form the bulk of the coastal plain terrace. Four other land types of lesser extent have also been recognized in the area, namely, the Ogwashiuku, the Sapoba, the Sapele and the Ubiane. In some cases, the latter are transitional in character.

*The Iyanomo land type (CP1).* This land type is developed on unconsolidated sands of the Benin Formation. On the SLAR imagery, these areas exhibit a generally featureless grey tone broken only by a coarse-textured dendritic drainage pattern (Fig. 8).

The Iyanomo land type is nearly level with long interfluve slopes that are inclined very gently towards the river valleys. The area is drained by three southward-flowing rivers, namely, the Ossiomo, Ogba and Osse, which reach the Benin river at the margin of the mangrove zone. Rainfall in the area is high, and the soils are acid (pH 4.7–5.4), with low exchangeable base levels. Bleached horizons are present in the soils of the southernmost areas. Most of the soils are Ultisols, falling into the subgroup Typic Paleudults (Ogunkunle *et al.*, 1980). On valley floors, profiles become poorly drained and organic surface layers are common. The thickness and degree of decomposition of the organic layers are variable, with generally thicker layers in the wetter valleys such as the Ossiomo. Around Sapele, Tropofibrists are reported (Ogunkunle *et al.*, 1980) (Fig. 9).

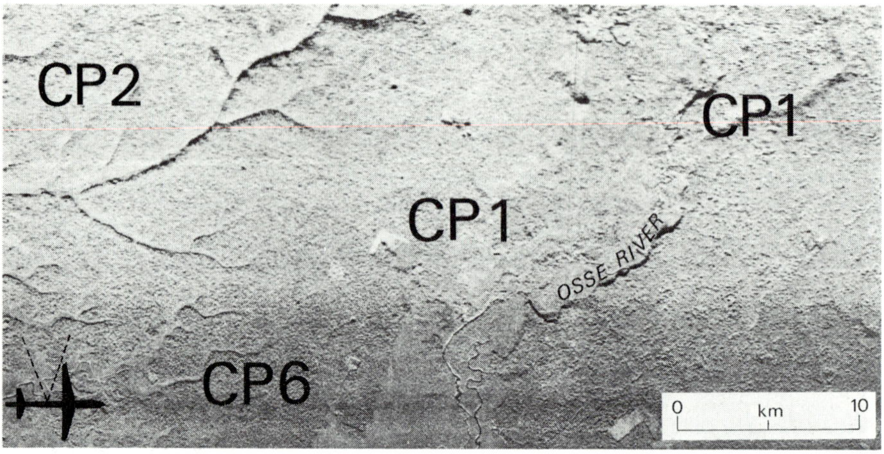


Fig. 8 Land types of the Coastal plain terrace (CP) in central western Bendel State. CP1: Iyanomo land type, CP2: Ekiadolor-Ehor land type, CP6: Sapele land type. AN/APS-94D SLAR image, February 1977. Approx. scale 1:250 000

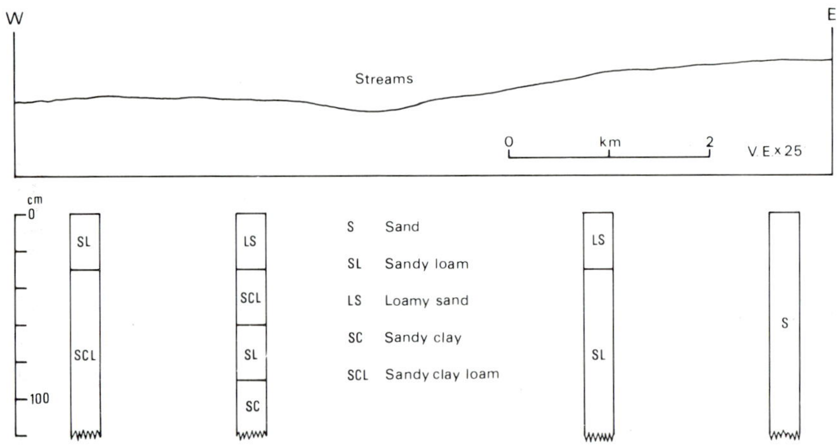


Fig. 9 Cross section through the Iyanomo land type (CP1).

*Ekiadolor-Ehor land type (CP2).* This land type is partly underlain by the sandy Benin Formation, but to the north it extends over the more consolidated Bende Ameki sandstone. The area is undulating and is dissected by deep gullies and incised, narrow streams. Interfluve slopes are short, and rivers are actively eroding the landscape. The area is thickly forested and is the major source of timber within Bendel State (Fig. 10).

On interfluve slopes, the soils are sandy loam at the surface, grading to sandy clay loam to sandy clay at depth. The clayey layer appears at shallower depths

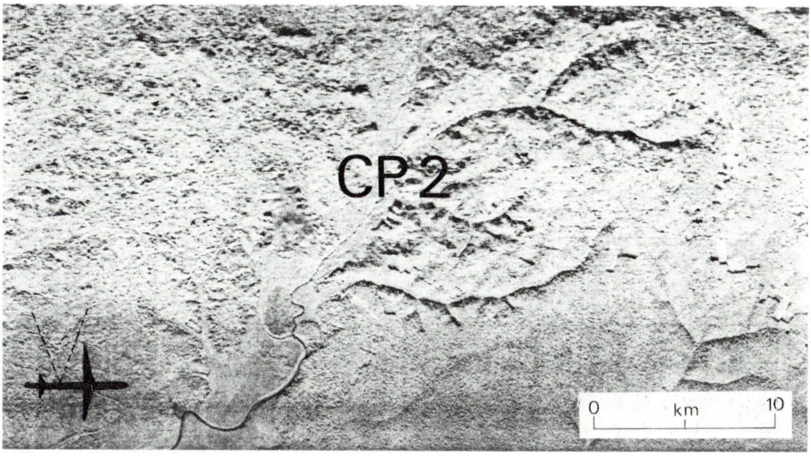


Fig. 10 The Ekiadolor-Ehor land type (CP2) in western Bendel State. AN/APS-94D SLAR image, February 1977. Approx. scale 1:250 000

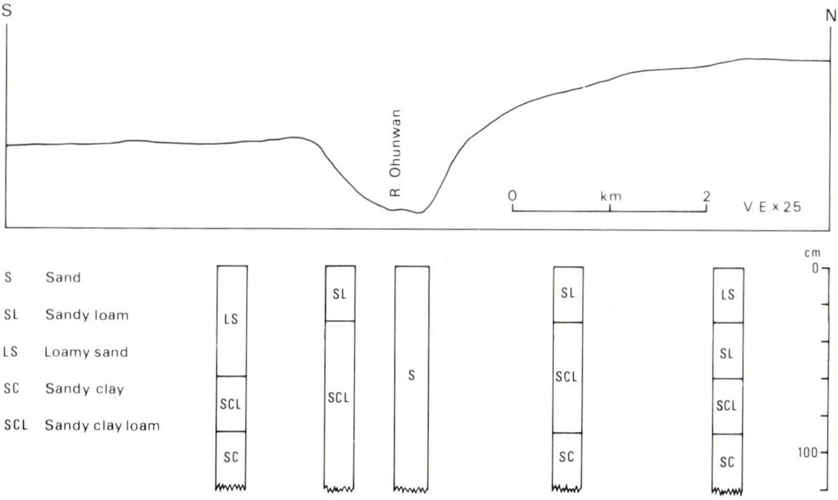


Fig. 11 Cross section through the Ekiadolor-Ehor land type (CP2).

on upper slopes. Soil colour is uniformly red. There are few poorly drained soils within the land type (Fig. 11).

The southern part of the Ekiadolor-Ehor land type is extensively cultivated for rubber and oil palm, and the Nigerian Institute for Oil Palm Research (NIFOR) is located there. A few cocoa and kolanut farms exist in the area, which also produces the bulk of plantains and bananas within Bendel State.

The soils support good crops of yams, cocoyams, cassava, and maize, but these occupy a smaller area than do the permanent tree crops.

*Agbor land type (CP3).* This land type has distinct patterns which serve to differentiate it from other units on the SLAR imagery. Rather subdued interfluves are characteristic; they cast little shadow on the imagery and their dome-like morphology is associated with distinctive radial drainage patterns. The streams and gullies are deeply incised. Towns like Agbor, Ogwashiuku, and Ubiaja are located on the summits of interfluve domes. The slopes between summit areas and the valleys of major streams are up to 3 km or more in length, and have gradients of 3–5°.

The soils of the Agbor land type are distinguishable from those of the Iyanomo and Ekiadolor-Ehor land type in that they are generally much coarser-textured and have a thick organic or organic-stained surface layer. The soils have been provisionally classified as Ultisols, although in places they appear to display some oxic characteristics. They are susceptible to gully erosion.

Floodplains are generally narrow, and so areas of poor internal drainage are very restricted. Valley floors are generally covered by raffia and coconut palms and the transition from the floodplain to the better-drained interfluve slopes is abrupt (Fig. 12).

*Ogwashiuku land type (CP4).* This land type is transitional between the Sombreiro-Warri deltaic plain to the east and the coastal plain terrace to the west. The unit is nearly flat and displays a coarse, sub-parallel pattern of drainage that is tributary to the river Niger. The original climax vegetation of oil palm forest has been cleared to make way for cultivation and the only remaining forest

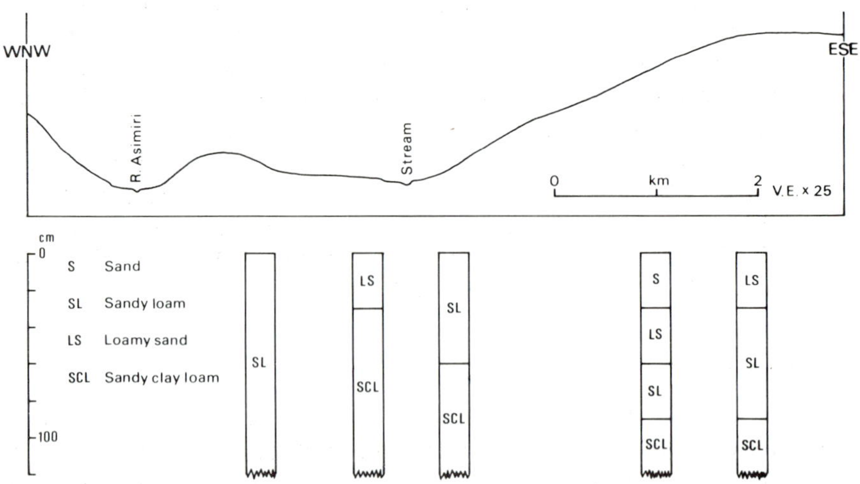


Fig. 12 Cross section through the Agbor land type (CP3)

areas are secondary growth. Most of the land (70–80% of the area) is used intensively for arable cultivation, although in some areas scattered grasslands have resulted from continuous cropping and land abandonment. Cassava is the main crop of the area. Maize, melons, vegetables, and rubber are also produced.

The soils are dark-coloured loamy sands, becoming finer textured and reddish in colour at depth (Fig. 4). Internal drainage is good and soils have been provisionally classified as Ultisols or Alfisols. In many areas continuous cultivation has led to deterioration of the structure of the topsoil.

*Sapoba land type (CP5).* This land type is underlain by the Benin Formation. On the radar imagery the Sapoba land type appears very similar to both the Iyanomo and the Ekiadolor-Ehor land types, the only distinctive feature being the somewhat greater width of the valley floors. Additional fieldwork is required to relate the soil characteristics of the Sapoba land type to those of other soils developed over the Benin Formation. Preliminary investigations suggest that in the south the Sapoba soils are similar to those of the Iyanomo land type and in the north to those of the Ekiadolor-Ehor land type.

*Sapele land type (CP6).* This land type is located on the margin of the mangrove mudflat. On the geological map of Nigeria the area is included within the Sombreiro-Warri deltaic plain, but it displays distinctive features on the radar imagery that distinguish it from the delta land types. Much of the area is permanent swamp land and the soils are classified as Tropaquents.

Rubber is the major crop in the area. The potential for rice is apparently high, but further soil investigations of possible aluminium toxicity are necessary.

*Ubiane land type (CP7).* This land type occupies the northern margin of the coastal plain terrace. The river valleys are narrow, and the land type is less dissected than the adjoining Agbor, Sapele, and Ekiadolor-Ehor land types. The parent material is the Imo clay shale which is more resistant to gully erosion than the sandstone of the adjacent units because of its cohesiveness.

The vegetation is a mixture of secondary forest and derived savanna. The most important feature of land use in the area is the cultivation of cocoa, apparently reflecting the high clay content of the soils. In places cashew is also cultivated. Most soils are classified as Alfisols, including Aqualfs, Plinthaqualfs and Ochraqualfs (Ogunkunle *et al.*, 1980).

### *Crystalline Basement Complex (BCH)*

The Kukuruku hills occupy the northern part of Bendel State. They represent the easternmost extension of the dissected upland of southwest Nigeria — a gently undulating to rolling terrain with narrow, V-shaped valleys and rock exposures in the stream channels.

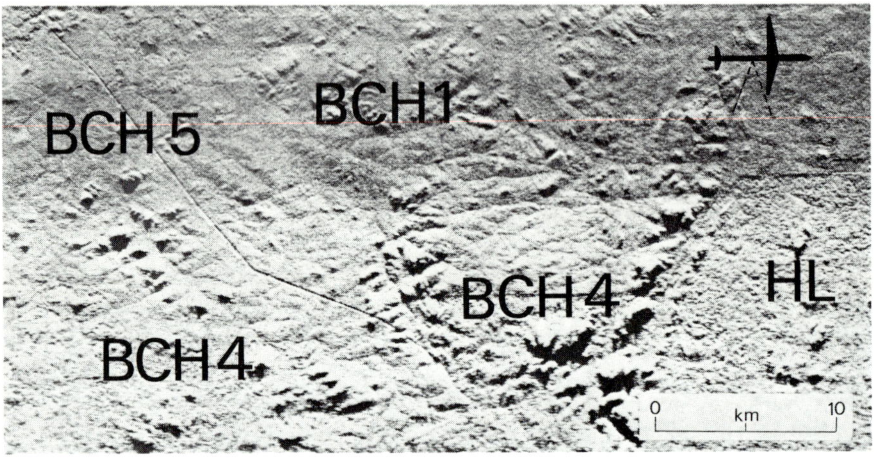


Fig. 13 Land types of the Crystalline Basement Complex (BCH) and Transitional landscape (HL) in northern Bendel State. BCH1: Ore land type, BCH4: Auchi land type, BCH5: Ibillo land type, HL: Transitional landscape. AN/APS-94D SLAR image, February 1977. Approx. scale 1:250 000

The igneous and metamorphic rocks of the area outcrop as massive ridges, ranges of hills, and inselbergs. The scattered inselbergs are commonly surrounded by level to gently undulating plains (Fig. 13). Complex slopes are characteristic of the landscape, and local soil variations frequently occur. The following land types have been recognized on the SLAR imagery:

*Ore land type (BCH1).* This land type is underlain by fractured rocks, with mappable lineaments beneath a thick forest cover. The area is strongly undulating. Soils are deep Alfisols or Ultisols.

*Okene land type (BCH2).* This land type shows similar lineations to the above, but is covered by savanna vegetation.

*Ifon land type (BCH3).* This is a non-lineated land type under a cover of forest. The soils are similar to those of the Ore land type, but some of the parent materials may in fact be sedimentary in character, derived from the Imo clay-shale and false-bedded sandstones (Fig. 1).

*Auchi land type (BCH4).* This land type, which is also non-lineated and has similar geological complexity to the Ifon land type, is under savanna vegetation. Colluvial soils occurring along streams can be delineated on the imagery.

*Ibillo land type (BCH5).* This land type is characterized by an absence of lineaments and by medium to fine dendritic drainage. The area is under savanna, and inselbergs and tors are present. Colluvial soils are present along streams and river valleys.

In general, limited data are currently available on the rather variable soils of the Basement Complex landscape, but some preliminary observations can be made. Shallow, stony, and gravelly soils are generally associated with rock outcrops. In level to gently undulating areas, lateritic soils are very common, often with shallow, hard iron pans. Where the interfluve slopes are sufficiently long, the soils are well-drained, with sandy loam materials overlying sandy clay loam to sandy clay. Floodplains are very narrow or absent, and poorly drained soils are of limited extent. A typical downslope toposequence comprises:

1. a shallow to very shallow soil over bedrock (Lithic Ustorthents);
2. a well-drained sandy loam topsoil, overlying sandy clay loam to sandy clay subsoil (Oxic Paleustalfs);
3. a poorly drained loamy sand over sandy loam subsoil (Typic Tropaquepts/Abruptic Tropaqualfs).

The Ore (BCH1) and Ifon (BCH3) land types provide some of the best cocoa soils of southwest Nigeria. Although characterized by the presence of stones and gravels (Nye, 1954; Ojanuga and Wirth, 1977; Fagbami and Olaniyi, 1982), these soils are better able to support cocoa than the deeper, stone-free sedimentary soils, because of their higher water-holding capacity and their higher reserve of nutrients in the underlying crystalline rocks (Smyth and Montgomery, 1962). Rubber and oil palm thrive on these soils, although yields are lower than on soils developed from sedimentary materials; a similar situation prevails in respect of arable crops, especially the root crops. The Okene (BCH2) and Ibillo (BCH5) land types, which are under savanna, are mainly used for arable farming, including yams, cassava, and maize, but small areas of cocoa farming are also present. These are the only land types in Bendel State on which grazing and ranching can be undertaken.

*Transitional landscape (Basement Complex-sediments) (HL)*

In the area between Agenebode on the Niger river and Auchi in the west there is a zone of transition—the land type displays some of the characteristics of the Basement Complex and some features of the areas of sedimentary rocks.

## Land potential and regional land use planning

Data on the potential of soil and land types in Bendel State can be summarized as follows. The Sombreiro-Warri deltaic plain is best suited, in order of importance, to cropping for tubers (yams, cassava, cocoyams), maize, rubber, and oil palm. The Niger delta floodplain can sustain the same arable and tree crops as the Sombreiro-Warri land type, but it suffers the risk of annual flooding and also the problem of permanently-high water tables which reduce the crop potential. Lumbering, rice cultivation, and fishing appear to be the most profitable land uses for the area.

The soils of the mangrove mudflat, which are more or less saturated throughout the year, cannot be used for cropping, unless major land reclamation is undertaken. Parts of the area may be devoted to wet rice cultivation, but further studies are required to identify suitable sites. The extent of arable soils in the mangrove zone is so small and the people so averse to farming that the entire zone should, for the present, be reserved for fishing, for the tapping of palm wine from *Raphia* sp., and for lumbering. The beach ridges and some adjacent river levees, which have seasonally-deep water tables, should continue to be cultivated for arable subsistence crops and also for coconuts and plantains.

The coastal plain terrace is the largest landform unit in Bendel State, occupying 55.8% of the land area. The Iyanomo land type is the best rubber-producing area, followed by the Ekiadolor-Ehor and Sapoba land types. The Ekiadolor-Ehor land type is the most suitable area for oil palm. The Ekiadolor-Ehor and Sapoba land types are also well-suited to arable crops, especially plantains and bananas. The Agbor land type is mainly devoted to arable cropping, and root crops, in particular, appear most appropriate for the area. The soils are generally too coarse textured for tree-cropping, which is limited by low levels of soil moisture. The Ogwashiuku land type, whose soils are somewhat finer-textured, can support productive growth of tubers, maize and vegetables, as well as rubber and, to some extent, oil palm. The Sapele land type, on the margin of the mangrove zone, may be the best area for wet rice cultivation, at least in the absence of land reclamation. Finally, there is the Ubiane land type, which is the only part of the coastal plain terrace that is favourable for both cocoa and arable crops.

The Basement Complex soils need to be considered in relation to their vegetation cover. Forest soils in this area are the most favourable for cocoa in Bendel State, while the savanna soils are suitable for grazing and ranching activities.

Land use suitabilities for all the land types are indicated in Table 2, based on current levels of performance and on estimated land potential.

## CONCLUSION

Using SLAR imagery of Bendel State, it has been possible to identify and demarcate different land types each with distinctive soil characteristics. Many aspects of the lithology, relief, drainage pattern, moisture regime, and vegetation cover can be directly interpreted or derived from the radar imagery. Within land types, major soil groupings are recognizable and these can be used as an aid to regional land planning at the reconnaissance level.

Table 2 Land types, landscape characteristics, and land use suitability of Bendel State.

| | Land type | Landscape characteristics | Land use suitability |
|---|---|---|---|
| FP | Niger floodplain | Sandbars, braiding river system | Paddy rice, arable cropping |
| D | Niger delta floodplain | Network of meandering distributaries | Paddy rice, fishing, lumbering |
| SW | Sombreiro-Warri deltaic plain | Dense network of fossil stream channels | Root cropping, rubber |
| M | Mangrove mudflat | Anastomosing drainage pattern | Fishing, lumbering |
| BR | Beach ridges | Subparallel series of swales and ridges | Coconuts, arable cropping |
| CP | Coastal plain terrace | | |
| | CP1 Iyanomo land type | Coarse-textured dendritic drainage pattern; forested | Rubber, arable cropping |
| | CP2 Ekiadolor-Ehor land type | Undulating relief; gully dissection; long slopes; forested | Oil palm, rubber, plantains, pineapple, root crops, maize |
| | CP3 Agbor land type | Dome-shaped interfluves, with radial drainage patterns | Arable cropping, especially root crops, oil palm |
| | CP4 Ogwashiuku land type | Transitional to Sombreiro-Warri land type; subparallel drainage pattern; forested | Root crops, rubber |
| | CP5 Sapoba land type | Forested | Arable crops, rubber |
| | CP6 Sapele land type | Swamp forest | Rubber, paddy rice |
| | CP7 Ubiane land type | Northern margin of Coastal plain terrace | Cocoa, arable cropping |
| BCH | Crystalline Basement Complex | | |
| | BCH1 Ore land type | Lineated rocks; forested | Cocoa, plantains, arable cropping, lumbering |
| | BCH2 Okene land type | Lineated rocks; savanna | Arable cropping, ranching |
| | BCH3 Ifon land type | Non-lineated rocks; coarse dendritic drainage pattern; forested | Cocoa, arable cropping |

*continued*

Table 2 *continued*

| | Land type | Landscape characteristics | Land use suitability |
|---|---|---|---|
| | BCH4 Aucni land type | Non-lineated rocks, with coarse dendritic drainage pattern; savanna | Arable cropping, ranching |
| | BCH5 Ibillo land type | Non-lineated rocks, with dendritic drainage pattern; inselbergs/tors; savanna | Arable cropping, ranching |
| HL | Transitional landscape (Basement Complex to sediments) | | Arable cropping, ranching |

## ACKNOWLEDGEMENTS

We wish to thank the Director of the Nigerian Federal Department of Forestry for giving us access to the imagery and to facilities for interpretation at the Department's offices in Ibadan; also for permission to use the imagery displayed in this paper. Our thanks go to the staff of the Department in Ibadan for their co-operation, in particular Mr. Narayan Shinde. We also wish to thank the Director of the Federal Department of Agricultural Land Resources in Kaduna for involving us in the project and for related funding.

## REFERENCES

Anderson, B., 1967. *Report on the Soils of the Niger Delta Special Area*. Port Harcourt: Niger Delta Development Board.

Bleeker, P., 1978. *The Application of Landsat Imagery to Soil Degradation Mapping at 1:500,000: Gambia, Guinea, Sierra Leone, and parts of Senegal, Liberia and Ivory Coast*. Rome: Remote Sensing Unit, Food and Agriculture Organization.

Buringh, P., 1960. Application of aerial photographs in soil surveys. In Colwell, R. N. (Ed.), *Manual of Photographic Interpretation*. Washington: American Society of Photogrammetry, pp. 633–666.

Fagbami, A., 1980. Interpretation of Landsat MSS imagery of the Nigerian Federal Capital Territory for soil survey purposes. *Geoforum*, **11**, 71–83.

Fagbami, A., 1981a. SLAR imagery interpretation for eastern Niger delta for soil survey purposes. *Tech. Pap., American Society of Photogrammetry, San Francisco, September 1981*, pp. 191–202.

Fagbami, A., 1981b. Soil formation processes in the subhumid tropics on Basement Complex. *Niger. Soil Sci. J.*, **2**, 131–146.

Fagbami, A. and Olaniyi, K., 1982. Reconnaissance soil mapping from SLAR imageries in the cloud-covered zones of Nigeria: the Central Southwestern Nigeria example. *Proc. 11 Congreso VII Pan Americano Nacional Fotogrametria, Fotointerpretacion y Geodesia, Mexico, 1982*, ***II***, 1, 11–23

FAO, 1974. *Guidelines for Soil Profile Description*. Rome: Food and Agriculture Organization.

FDF, 1978. *NIRAD Project Report by Motorola Aerial Remote Sensing Inc., U.S.A. Final Report of the Interpretation Phase*. Ibadan: Federal Department of Forestry.

Goosen, D., 1967. Aerial photo interpretation in soil survey. *FAO Soils Bulletin*, **6**, 1–55.

Kogbe, G. A. (Ed.), 1976. *Geology of Nigeria*. Ibadan: Abiprint and Pak.

Mainguet, M. M., Spiers, B. and Canon, L. B., 1978. *Application of Landsat Imagery to Soil Degradation Assessment and Mapping at 1:500 000: Niger, Upper Volta, Eastern Mali, Northern Nigeria, Chad and Central African Empire*. Rome: Remote Sensing Unit, Food and Agriculture Organization.

Motorola Inc., 1970. An introduction to radar surveillance set AN/APS 94D equipment and imagery. *Motorola Aerial Remote Sensing Inc. Rept. MAE 4/2.*

Nye, P. H., 1954. Some soil forming processes in the humid tropics. Part 1. A field study of a catena in the West African forest. *J. Soil Sci*, **5**, 7–21.

Ogunkunle, A. O., Omoti, U., Ataga, D. O. and Eseiche, A., 1980. *Field Tour Guide, Soil Science Society of Nigeria Conference, Benin, November-December 1980.*

Ojanuga, A. G. and Wirth, K., 1977. Threefold stonelines in south-western Nigeria: evidence of cyclic soil and landscape development. *Soil Sci.*, **123**, 249–257.

Smyth, A. J. and Montgomery, R. F., 1962. *Soils and Land Use in Central Western Nigeria*. Ibadan: Government Press of Western Nigeria.

Vink, A. P. A., 1968. Aerial photographs and the soil sciences. In UNESCO, *Aerial Surveys and Integrated Studies*. Paris: United Nations Educational, Scientific and Cultural Organization, pp. 81–141.

Remote Sensing and Tropical Land Management
Edited by M. J. Eden and J. T. Parry


# CHAPTER 5

# *Radar Surveys for Resource Evaluation in Brazil: an Illustration from Rondônia*

PETER A. FURLEY
*Department of Geography, University of Edinburgh, U.K.*

Resource surveys in the tropical interior of Brazil present considerable problems, particularly in forest areas that are seasonally or perennially wet, cloud-covered, and inaccessible. Conventional aerial photography is only useful for a limited part of the region. Existing satellite imagery has potential at the reconnaissance level, but in the late 1960s, when Brazilian pressure for resource exploration was at a peak, satellites were unable to offer an adequate basis for the survey of forest areas.

Projeto Radambrasil was set up in the early 1970s, using side-looking airborne radar (SLAR), with the object of producing a rapid reconnaissance of Amazonian natural resources. The GEMS 1000 SLAR system used for Projeto Radambrasil is an HH system operating at X-band frequencies (9375 MHz—3.2 cm). It offers a rapid, large-area coverage with adequate resolution (16 m) for reconnaissance surveys (Azevedo, 1971; van Roessel and Godoy, 1974). The system is mounted in a Caravelle jet aircraft which has a cruising speed of 850 km $h^{-1}$. From an altitude of 11 000 m the radar swath width is 37 km. The flight lines generally run north-south, 15 minutes of longitude apart, allowing 25% sidelap. The swath coverage rate is 31 450 $km^2 h^{-1}$. Radar is an active system and so radar surveys can be carried out by day or by night virtually independent of weather conditions.

The radar imagery was initially generated at a scale of 1:400 000. This was subsequently enlarged, with minimum loss of detail, to produce semi-controlled mosaics at 1:250 000, each of which covered 1° of latitude by 1°30′ of longitude. The mosaics were then compiled into radar mosiac map sheets at 1:1 000 000, which covered 4° of latitude by 6° of longitude (van Roessel and Godoy, 1974).

The advantages of SLAR for the Amazon survey were considered to be its synoptic coverage of geological structures and drainage systems, its ability to detect many small-scale features, and its excellent definition of land-water boundaries. Despite the high cost of setting up and undertaking the survey, the overall size of the project was considered to justify the initial expense, quite apart from the technical argument that only a radar system could have achieved the desired result. Moreover, radar survey had already been successfully demonstrated in the Darien Province of Panama (MacDonald, 1969); under conditions similar to those of the Amazon, 6 hours flying had provided comprehensive radar coverage, whereas 25 years of conventional aerial photography had failed to complete the equivalent task.

The object of this account is to review the application of the Radam survey in the Rondônia region of the western Amazon. Rondônia has recently become a full State within the Brazilian Federation and has witnessed some of the most rapid development in the Amazon region. It is arguable that its growth was partly provoked by the initial results of the Radam survey, but the area represents a microcosm of the larger Amazon situation and provides an excellent case study illustrating the application of SLAR surveys.

## PROJETO RADAMBRASIL

The Radam project evolved in the early 1970s, as a result of the Brazilian government's desire to integrate the undeveloped North and the backward Northeast of the country into the national economy. One specific factor behind the project was a growing appreciation of the need for a survey of the renewable and non-renewable resources in the Amazon region. This need increased dramatically with the events that led to the Transamazon Highway scheme and with the construction of the Highway itself (Bourne, 1978). A second factor behind the project was the availability of non-military radar systems owned and operated by U.S. companies. Thus, an appropriate technology was available to meet Brazil's needs for rapid resource inventory at the time when such surveys were required.

The transfer of radar technology to Brazil began in 1969. The United States National Aeronautics and Space Administration (NASA) initially undertook a demonstration survey over 5000 $km^2$ of iron-rich territory in Minas Gerais State. This proved sufficiently persuasive for the National Department of Mineral Production (DNPM) of the Brazilian Ministry of Mines and Energy to attempt a larger survey, covering 44 000 $km^2$, in the Tapajós region of Pará State. The survey was undertaken to determine if SLAR could provide useful imagery in spite of the region's persistent cloud and heavy rain which severely limited aerial photography. The experiment demonstrated the effectiveness of the technology and, in November 1970, the Ministry created a Commission for the radar survey of Amazônia (CRADAM). The specific aim of CRADAM, within the framework of the newly established Programme for National Development (PIN), was to undertake a rapid survey of the natural resource potential of an area of 1.5 m $km^2$ along the Transamazon Highway system. In 1971 the original contract was extended to cover an additional 1.8 m $km^2$, and in 1973 a further 1.45 m $km^2$ were added. Finally, in 1975, the survey was expanded by 3.76 m $km^2$ to cover the whole of the Brazilian territory giving a total area of 8.51 m $km^2$.

The technical details of the survey and the methods of processing and interpreting the radar images have been reviewed elsewhere (see for example: Moreira, 1973; Comissão Executiva, 1983), and the emphasis in this study is upon the objectives and achievements of Projeto Radambrasil. Its primary objective was to carry out an integrated survey of natural resources at the reconnaissance level, using SLAR imagery and other remote sensing data, as appropriate. The resources were to be investigated by survey groups concerned with geology (especially mineral resources), hydrology, geomorphology, soils, vegetation, land suitability for agriculture, pasture, and forestry, and land use potential. Socio-economic data necessary for planning were to be considered. Additional objectives were the identification and selection of favourable areas for settlement and the evaluation of areas already settled, as a basis for determining ways to increase production. Finally, it was intended to establish a national data bank to make the results of the various surveys available for planning projects.

Although the Radam project has covered the Amazon region, and indeed extended across the entire national territory, the analysis and publication of the results are not yet complete. However, achievements at the present time include:

1. The coverage of the whole of Brazil by semi-controlled radar mosaics (551 sheets at a scale of 1:250 000), subsequently compiled into radar map sheets (41 sheets at 1:1 000 000). These are supplemented by other types of imagery for selected areas:
(a) High altitude colour infrared aerial photography at a scale of approximately 1:130 000 (60% forward overlap, 10% sidelap); photoindices of the above at a scale of 1:500 000; also some black and white infrared photography at a scale of approximately 1:75 000.
(b) Multispectral photography in four channels (blue, green, red, and infrared) at a scale of approximately 1:70 000; these were acquired at the same time as (a) above, but are only available for the Amazon region.
(c) Video imagery at a scale of approximately 1:23 000; again this coverage is only available for the Amazon region.
(d) Altimetric profiles at 27 km intervals along the flight lines, plotted at a horizontal scale of approximately 1:400 000.
2. Integrated resource surveys covering five principle themes: geology, geomorphology, soils, vegetation, and land use potential, each yielding a

Fig 1. Radar map sheets (1:1 000 000 scale) and volume reference numbers for the Projeto Radambrasil resource series *Levantamentos de Recursos Naturais* (after Sampaio de Almeida, 1984).

report and maps at 1:1 000 000 scale. The reports form the series known as *Levantamentos de Recursos Naturais* (Surveys of Natural Resources), of which thirty volumes have so far been published, with a further eight nearing completion (Fig. 1). The published volumes cover approximately 80% of the national territory.

3. The production of planimetric, topographic, and radar image maps at a scale of 1:250 000.

4. A map of *Amazônia Legal* (the official Amazon region) at a scale of 1:2 500 000 showing details of the regional infrastructure, mineral workings, vegetation, National Parks, and Indian and Biological Reserves.

5. Special-purpose maps of promising metalliferous areas and hydrological resources in Northeast Brazil.
6. Special-purpose maps of the Grand Carajás development area in Pará State.
7. Technical bulletins on geology and vegetation.

A great variety of subsidiary projects and ancillary contracts has stemmed from the principle surveys, and a recent review of work in progress is given in Sampaio de Almeida (1984).

## RESOURCE SURVEYS AND DEVELOPMENT ON RONDÔNIA

The Federal State of Rondônia covers approximately 243 000 km$^2$ and represents about 5% of the Brazilian Amazon. Until the 1970s the area was largely undeveloped, with a population averaging less than one inhabitant per km$^2$. The general migration of landless squatters into the Amazon region and the growth of official colonization schemes date from a period of intense highway construction during the early 1970s. In the case of Rondônia, such migration and colonization were particularly encouraged by the modernization of the BR 364 Highway, the main access route from the centre and south of Brazil (Fig. 2). The migrants themselves came principally from Mato Grosso, Paraná, and other southern States. Mineral deposits in Rondônia, some of which were already being exploited, were more accurately located by the Radam survey, and this further stimulated development. In addition, a land boom was provoked by tax incentives offered by the Federal Government.

Although Landsat coverage of Rondônia was available in the 1970s, little use has been made of this imagery. In contrast, the SLAR imagery acquired in the period 1971–1974 has provided a major source of information for resources inventory (Cruz, 1974). The results of the radar image analysis have been published in the volumes of the *Levantamentos* series, each volume covering a particular map sheet area: vol. 12 Rio Branco map SC 19 (Brasil, 1976); vol. 16 Porto Velho map SC 20 (Brasil, 1978a); vol. 19 Guaporé map SD 20 (Brasil, 1979); vol. 20 Juruena map SC 21 (Brasil, 1980); vol. 26 Cuiabá map SD 21 (Brasil, 1982). Each of the volumes has a similar organization with sections devoted to geology, geomorphology, soils, vegetation, and land use potential. In the following pages an attempt will be made to provide a summary appraisal of the value of the data provided under each of these headings in the volumes covering the State of Rondônia.

### Geology

In view of the organization of Projeto Radambrasil through the Ministry of Mines and Energy, and specifically by the National Department of Mineral

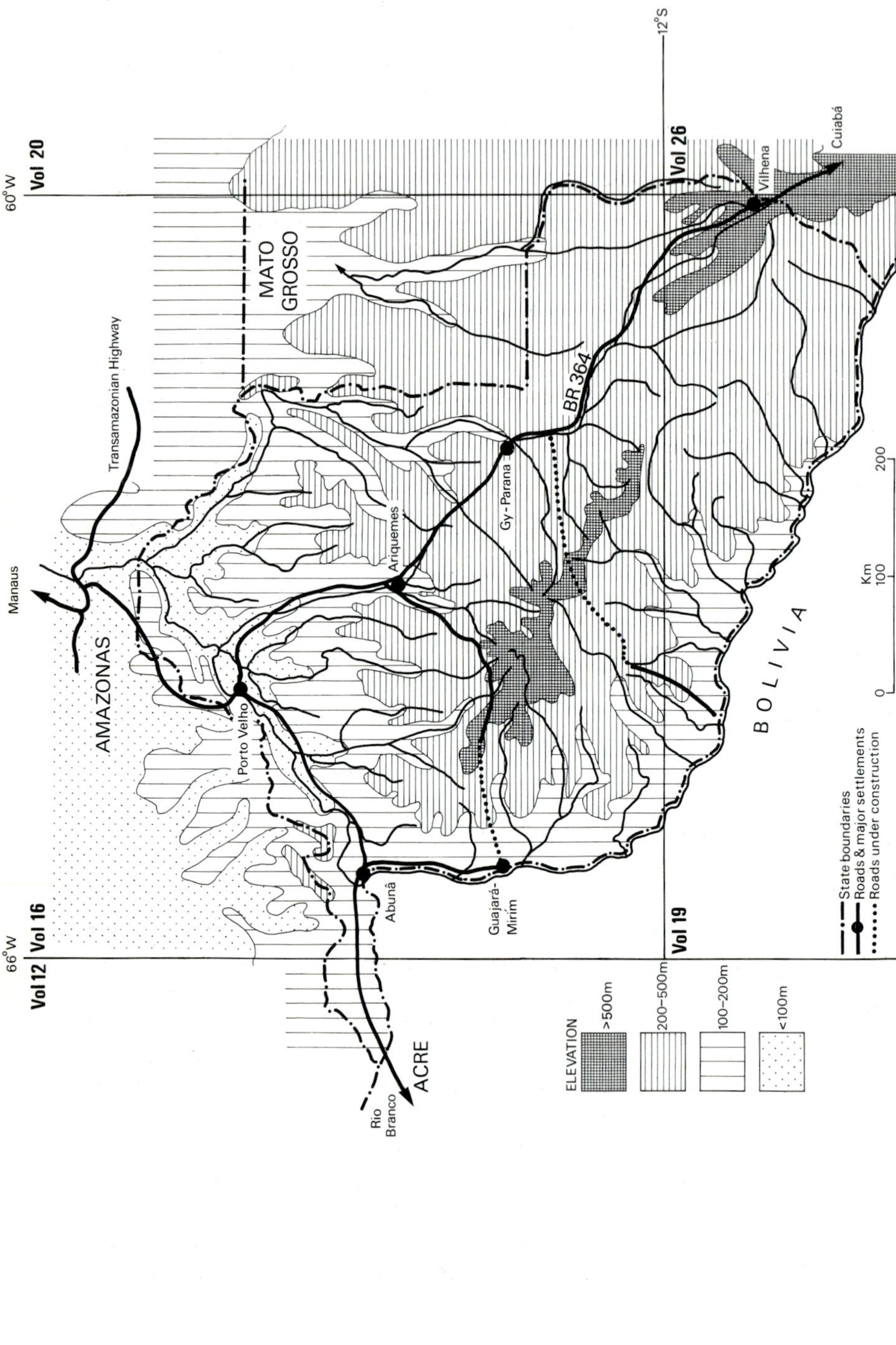


Fig. 2 Radar map sheets (1:1 000 000 scale) and volume reference numbers for the Projeto Radambrasil resource series *Levantamentos de Recursos Naturais* for Rondônia State.

Production (DNPM), it is not surprising that the major emphasis should have been placed on the description and analysis of mineral resources. Geologists are the largest group in the DNPM and in each volume of the Radam reports geology receives priority treatment. The pattern of each report is similar. After an initial outline of the area and methodology, there are sections on stratigraphy, structure, geological history, and economic geology, followed by conclusions and recommendations. However, as the survey progressed during the 1970s, the methodology itself was modified. As a result, it is not always possible to compare directly parts of an earlier volume with those of a later one.

Recent reviews of radar remote sensing have stressed the value of SLAR imagery for geological surveys (de Loor, 1983). In the Radam project the SLAR mosaics were initially examined at the 1:1 000 000 scale in a search for major structural patterns and signs of mineralization. Subsequently, more detailed analyses were carried out using the 1:250 000 scale radar maps, and the infrared and multispectral photography. Once an outline of the principal structural units had been obtained and a base map prepared, low altitude flights were made to check the preliminary results and field sampling was undertaken at selected points. Helicopters were used to land the field teams in remote areas; otherwise access was by river and, where practical, by track. Field samples were obtained and subjected to laboratory analysis using radiometric, palaeontological, and palynological techniques to establish the geochronology of the area. The information was then collated and prepared for publication.

Much of the economic value of the reports derives from the fieldwork and subsequent analysis of samples rather than directly from the interpretation of remotely sensed imagery. In the northern part of Rondônia, three new areas of promising mineral deposits were identified, namely, the intrusive area of Serra da Providencia, basins associated with volcanic activity, and sedimentary zones containing deposits of limestone, gypsum, and anhydrite. New deposits of cassiterite, which was already being exploited on a large scale in the area, were found and significant occurrences of gold and many other minerals were indicated (Fig. 3).

Although the main objective of the surveys was the identification of areas containing minerals of economic importance, other significant results were achieved. Stratigraphic and structural patterns are particularly well displayed on SLAR imagery and so it was possible to delimit effectively on the radar mosaics the three geological provinces recognized in the Porto Velho area and the five provinces of the Guaporé area. In addition, specific elements such as megalineaments and circular structures were readily identified (Fig. 4(a),(b)).

Additional studies based on interpretation of the SLAR imagery included a geological chronology for the area, an analysis of the processes responsible for mineral differentiation in ore bodies and preliminary predictions on the relationships between specific tectonic and structural features and mineral associations. The surveys understandably focused on the areas most favourable

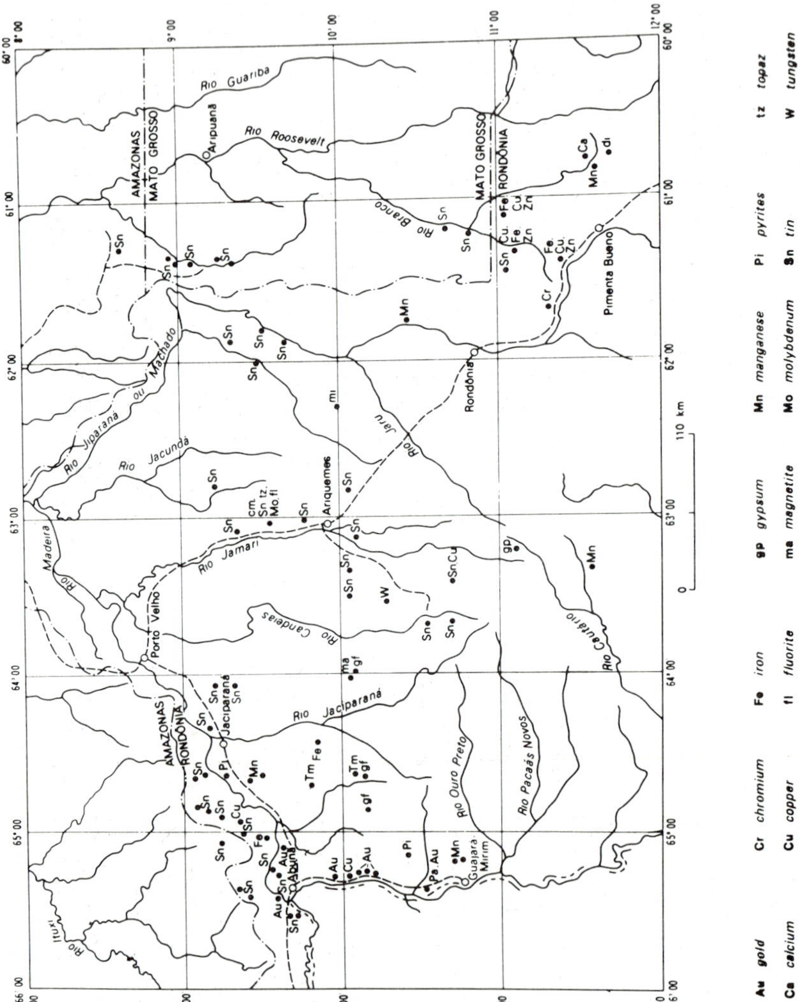


Fig. 3 Mineral discoveries reported over the northern part of Rondônia by Projeto Radambrasil (Brasil, 1978a).

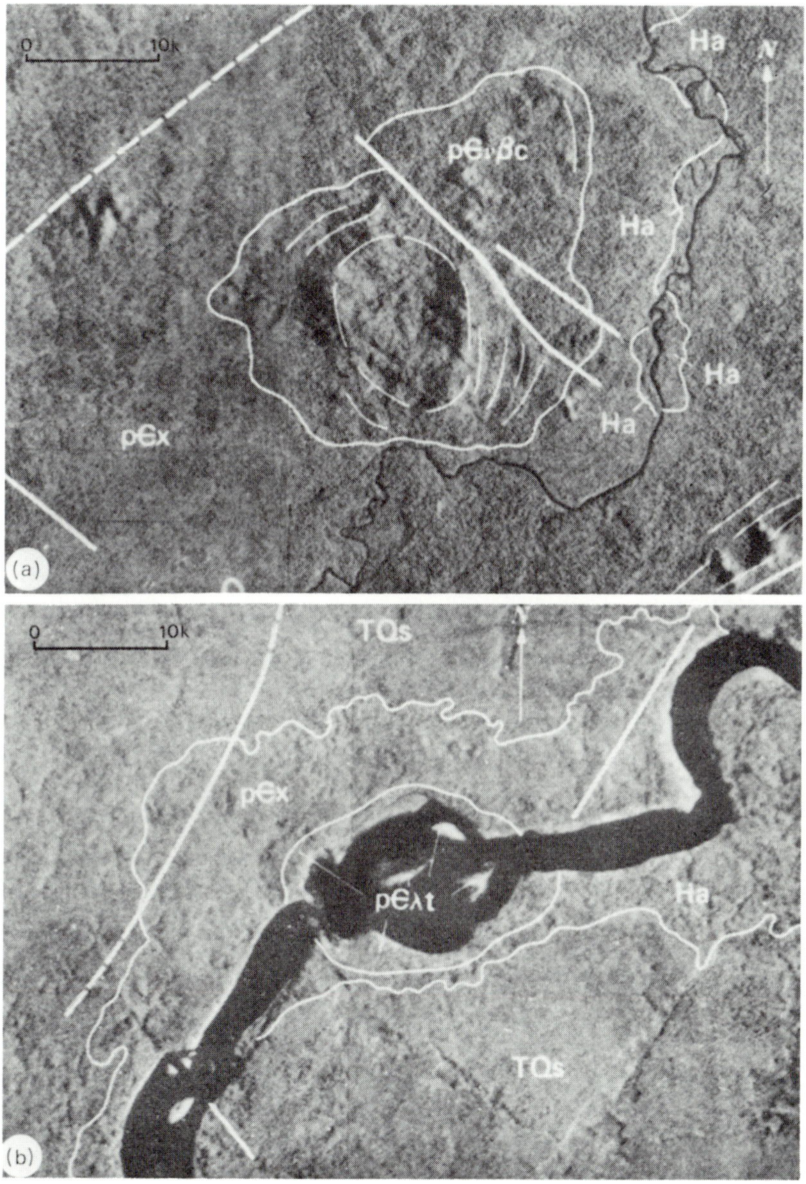


Fig. 4 Circular structures identified in Projeto Radambrasil X-band SLAR imagery. (*Reproduced by permission of Projeto Radambrasil.*) (a) Rio Branco circular structure: Ha recent alluvium, pЄx Xingú complex, pЄvβc basic and ultrabasic rocks of the Ciriquiqui complex, —— contact zones, — — structural alignments, - - — fault lines. (b) Teotonio circular structure on the Rio Madeira, Ha recent alluvium, TQs Solimões Formation, pЄλt Teotonio alkaline rocks, pЄx Xingú complex, —— contact zones, - - — fault lines.

for mineral exploitation, but they contributed to the work of the land use potential and other survey teams and to the formulation of the integrated development plans. In all this, the role of the initial radar survey was crucial.

**Geomorphology**

Since geological structures are well displayed on the radar images, it follows that geomorphological features are likely to be well represented. In fact, the geomorphological teams in Rondônia found the SLAR images to be of exceptional value, although they did encounter some difficulties in their use and interpretation. These are fully recognized in the reports. The general methodology adopted was similar to that of the geological surveys. At the outset, a bibliographic review was undertaken, together with a preliminary inspection of the 1:1 000 000 mosaics. The drainage pattern was outlined using the scheme of Howard (1967), landscape texture classes were identified (Fig. 5 (a), (b)), and physiographic units were delimited (Fig. 6). Details were plotted on the 1:250 000 scale radar maps and particular attention was given to the matching of features from one map sheet to another. Field checking and low level flights, providing vertical and oblique photographs, were used to correct the initial interpretations. A useful innovation was the creation of a checklist of reference features as an aid to interpretation. The standard black and white, infrared, and multispectral photographs were also examined, although their usefulness was frequently diminished by cloud cover. Some comparative use was made of Landsat imagery in the later stages of the surveys.

The main problems encountered in the surveys were, firstly, the difficulty of arriving at an appropriate system of landscape classification for use in the soil and vegetation surveys. In addition, there were problems in the local use of radar images in the field on account of their apparent similarity to standard stereo pairs of black and white photographs (Lewis, 1974). Interpreters found difficulty in assessing the effects of foreshortening and layover, two fundamental characteristics of the slant-range geometry of SLAR imaging, and this hindered the development of a workable slope classification. The contribution of the radar imagery to the mapping of surficial materials and the analysis of morpho-dynamic processes was found to be quite limited, while the scale of the imagery (1:40 000) and its effective resolution (16 m) precluded the identification of minor landforms. Even so, in earlier volumes, some problems arose from the complexity of the morphological detail on maps. It is noteworthy that when the decision was taken to extend the Radam surveys beyond the Amazon region, the original morpho-climatic and morpho-structural units of classification (Ab'Saber, 1967, 1970, 1971), derived from Cailleux and Tricart (1957), were replaced by simple 'geomorphological units' in order to avoid the complications of genetic interpretation.

The geomorphological reports extend beyond the basic units of study, namely, drainage patterns and landforms. Following a discussion of methodology, the reports review the history of geomorphological research in the area and provide

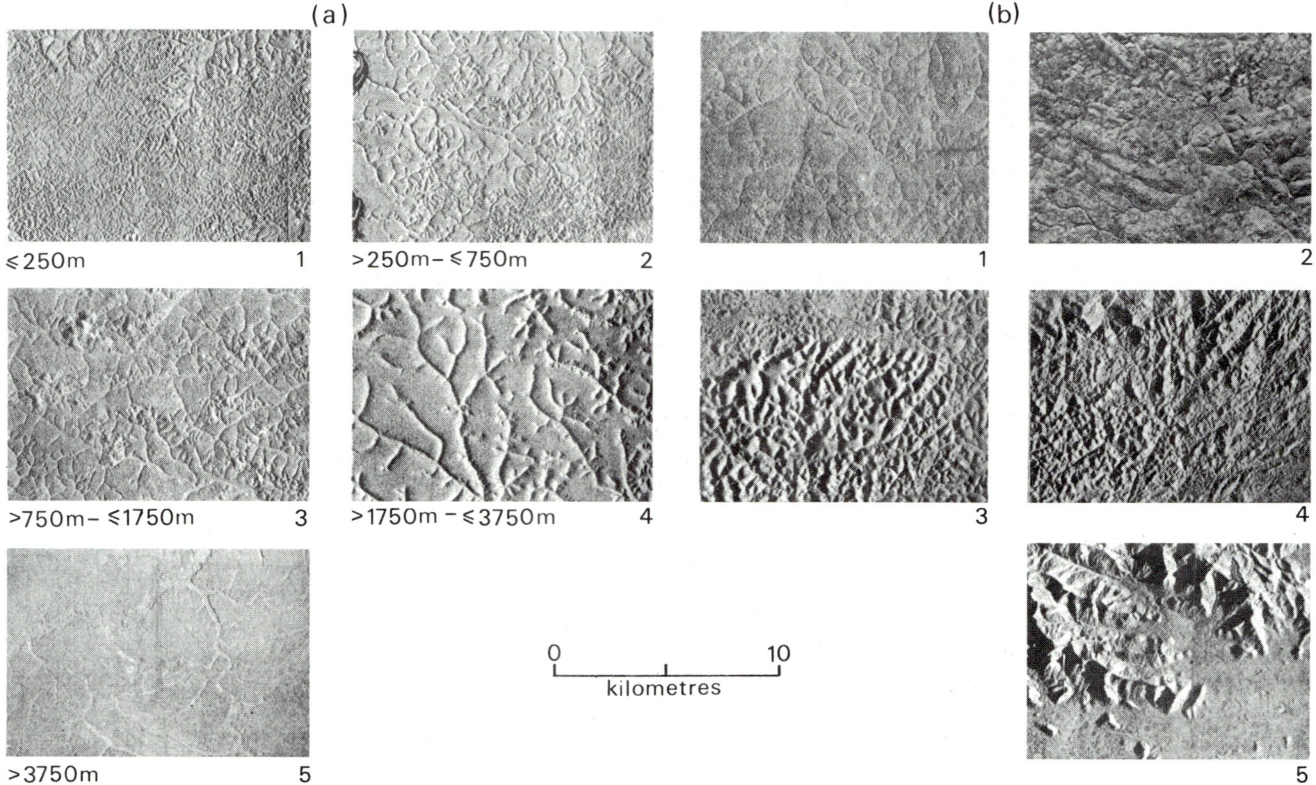


Fig. 5 Landscape texture and dissection classes: SLAR imagery chips. (*Reproduced by permission of Projeto Radambrasil.*) (a) Drainage line spacing classes 1–5; (b) depth of dissection classes 1–5.

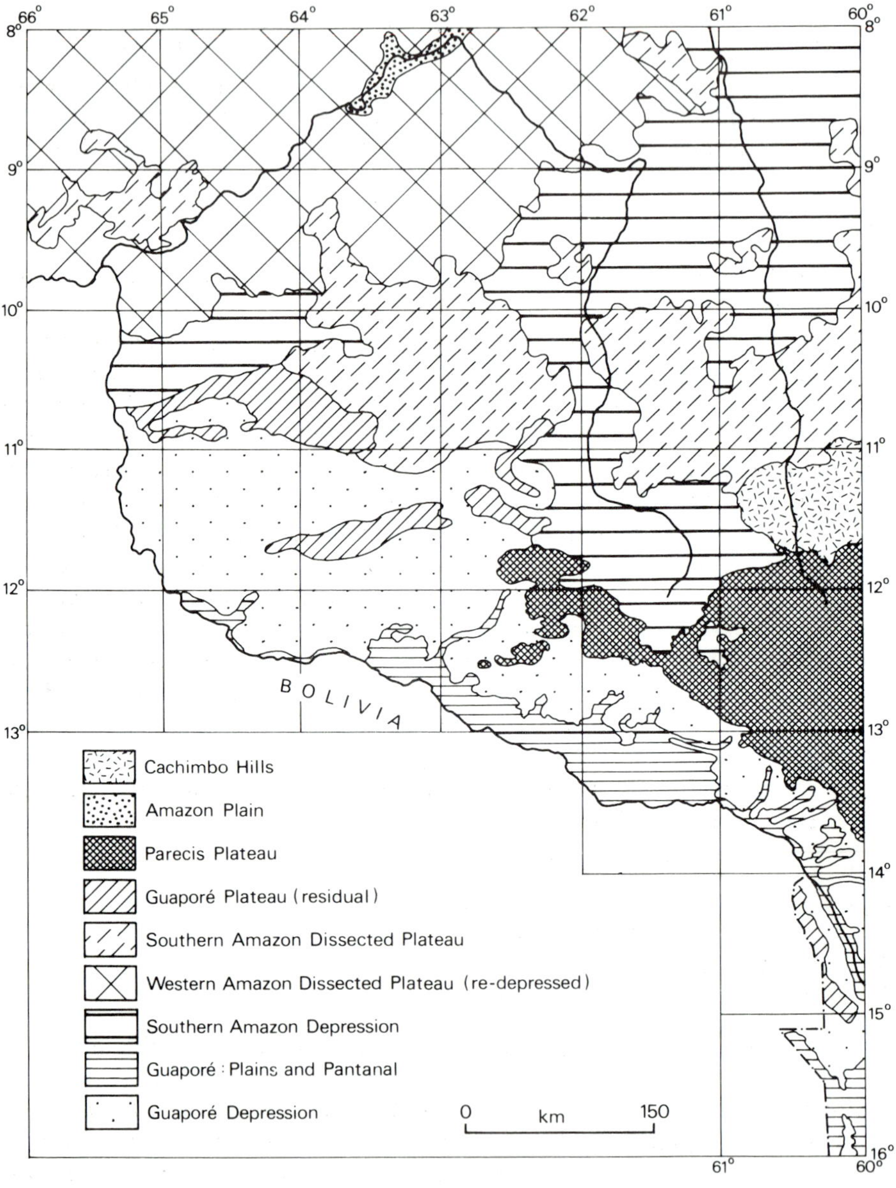


Fig. 6 Physiographic units of Rondônia State and adjacent areas interpreted from SLAR imagery in Projeto Radambrasil.

an outline of the structural and morphological evolution of the landscape. Practical applications are covered in a section dealing with road construction, river navigation, hydroelectric potential, and erosion hazard.

The geomorphological surveys, together with the geological mapping, provided a basis for the soil and vegetation studies and also furnished information directly relevant to the assessment of land use potential. Despite the initial problems and some persisting methodological criticism, the geomorphological surveys have been greatly facilitated by the availability of SLAR imagery.

**Soils**

The radar imagery provides a synoptic view of drainage and relief. As noted by White (1977, p. 55), such imagery offers a useful framework for soil survey, particularly in areas where topography is a dominant factor in soil genesis. In Rondônia, SLAR imagery mosaics have been used as a basis for reconnaissance soil survey and in the assessment of the suitability of different areas for particular types of agricultural development. Preliminary work was based on the semi-controlled radar mosaics at 1:250 000. These allowed a provisional interpretation of soil mapping units and provided a framework for subsequent field sampling. Available infrared and multispectral imagery was used to clarify problem areas.

The fieldwork consisted of the identification and confirmation of the provisional soil units, the description of representative soil profiles, and the collection of soil samples for laboratory analysis. The field descriptions followed the standard guide of the Brazilian Soil Science Society (Sociedade Brasileira de Ciência do Solo, 1973). The soil analytical procedures followed the recommendations of Vettori (1969). The soil classification system adopted was that of Bennema (1966), updated on numerous occasions to match the United States Soil Taxonomy (Soil Survey Staff, 1975). The following first-order soil categories were used (Furley, 1979):

1. non-hydromorphic latosols,
2. non-hydromorphic soils with a textural B horizon,
3. hydromorphic soils with gley and plinthite horizons,
4. non-hydromorphic sandy soils,
5. poorly developed shallow soils.

On completion of the survey, the final soil maps were produced at a scale of 1:1 000 000.

The reconnaissance survey was followed by an assessment of the potential of the soils for agricultural development. The interpretation was based on soil properties, following the guidelines of Bennema *et al.* (1965) and Beek (1975), and on assessment of climatic conditions, especially the water balance (Thornthwaite and Mather, 1955; Bagnouls and Gaussen, 1963). Water balance calculations were made for two soil depths: for annual crops the calculation was for the 0–60 cm layer and for deeper-rooted perennial crops for the 0–120 cm layer. The radar contribution to this section of the work was obviously limited, although other remote sensing techniques helped improve the accuracy of climatic information during the survey period.

The first step in the interpretation of the soils for agricultural development was a description of the contemporary land use (Furley, 1980). In an area such as Rondônia with very limited development, Landsat images might have been employed in this baseline survey, but they were either unavailable or not used when the work was undertaken. Instead, brief descriptions of land use were obtained from field observations and a few published sources. Subsequently, the soil types were evaluated on the basis of their limitations for agricultural development. As in previous Radam surveys, five characteristics were considered: (1) nutrient deficiencies, (2) excess water, (3) water deficiency, (4) susceptibility to erosion, and (5) suitability for mechanization. The soil types were then classified as good, standard, restricted or unsuitable with respect to agricultural, pastoral, and silvicultural usage. At the time, no consideration was given to possible drainage or irrigation improvements. In the case of agriculture, three management systems were postulated for annual or perennial crops: (1) traditional or largely subsistence management, (2) simple management practices, involving modest capital and technical inputs, and (3) advanced management practices, operating at a high technical level.

On the basis of this analysis, maps were prepared showing the land aptitude for agriculture, pasture, and forestry. The most significant finding of the surveys in Rondônia was the relatively high proportion of fertile soils that were encountered in comparison with the Amazon basin as a whole. It may be concluded that, although the survey and interpretation of soils extend well beyond the remote sensing base, it would have been impossible to achieve the detailed land assessments without access to the SLAR imagery.

## Vegetation

The notion that a large proportion of the Amazon basin and its surrounds is clothed in a uniform evergreen rain forest was effectively dispelled by the Radam surveys. Differing vegetation types were shown to be associated with broad topographic and drainage units and to be identifiable on the radar imagery at a reconnaissance level.

Table 1. SLAR image tone and texture classes for vegetation types (Brasil, 1979).

| Phyto-ecological region | SLAR image grey tone | SLAR image texture |
|---|---|---|
| Savanna | Light to medium | Smooth to medium |
| Dense forest | Medium | Rough to medium |
| Open forest | Medium | Rough to medium, mottled |
| Semi-deciduous forest | Medium | Medium to rough, mottled |

The methodology used for the vegetation survey was broadly similar to that of the other resource surveys. The radar mosaics at 1:250 000 and the reduced sheets at 1:1 000 000 provided the basis for a preliminary map of the principal vegetation formations. Six major units were identified, as follows: savanna, pioneer formations (seral communities), dense tropical forest, open tropical forest, ecological tension zones (transitional areas), and anthropically disturbed areas. The initial division of the vegetation was based on relief and drainage differences. As indicated in Table 1, it was possible to distinguish vegetation types on the basis of the tone and texture of the radar images. Key illustrations are provided in the Radam reports, but these do not reproduce convincingly and are not included here. In general, vegetation units covering areas of less than approximately 6 $km^2$ (1 $cm^2$ at the 1:250 000 image scale) could not be discriminated and so the final map legend included some combined vegetation classes.

Once the preliminary vegetation maps had been prepared, low-level flights, helicopter lifts to inaccessible sites, and road and river traverses were made to check the initial interpretation. Forest inventories were prepared and these formed the basis of a separate report series covering each 1:250 000 map sheet. The vegetation description in these reports is based on the phytogeographic system of Ellenberg and Mueller-Dombois (1965–66) modified for large-scale regional surveys (UNESCO, 1973; Veloso *et al.*, 1973; Brasil, 1978b). The reports provide an assessment of the timber reserves and a commentary on the feasibility of exploitation.

As with the other resource surveys, the vegetation survey has evolved since the first Radam publications, but there is a logical and comparable order of presentation throughout the series. In the introduction there is a review of the principal factors affecting the vegetation, with an emphasis on morpho-structural controls, and the main vegetation formations and subtypes are outlined. The methodology section highlights the advantages of radar remote sensing for the study of vegetation, and outlines the stages of the survey. Each vegetation formation and subtype is reviewed and the distribution is mapped. Particular attention was paid to the anthropically disturbed areas, which were surveyed in both January 1972 and January 1976 and assessed in the light of the very rapid migration into Rondônia at that time. A final section outlines bioclimatic controls, which receive increasing attention through the series of reports.

Once again radar imagery contributed to the overall effectiveness of the survey by providing a base map. Information from other airborne sensors was used to clarify problem areas, but the preliminary vegetation map and the survey framework were effectively prepared from the semicontrolled radar mosaics. A particular feature of the survey is the impressive collation of floristic data and the valuable prediction of timber resources. Although these results were not directly derived from the radar images, it is unlikely that any other method of reconnaissance available at the time would have provided as satisfactory a starting point for the survey.

**Land use potential**

In preparing the reports in this section, the team members effectively synthesized the results of the geological, geomorphological, soil, and vegetation surveys, assessed this information in terms of the relative advantages and constraints on development, and provided guidelines for a development strategy. The work did not involve direct radar interpretation, but utilized information therefrom and provided the crucial analysis for development planning. The radar survey was thus of fundamental importance through to the final planning stage.

The land use potential reports firstly review the principal advantages and limitations created by environmental conditions in Rondônia. The reports emphasize the variety of natural resources and the range of conditions. From this review there emerges a definition of areas that have potential for primary development, with emphasis being placed on the most promising areas which are evaluated on the basis of their soil, climate, and relief conditions and on their economic potential. The reports also identified areas that should be protected under Article 2 of the Forest Laws (*Codigo Florestal*), which gives a mandate for the preservation of the diversity of natural vegetation formations within Brazil. They also indicated areas with specific attributes worth conserving, including scenic landscapes of possible tourist value. In Rondônia, the recommendations concerned with conservation and wildlife were supported by work undertaken by the Brazilian Institute of Forest Development (Wetterberg *et al.*, 1976; Jorge Padua, undated). The general methodology of this section, which was established in the first Radam report (Azevedo *et al.*, 1973), was dependent on the individual resource surveys, with appropriate weightings being developed for each resource. The result was a calculation of the 'average natural capacity' for each potential land use (Fig. 7).

As well as synthesizing the original resource surveys, the land use potential reports contain socio-economic analyses. These seek to provide basic information for planners and investors and to alert 'competent agencies' to actual and potential problems in the survey areas. This daunting task was not immediately dependent on the radar surveys, although some use was made of available, larger-scale aerial photography. The socio-economic analyses depended mainly on questionnaires and interviews with government and business personnel. In Rondônia, they focused on the dramatic colonization processes, the educational, sanitary and tenurial conditions, the agricultural and mining activities, and the commercial infrastructure. The analyses were based on detailed studies undertaken in six municipalities.

The final part of the land use potential reports deals with aids to planning decisions, based on the methodology established by Rossy-Gralato *et al.* (1976). From the information available in the resource surveys and from the socio-economic analyses, the reports distinguished areas that might be occupied over the short and medium term. This was mainly based on soil and drainage

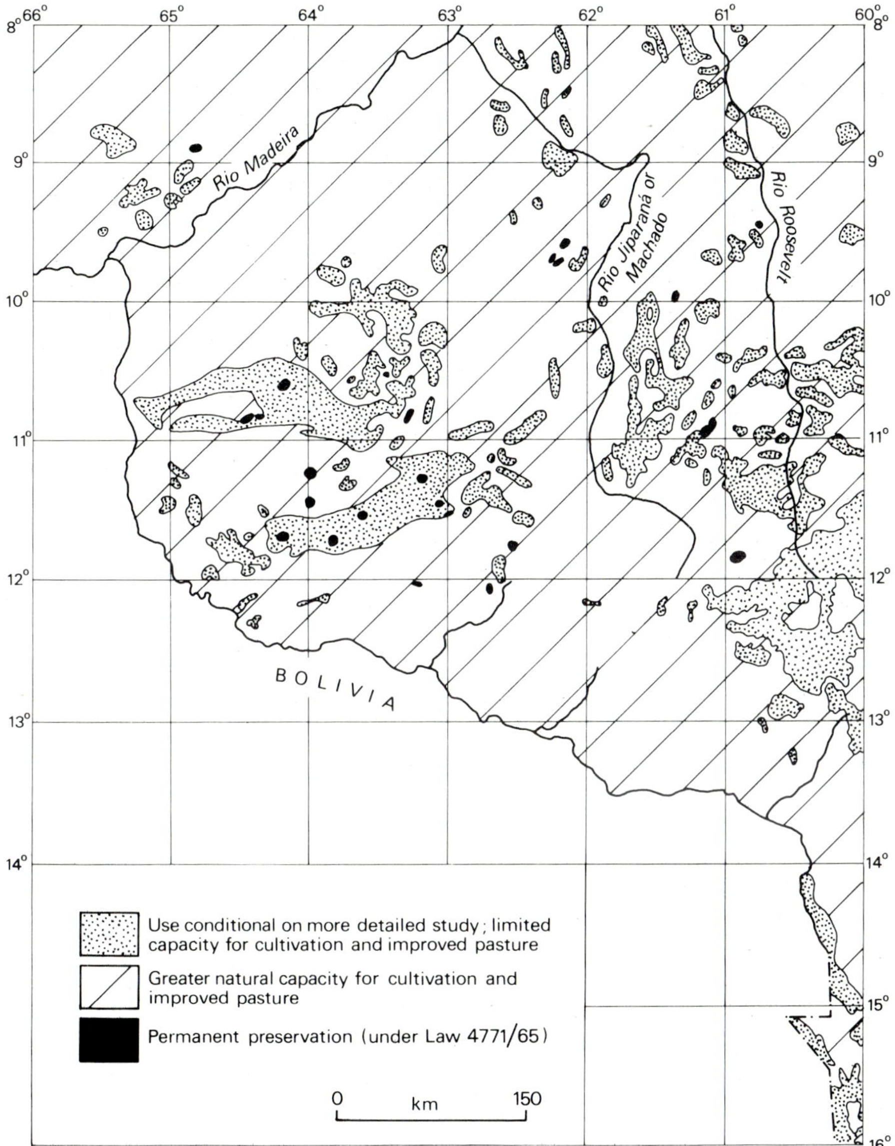


Fig. 7 Land development potential in Rondônia State and adjacent areas determined by Projeto Radambrasil.

characteristics (Table 2). Forestry potential was indicated on the basis of extractable timber estimates, but little attention was given to the problems of sustaining forestry or agricultural enterprises. Finally, the areas selected for different land uses were specified, together with the recommended crops.

Table 2. Recommended land development in Rondônia based on land use potential (after Radam reports).

| Designated area | Definition (simplified) | Recommendation | Per cent total area (approx.) |
|---|---|---|---|
| A | Naturally fertile, suitable climate; no restricting conditions; best opportunities for agro-pastoral development, with least capital investment | Suitable for colonization; small to medium enterprises; short and long cycle crops; diverse range of plants *Controlled occupation* | 1.5 |
| B | Fertile areas subdominant or codominant; no major limitations | Suitable for colonization; small to medium enterprises; short and long cycle crops; intensive dairying *Controlled occupation* | 8.5 |
| C | Inundated for part of the year; usable in the dry season | Subsistence agriculture and small commercial enterprises *Spontaneous occupation* | 4.0 |
| D | Poor soils (chemical fertility), but satisfactory physical properties; no drastic limitations, especially relief; good opportunities with management and technical skills | Medium and large commercial enterprises: timber and beef cattle; small enterprises with cooperative systems *Incentives for occupation* | 40.0 |
| E | Areas with poor soils and limiting physical factors | — | 0 |
| F | Areas with drastic soil, climatic and morphological limitations | Not recommended for any form of development at present *No occupation* | 46.0 |

## CONCLUSIONS

In the early 1970s there were no conventional techniques available to provide effective reconnaissance surveys and base maps for an area such as Rondônia. The SLAR survey, which was the prime data source of the Projeto Radambrasil, was a remarkable achievement. It provided a relatively consistent and informative reconnaissance of geology, geomorphology, soils, and vegetation and supplied the basis for the land use evaluation and firm planning proposals. It was particularly effective for identifying physiograpic units with distinct topographic and drainage characteristics. In Rondônia and the western Amazon in general, the broad patterns of soil and vegetation are controlled by the relief and drainage and so the radar surveys proved to be effective in the analysis of both topographic and soil-vegetation patterns.

It is to be hoped that the Radam project will not be terminated with the completion of the 1:1 000 000 maps and the associated reports. The areas identified as potentially useful for development need to be surveyed at scales of 1:10 000 or 1:20 000 in order to ensure a rational pattern of clearing and cropping in areas of variable soil fertility. The expertise and organization of the Radam teams is needed in the development process, as there are currently few specialists active in the field, as opposed to the office or laboratory. Moreover, there is considerable urgency in this matter because of the rapid rate of land colonization which has in places already pre-empted the original survey recommendations. It seems likely that the teams will be employed for special-purpose surveys, of which some have already been commissioned, but the future of the organization is not yet clear.

In the 1980s the SLAR image coverage can be integrated with other sensor products, such as Landsat MSS and SPOT imagery, as these become available for particular areas. Already the Landsat multispectral (MSS) system is used extensively in Brazil, especially for the Monitoring Programme of the Brazilian Institute of Forest Development (IBDF). This project monitors deforestation, especially of the tropical evergreen forest, and reafforestation, involving mainly eucalypt and pine species. It is also responsible for the 'multidisciplinary mapping' of the National Parks and Reserves. It is likely that natural resource surveys, at progressively larger scales, will be carried out by the new generation of satellites, with their improved levels of ground resolution. In areas of dense forest and thick cloud cover, however, radar imagery will continue to be a useful, investigative, and mapping tool.

## REFERENCES


Ab'Saber, A. N., 1967. Domínios morfoclimáticos e províncias fitogeográficas do Brasil. *Orientação*, **3**, 45–48.

Ab'Saber, A. N., 1970. Províncias geológicas e domínios morfoclimáticos no Brasil. *Geomorfologia*, **20**, 1–26.

Ab'Saber, A. N., 1971. A organização natural das paisagens inter e subtropicais brasileiras. In *III Simpôsio Sobre o Cerrado*, São Paulo: USP e Edgard Blücher, pp. 1–14.

Azevedo, L. H. de, 1971. Radar in the Amazon. *Proc. 7th int. Symp. Remote Sensing Environ.*, **3**, 2303–2306.

Azevedo, L. G. de, Cunha, J. C. T. da, and Tuyama, V., 1973. Uso potencial da terra, avaliação media da capacidade natural do uso da terra. In Brasil, *Levantamento de Recursos Naturais*, vol. 1, *Parte das Folhas SC 23 Rio São Francisco e SC 24 Aracaju*. Rio de Janeiro: Departamento Nacional da Produção Mineral, Ministério das Minas e Energia.

Bagnouls, F. and Gaussen, H., 1963. Os climas biológicas e sua classificação. *Bull. Geogr., Rio de Janeiro*, **22**, 545–566.

Beek, K. J., 1975. *Recursos Naturais e Estudos Perspectivos a Longo Prazo; Notas Metodologicas*. Brasília: SUPLAN.

Bennema, J., 1966. *Report to the Government of Brazil on the Classification of Brazilian Soils*. Rome: Food and Agriculture Organization.

Bennema, J., Beek, K. J., and Camargo, M. N., 1965. *Interpretação de Levantamento de Solos no Brasil; Primeiro Esboço*. Rio de Janeiro: Divisão de Pedologia e Fertilidade de Solo.

Bourne, R., 1978. *Assault on the Amazon*. London: Gollancz.

Brasil, 1976. *Levantamento de Recursos Naturais, Folha SC 19,* vol. 12, *Rio Branco*. Rio de Janeiro: Departamento Nacional da Produção Mineral, Ministério das Minas e Energia.

Brasil, 1978a. *Levantamento de Recursos Naturais, Folha SC 20*, vol. 16, *Porto Velho*. Rio de Janeiro: Departamento Nacional da Produção Mineral, Ministério das Minas e Energia.

Brasil, 1978b. *Documento Preliminar de Uniformização da Nomenclatura Fisionômico-ecológico*. Relatório Interno Radambrasil 5-V, Departamento Nacional da Produção Mineral, Rio de Janeiro.

Brasil, 1979. *Levantamento de Recursos Naturais, Folha SD 20*, vol. 19, *Guaporé*. Rio de Janeiro: Departamento Nacional da Produção Mineral, Ministério das Minas e Energia.

Brasil, 1980. *Levantamento de Recursos Naturais, Folha SC 21*, vol. 20, *Juruena*. Rio de Janeiro: Departamento Nacional da Produção Mineral, Ministério das Minas e Energia.

Brasil, 1982. *Levantamento de Recursos Naturais, Folha SD 21*, vol. 26, *Cuiabá*. Rio de Janeiro: Departamento Nacional da Produção Mineral, Ministério das Minas e Energia.

Cailleux, A. and Tricart, J., 1957. Zones phytogeographiques et morphoclimatiques au Quaternaire au Brésil. *C.r.somm. Séanc. Soc. Biogéogr.*, **34**, 293, 7–13.

Commissão Executiva do Projeto Radambrasil, 1983. *Informações Básicas*. Brasília: Secretaria Geral, Ministério das Minas e Energia.

Cruz, O., 1974. *Sobrevôos no Território de Rondônia, Norte de Mato Grosso e Sul do Pará e Amazonas*. Relatório Interno Radam 1-GM, Rio de Janeiro.

Ellenberg, H. and Mueller-Dombois D., 1965–66. A tentative physiognomic and ecological classification of plant formations of the earth. *Ber.geobot. ForschInst. Rübel*, **37**, 21–55.

Furley, P. A., 1979. Levantamento do solo e uso potencial de terra. In SUDECO, *Diagnóstico e Estudo de Perspectivas para o Território Federal*. Brasília: Superintência do Desenvolvimento da Região Centre-Oeste, Ministério do Interior.

Furley, P. A., 1980. Development planning in Rondônia based on naturally renewable resource surveys. In Barbira-Scazzochio, F. (Ed.), *Land, People and Planning in Contemporary Amazonia*. Cambridge: Centre for Latin American Studies.

Howard, A. D., 1967. Drainage analysis in geologic interpretation: a summation. *Bull. Am. Ass. Petrol. Geol.*, **51**, 2246–2259.

Jorge Padua, M. T., undated. *Parques Nacionais e Reservas Equivalentes*. Brasília: Departamento de Parques Nacionais e Reservas Equivalentes, Instituto Brasileiro de Desenvolvimento Florestal.

Lewis, A. J., 1974. Geomorphic-geologic mapping from remote sensors. In Estes, J. E. and Senger, L. W. (Eds), *Remote Sensing*. Santa Barbara: Hamilton, pp. 105–126.

de Loor, G. P., 1983. *Radar Remote Sensing. Remote Sensing Reviews*, vol. 1, no. 1. New York: Harcourt.

MacDonald, H. C., 1969. Geologic evaluation of radar imagery from Darien province, Panama. *Mod. Geol.*, **1**, 1–63.

Moreira, H. F., 1973. *Controle de Qualidade de Imagens de Radar; Projeto Radam.* Relatório Interno Radam 9-T, Rio de Janeiro.

Roessel, J. W. van and Godoy, R. C. de, 1974. SLAR mosaics for Project Radam. *Photogramm. Engng,* **40**, 583–595.

Rossy-Gralato, I. R. *et al.*, 1976. Uso potencial da terra. In Brasil, *Levantamento de Recursos Naturais, Folha SA 21*, vol. 10, *Santarém*. Rio de Janeiro: Departamento Nacional da Produção Mineral, Ministério das Minas e Energia, pp. 415–522.

Sampaio de Almeida, A. L. de, 1984. *Realizacões*. Brasília: Projeto Radambrasil, Ministério das Minas e Energia.

Sociedade Brasileira de Ciência do Solo, 1973. *Comissão de Metodo de Trabalho de Campo*. Rio de Janeiro: Sociedade Brasileira de Ciência do Solo.

Soil Survey Staff, 1975. *Soil Taxonomy*. Agricultural Handbook 436. Washington: Soil Conservation Service.

Thornthwaite, C. W. and Mather, J. R., 1955. The water balance. *Publs Clim. Drexel Inst. Technol.*, **8**, 1–104.

UNESCO, 1973. *International Classification and Mapping of Vegetation. Ecology and Conservation* 6. Paris: United Nations Educational, Scientific and Cultural Organization.

Veloso, H. P., Japiassu, A. M. and Goes-Filho, S. L., 1973. As regiões fitoecológicas, sua natureza e seus recursos econômicos; estudo fitogeográfico. In Brasil, *Levantamento de Recursos Naturais*, vol. 1, *Parte das Folhas SC 23 Rio São Francisco e SC 24 Aracaju*. Rio de Janeiro: Departamento Nacional da Produção Mineral, Ministério das Minas e Energia.

Vettori, L., 1969. Método de analise de solo; equipe de pedologia e fertilidade do solo. *Boln téc., Rio de Janeiro*, **7**, 1–24.

Wetterberg, G. B. *et al.*, 1976. Uma analise de prioridades em conservação na Amazônia. *Serie Técnica, IBDF/PROPEPEF*, Brasília **8**, 1–62.

White, L. P., 1977. *Aerial Photography and Remote Sensing for Soil Survey*. Oxford: Clarendon Press.

Remote Sensing and Tropical Land Management
Edited by M. J. Eden and J. T. Parry


# CHAPTER 6

# *Landsat and the Detectability of Land Systems in Northern Kenya*

JOHN T. PARRY and MARK G. WILLIAMS
*Department of Geography, McGill University, Canada and*
*Department of Geography, University of New South Wales, Australia*

During the last fifty years increased population pressure on many countries has necessitated a serious examination of available resources and resource management methodology. At the national level there is a need for resource data banks to store existing information, incorporate new data that are being made available by a variety of techniques, and accommodate future revision and updating. Clearly such programmes require a geographical data base that allows the collection, reduction, and organization of data in a form that is readily accessible, easily understandable at the local level, and applicable at both macro and micro scales. It has been recognized for more than thirty years that land systems constitute a valid framework for mapping and classifying a wide range of resource data and also provide a basis for extrapolation from known to relatively unknown areas (Christian and Stewart, 1968). With the launch of the ERTS (Landsat) in July 1972 a new perspective on land resources was provided. For the first time it was possible to examine a landscape at the synoptic level and to see whole land systems in their regional context. Imagery from orbital altitudes is particularly useful in the analysis of the land systems of arid and semi-arid lands, such as those of northern Kenya, because of the clarity with which surface features can be interpreted in the relative absence of cloud and dense vegetation (Parry, 1983).

## LAND SYSTEMS AND LANDSAT: AN INTEGRATED APPROACH

In most regional surveys using the land system methodology, information about terrain conditions is provided by field survey, reference to existing data sources, such as regional geological surveys, soil reports, vegetation surveys, etc., and by air photo interpretation. In better-known areas, thematic data are generally available, although in remote, little-known areas the regional planner may be forced to rely almost exclusively on information obtained by remote sensing. In this situation Landsat imagery would appear to offer considerable potential as a data source. Some recent studies in Kenya illustrate how information on rangeland conditions, land degradation, and hydrology can be acquired from Landsat utilizing a variety of techniques (Gwynne, 1977; Odenyo, 1979; Hellden, 1980; Rango, 1980; Griffiths and Collins, 1981).

The Landsat user is faced with only minimal costs in carrying out reconnaissance-level resource inventory using analogue techniques. The imagery is multispectral and so the examination of specific features within discrete spectral bands is possible. The coverage is multi-temporal and so change detection can be attempted within the limits set by the scale of the imagery. However, before effective use of Landsat is possible it is necessary to address some fundamental questions concerning the rendition of land systems on the imagery.

Semi-arid regions are subject to seasonal climatic variations resulting in marked changes in certain landscape elements. In addition, there are longer-term climatic fluctuations which have brought serious droughts to much of the Sahel region. These changes affect the degree of detectability of many of the parameters used in the definition of land systems. For example, a brief rain spell will generally enhance the drainage pattern, improving its detectability because of the vegetation 'flush' along the channels; however, seasonal rains promote regional vegetation growth which often leads to a marked reduction in the visibility of structural and lithological detail. Landscape patterns are of fundamental significance in the delineation of land systems and so any changes in their level of detectability will affect the accuracy of the land system classification.

In this study an attempt is made to identify the extent to which there are changes in the detectability of land systems from season to season. The Maralal area of northern Kenya is taken as a representative sample of conditions experienced in the semi-arid zone of east Africa. The Maralal transect provides an opportunity to address the following interrelated problems:

1. The ease with which land system boundaries can be detected and delineated on Landsat imagery.
2. The extent to which the detectability of land system boundaries changes as a result of seasonal variations in climate.
3. The relative accuracy of the land system boundaries obtained from Landsat imagery compared with those obtained using conventional methods.
4. The relative detectability of land units at different levels in the land system hierarchy.
5. The extent to which relative detectability within the land system hierarchy changes as a result of seasonal variations in climate.

In essence, this study attempts an analysis of the temporal variability in the visibility of the land systems of the Maralal area as seen by the four-channel multispectral scanning system of Landsat. The conclusions are directly relevant for any researcher concerned with land systems in semi-arid areas.

## THE MARALAL STUDY AREA

Maralal is located in Samburu District, northern Kenya, approximately 250 km north of Nairobi. A transect of 7000 $km^2$ (180 km × 40 km) was selected for analysis by the McGill-Nairobi joint university project on the ecosystems of semi-arid Kenya in 1981. The transect includes a variety of landscape types from recent volcanic forms to extensive planation surfaces on rocks of the Basement Series. Differences in the regional climate give rise to a range of ecosystems from arid to subhumid. As indicated in Fig. 1, there is an ecoclimatic gradient from the subhumid conditions of the highland areas (Karisia Hills and Mathews Ranges) to the arid conditions of the Rift Valley in the west and the Laijok Plains in the east. In the rangeland classification of Pratt *et al.* (1966) this corresponds to a change from ecoclimatic zone II to ecoclimatic zones V and VI.

### Geological history

Geology and erosional history provide the main controls on the pattern of land systems in the Maralal area. The oldest lithological units in the area belong to the Basement Series (Shackleton, 1946). The main highland masses, such as the Karisia Hills and the Mathews Ranges, are composed of resistant granite gneisses of the Basement Series, the product of regional metamorphism in the Cambrian Period. After uplift in late Cambrian time, the area remained stable, and was subjected to subaerial erosion for an extended period of time. The end-product was the Cretaceous erosion surface which is represented by accordant summit levels at 2400 m elevation in the Karisia Hills and Mathews Ranges. Continued erosion in Eo-Oligocene times accompanied by isostatic readjustment resulted in incision of the drainage lines and the establishment of lower erosion levels in the classic pediplanation sequence (Pugh, 1955). As shown in Fig. 2, the sub-Miocene surface is developed over an extensive area. It forms a fairly uniform plateau at approximately 1400 m surrounding the more resistant massifs where remnants of the Cretaceous surface are preserved. A subsequent episode of isostatic readjustment initiated another erosion cycle resulting in the development of a lower, end-Tertiary surface which can be identified in the eastern part of the Maralal transect. The more irregular terrain of the southeastern part of the transect is characterized by the interpenetration of the sub-Miocene and end-Tertiary surfaces.

The landscape of old plateau surfaces has been completely obliterated in the western section of the transect by great thicknesses of lava emanating from fissure flows in the Miocene Period and from later central vent eruptions in the Rift Valley. The lavas are characteristically alkaline consisting of phonolites,

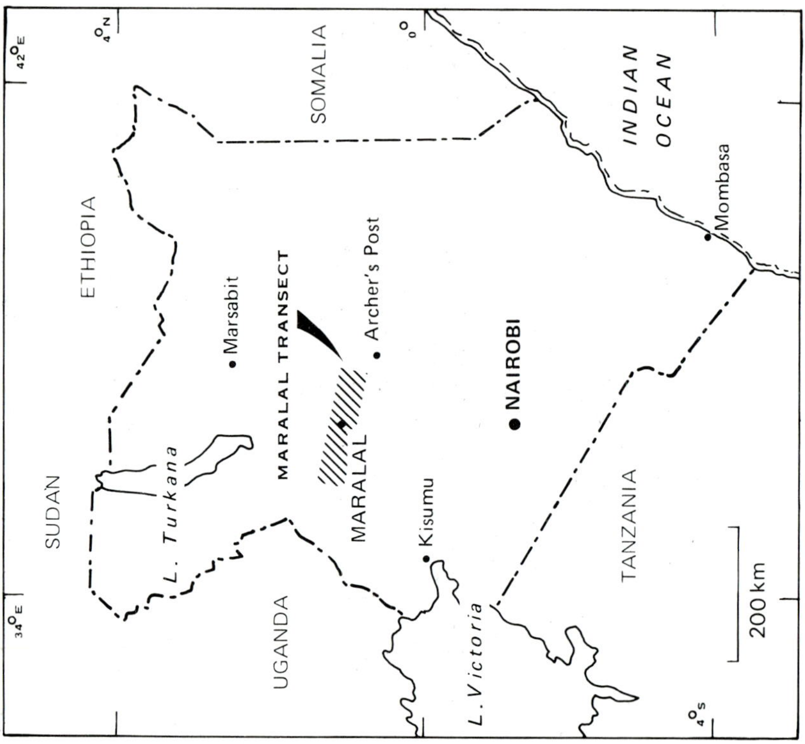


Fig. 1a Location of the Maralal transect.

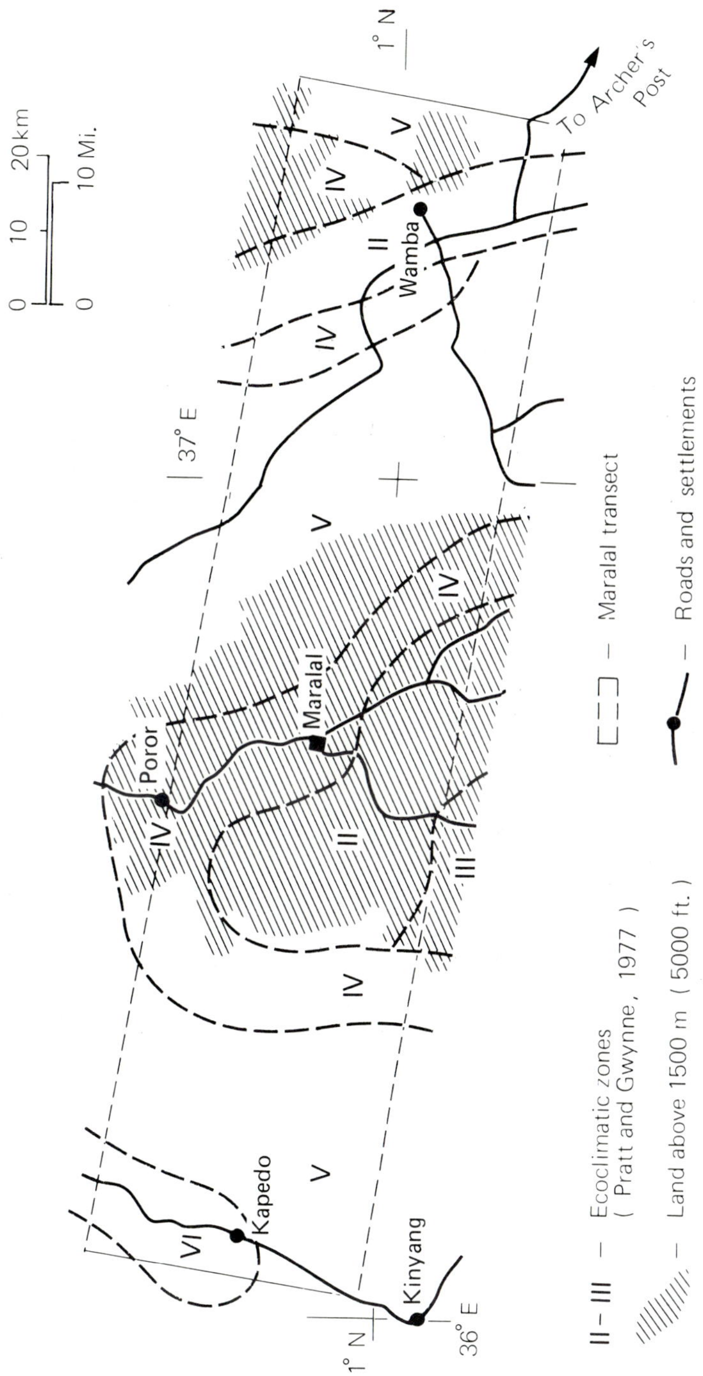


Fig. 1b Ecoclimatic zones of the Maralal transect (after Pratt *et al.*, 1966).

nephelinites, basalts, and tuffs. Two major groups were distinguished by Shackleton (1946): a lower series of basalts with associated red and purple tuffs, and an upper series consisting of phonolites which cover more than a third of the transect (Fig. 2). The underlying topography exercised some control on the spread of these lavas. The eastern scarps of the Rift Valley step faults mark the limit of the eastward spread of the Samburu Basalts, and the Karisia upland served to stem the flood of the Plateau Phonolites except in the south, where some of the lavas spilled through a lower gap and appear to have invaded the valley of the proto-Seya River. Remnants of this flow are preserved in the landscape today as a sinuous plateau outlier providing a nice example of inversion of relief.

The floor of the Rift Valley is occupied by younger volcanics, including the large strato vents of Pakka and Silali. The latter has experienced different phases of activity involving both tephra accumulation and lava flows. A very recent destructive episode has resulted in the formation of a large caldera bounded by a collapse rim.

## Climate

The critical parameter in the climate of the semi-arid areas of Kenya is rainfall. The main characteristics of the rainfall pattern of the Maralal transect can be summarized as follows:

1. Relatively low total amounts as indicated in Table 1 and Fig. 3. Rainfall data are available for six stations within or close to the transect (their locations are shown in Fig. 1). The mean annual precipitation for all but one of the stations is below 750 mm, the minimum limit identified by Nieuwolt (1978) for most types of agriculture. It should be noted, however, that rainfall totals from

Table 1. Stations maintaining weather records within or close to the Maralal transect.

| Station name | Situation in the transect | Record period | Mean annual precipitation (mm) |
|---|---|---|---|
| Kapedo Mission | West | 1966–79 | 446.6 |
| Kinyang Health Centre | | 1966–78 | 624.5 |
| Poror Police Post | Central | 1966–79 | 628.4 |
| Maralal District Office | | 1966–79 | 632.1 |
| Wamba | East | 1969–79 | 772.3 |
| Archer's Post | | 1966–79 | 355.9 |

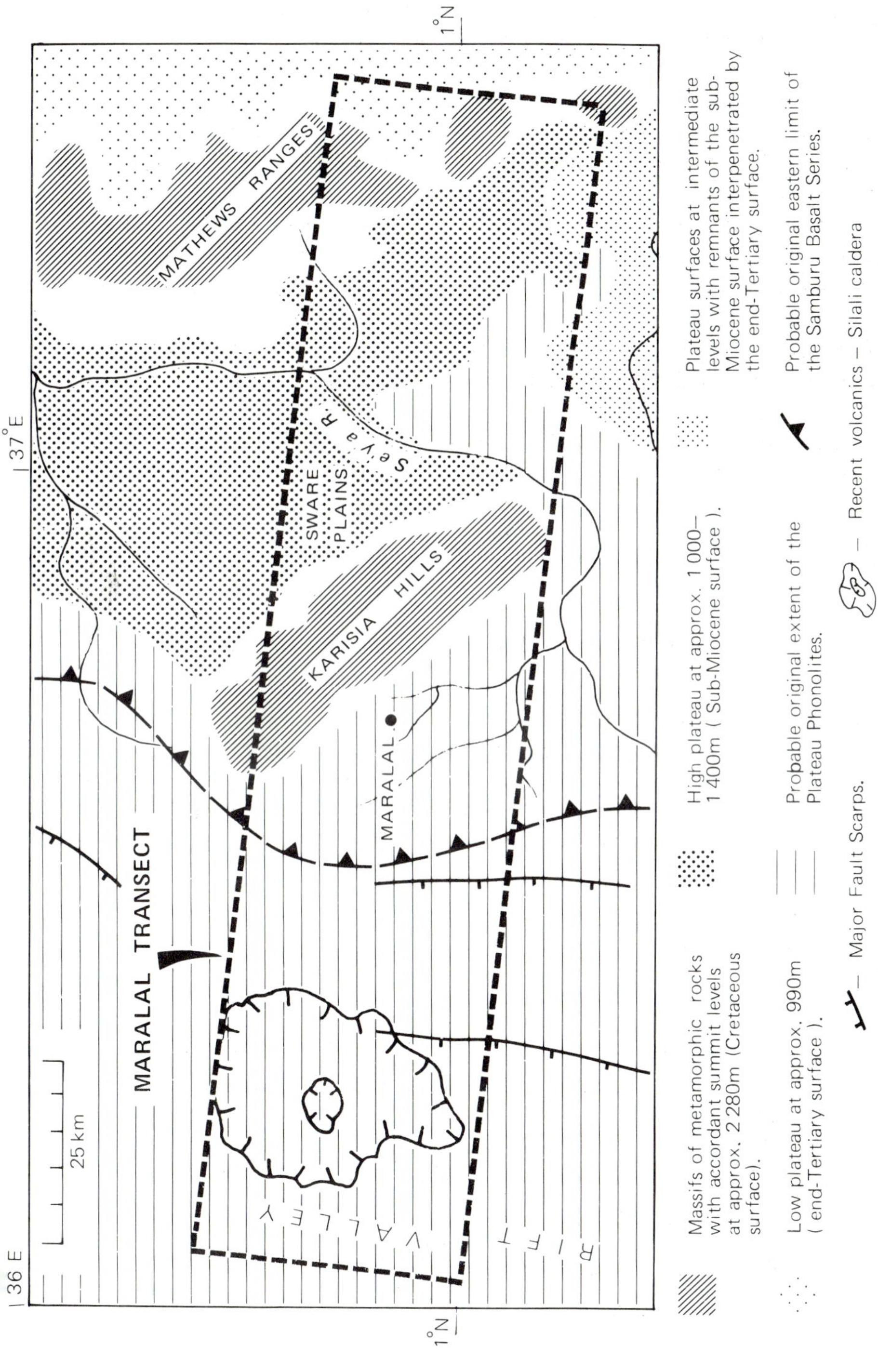


Fig. 2 Major physiographic units of the Maralal transect (adopted from Shackleton, 1946).

900 mm to more than 2000 mm are received in the Karisia Hills and Mathews Ranges as a direct result of orographic effects (Trapnell and Griffiths, 1960).
2. Great variability from year to year. An examination of the graphs of annual precipitation illustrates this variability (Fig. 3). Precipitation in the central portion of the transect is generally more reliable than in the eastern or western sections because of the orographic effect of the Karisia Hills and the same effect can be expected in the Mathews Ranges. The greatest variability in rainfall is experienced in the eastern portion of the transect, particularly the southeast section, which is characterized by the data from Archer's Post.
3. Disparity in the seasonal distribution of rain across the transect. Examination of the monthly precipitation data reveals a very different pattern between the east and west sections of the transect (Fig. 4). In the east, the rainfall distribution is markedly bimodal with one peak in the period February to April and another in the period September to November. These rain periods are the product of the easterly wind systems of the Inter-Tropical Convergence Zone; a northeasterly component, strongest in April, and a southeasterly component which assumes dominance by November, reaches farther inland, and generally provides heavier precipitation. However, it can be seen from Fig. 4 that these rains are not reliable; failures occurred in 1970 and again in 1974. Archer's Post recorded a four-year failure in the late season rains (1973–1976), resulting in a major drought. In the western and central portion of the transect, the rainfall pattern is unimodal. The moisture bearing winds are the Congo Westerlies, which fluctuate in strength in response to changes in the dominant easterlies. A major surge occurs in the period July to September, reaching well within the Maralal transect (Kenworthy and Glover, 1958; Hills, 1979). This airmass is very unstable as it is forced upslope resulting in orographic rainfall in the Karisia Hills. A minor westerly surge frequently occurs in April and May with associated orographic rainfall. This is a significant rainfall period for the western section of the transect, as shown by the data for Kapedo and Kinyang.

## Vegetation

The pattern of vegetation in the Maralal transect is closely related to the pattern of rainfall. The availability of moisture is probably controlled here, as in the rest of the semi-arid region, by the combined effects of rainfall and temperature (Pratt and Gwynne, 1977). Dense evergreen forests characterized by *Ocotea usambarensis* and *Podocarpus* sp. tend to dominate the cooler, high rainfall areas above 1500 m, such as the Karisia Hills and Mathews Ranges. In somewhat drier upland areas, more open woodlands occur with cedar (*Juniperus procera*) and *Olea* sp. predominating. On the open plains, various bushland and grassland communities occur. In some areas, edaphic and biotic factors override the climatic in determining the distribution of vegetation types. Fire is historically important in these areas and has stimulated the spread of red oatgrass (*Themeda*

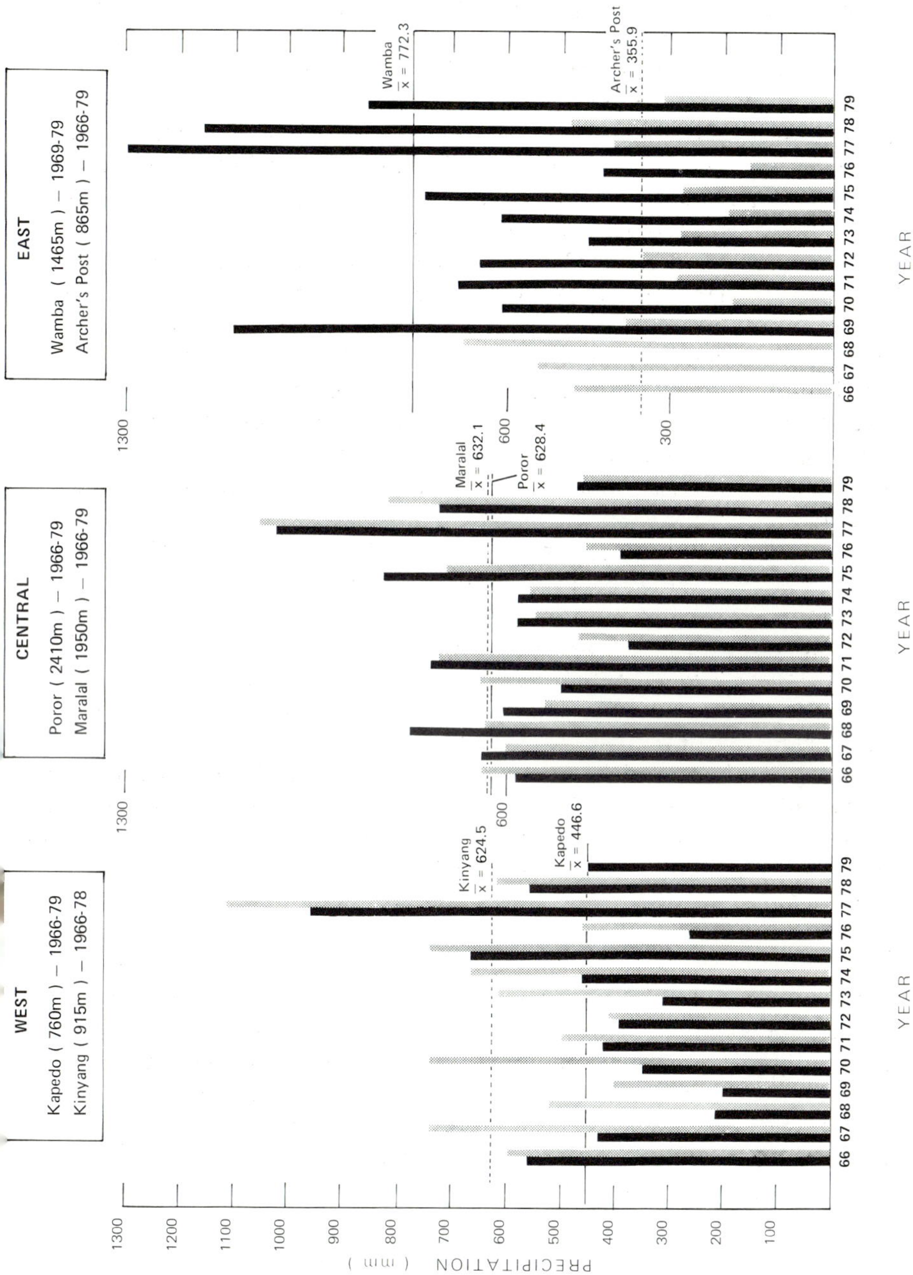


Fig. 3 Annual precipitation for stations within or close to the Maralal transect, 1966–1979.

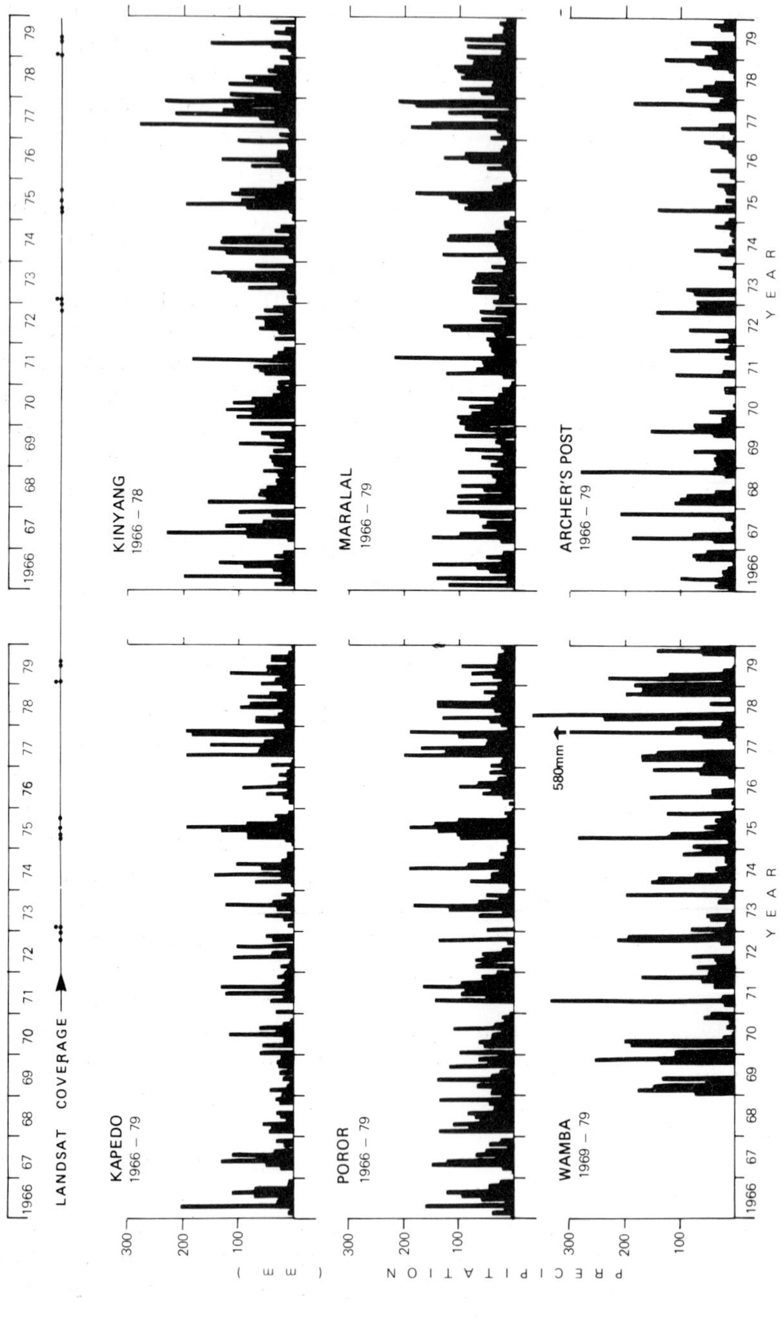


Fig. 4 Monthly precipitation for stations within or close to the Maralal transect, 1966–1979.

*triandra*) and the regeneration of various species of *Acacia*. Ligneous thorn plants and succulents with a sparse grass cover tend to dominate the drier plains, and in the lowest rainfall areas of the Rift Valley sparse semi-desert bush is characteristic. Bushland types tend to be deciduous and thorny consisting of *Acacia commiphora* with *Boscia* sp. (Lind and Morrison, 1974).

Over much of the transect the vegetation growth is essentially seasonal. There is a marked flush with the onset of a period of rains and dieback and dormancy in the following dry period. The appearance of the landscape changes quite dramatically from season to season and regional patterns as seen on aerial photographs or Landsat imagery vary accordingly. It is essential, for proper interpretation of the Landsat scene, to have information about the vegetation conditions in the period immediately preceding the overpass. In the Maralal transect the link between rainfall and vegetation growth is direct and immediate even in situations where edaphic factors are dominant. Thus, the rainfall data for a particular station can be used to determine the status of the vegetation in that area at the time of the Landsat orbit. Reference to the details of the antecedent rains in different parts of the transect, as revealed by the monthly precipitation data (Fig. 4), allows an assessment of the state of the ground and the vegetation growth stage—dormant, die-back, flush, etc.—at the time of each Landsat overpass, These details are summarized in Table 2.

## THE LAND SYSTEMS CONCEPT

In landscape science the term 'land' is defined as a complex of surface and near surface attributes that exhibit spatial variability and when expressed in combination give local character or individuality (Mabbutt, 1968). In order to define and describe the character and potential of particular tracts of 'land' it is useful to impose a framework of generalization. Land classification provides such a framework by identifying physiographic units in an hierarchical series. The physiographic approach to land classification was initially developed by Bourne (1931) and Veatch (1933) for forestry and agriculture purposes. It has been refined by a number of organizations in different parts of the world to meet different local requirements; however, the main elements are similar in all the systems, which include those of the Commonwealth Scientific and Industrial Research Organization, Australia (CSIRO), the Military Engineering Experimental Establishment, United Kingdom (MEXE), and the United Nations Food and Agriculture Organization (FAO).

The land systems of western Kenya were identified and mapped by a research group at Oxford University under contract to MEXE. The approach followed in preparing the *Land Systems Atlas of Western Kenya* (Scott *et al.*, 1971) was similar to that used by MEXE researchers in other areas (Beckett and Webster, 1965). A standard vertical air photo coverage (medium scale panchromatic 1:40 000) was used to identify a three-level hierarchical series consisting of land

Table 2. Landsat image data and ground conditions in the Maralal transect.

| Year | Date | Scene ID | Band quality 4 | Band quality 5 | Band quality 7 | % cloud | Sun elevation° | Sun azimuth° | Antecedent rain/stage of vegetation growth: West | Central[†] | East | Mean Ca value[‡] |
|---|---|---|---|---|---|---|---|---|---|---|---|---|
| 1972 | Sep 9 | 1048–07163 | 5 | 5 | 5 | 10 | 56 | 082 | E/B | Dy/D | Dy/D | 0.652 |
| | Nov 20 | 1120–07172 | 2 | 5 | 5 | 30 | 53 | 126 | L/D | H/A | H/A | 0.606 |
| 1973 | Jan 13 | 1174–07165 | 8 | 8 | 8 | 20 | 47 | 124 | Dy/D | Dy/B | L/D | 0.583 |
| | Jan 31 | 1192–07171 | 8 | 2 | 8 | 0 | 48 | 118 | Dy/D | Dy/B | H/F | 0.657 |
| 1975 | Mar 25* | 2062–07052 | 5 | 5 | 8 | 10 | 51 | 089 | Dy/D | Dy/D | Dy/D | 0.664 |
| | Sep 21* | 2242–07041 | 5 | 5 | 5 | 10 | 54 | 090 | H/A | H/A | S/F | 0.616 |
| 1979 | May 24 | 30445–07054 | 8 | 8 | 8 | 20 | 49 | 058 | E/A | E/B | E/B | 0.518 |

*State of ground vegetation*
A: active growth, B: brown off, D: dormant, Dy: dry, E: end of rain period, F: vegetation flush, H: heavy rain, L: light rain, S: start of rain period

*Band quality*
1 Poor → 8 excellent

* Images on which the ABD test was performed
[†] irrespective of orographic effects in mountain areas
[‡] Ca value: coefficient of areal correspondence

systems, land units or facets, and their constituent land elements or sites. Land systems are defined by the MEXE researchers as blocks of country, each having a common lithology and genesis giving rise to a recurrent pattern of recognizable features on an air photograph or in the field (Beckett and Webster, 1965). This is essentially similar to the definition adopted by CSIRO researchers in Australia who considered land systems as areas or groups of areas throughout which there is a recurring pattern of topography, soils, and vegetation (Christian and Stewart, 1968). Each land system consists of a number of land units or facets, each with its own distinctive combination of topography, lithology, and soil, and a soil water regime that is uniform or varies in a simple or predictable way. Land units are essentially homogeneous for practical purposes. They are combined in the landscape and form a recurrent pattern which gives distinctive character to each land system. At the base of the hierarchy are smaller constituents of terrain which are considered indivisible in most systems. These are the land elements or sites consisting of flats and slopes which are not divisible further on the basis of form.

Within the Maralal transect, 16 land systems have been identified each with its own distinctive recurring pattern of soil, topography, and vegetation. The *Land Systems Atlas of Western Kenya* (Scott *et al.*, 1971) provides a description of each land system including its climate, geology, relief, and altitude, its geomorphology, and its dominant vegetation and soil type. The component facets are defined and described individually with reference to the form, surface material (soil, rock or surficial deposits), water regime, and land cover, which may be either natural vegetation or the product of a particular land use. The distribution of the land systems in the Maralal transect is shown in Fig. 5(a). Several land systems have multiple occurrences so that in all 40 map units are encountered in the transect. Scott *et al.* (1971) did not attempt to map land units, but the Atlas provides a schematic block diagram and concise listing of the land units in each land system. As an example of the level of detail provided in the Atlas, the description of the Lopet land system and its component land units is reproduced in Fig. 6.

## LANDSAT COVERAGE

Spacecraft photography of Kenya has been available since the initial orbits of Landsat 1 in 1972 and now includes images acquired by Landsat 2 (launched in 1975) and Landsat 3 (launched in 1978). Unfortunately, the coverage has been irregular and of variable quality. Until recently all of East Africa lay beyond the range of the Landsat receiving stations, and so Kenya was dependent on the onboard tape recording system. Both Landsat 1 and Landsat 2 experienced tape recorder difficulties. The absence of images for most of 1974, for example, was the result of the near total failure of the recording system for Landsat 1. The limited life of the wide-band video tape used in the recorders resulted in

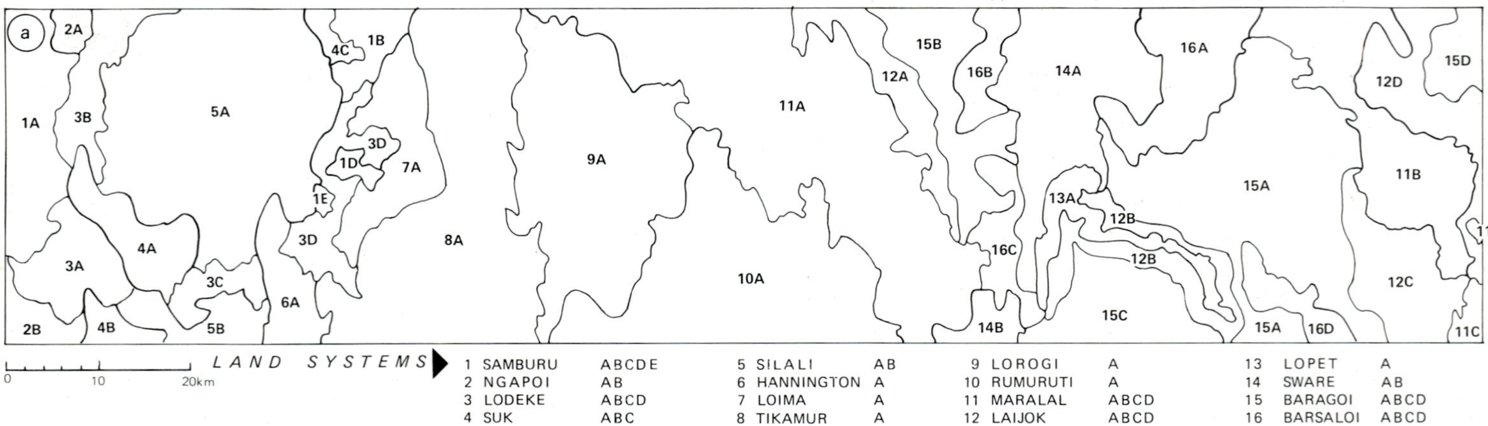


Fig. 5(a) The land systems of the Maralal transect as shown in the *Land Systems Atlas of Western Kenya* (Scott *et al.*, 1971).

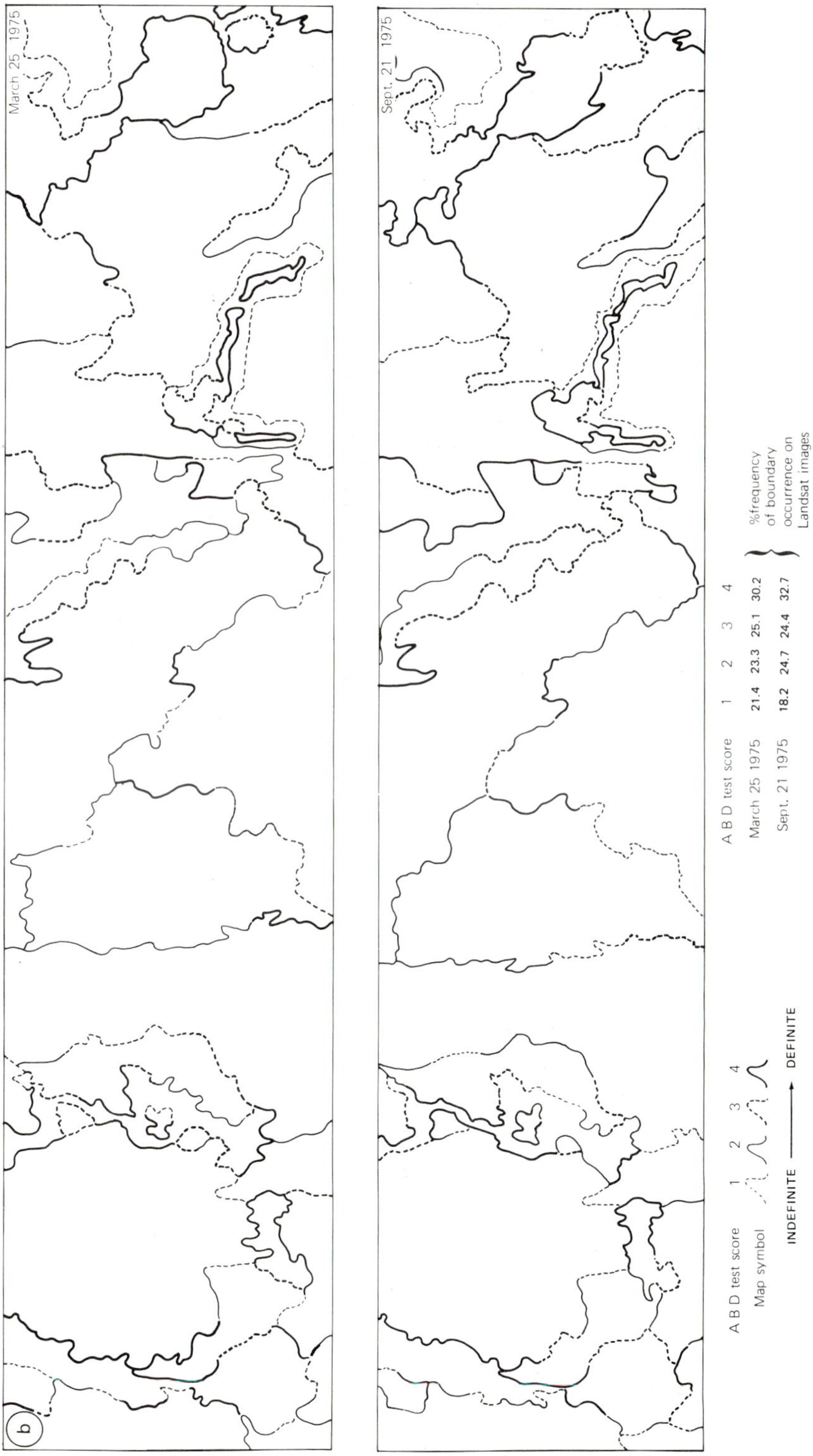


Fig. 5(b) The land systems of the Maralal transect as mapped from Landsat images at different seasons showing boundary definition (strength) on the bases of ABD test scores. March 25 1975 (dry) image ID 2062–07052. September 21 1975 (wet) image ID 2242–07041.

Table 3. Landsat multispectral scanner (MSS) bands.

| MSS band | Wavelength $\mu m$ | | Colour filter used in processing |
|---|---|---|---|
| 4 | 0.5–0.6 | green | blue |
| 5 | 0.6–0.7 | red | green |
| 6 | 0.7–0.8 | infrared | red |
| 7 | 0.8–1.1 | infrared | red |

other problems: firstly, deterioration in the quality of the images generated and, secondly, inadequate coverage, particularly sequential series because of the reluctance of NASA to expend the remaining hours of recording time on anything short of major disasters. These factors explain why the images available for analysis are so few in number and occur in such an irregular time series.

Landsat is in sun-synchronous orbit and the equatorial crossing occurs at approximately 0930 hours local time. Solar elevations for the Maralal area vary from 47° to 56° giving good relief enhancement effects as a result of shadowing. The Landsat multispectral scanner (MSS) is a four-channel system with an instantaneous field of view of 0.086 mrad (Table 3). This generates false colour imagery with a nominal ground resolution of $79 \times 79$ m which means that each image pixel is integrating the reflection spectra from the various landscape elements occurring within an area of 6241 $m^2$. In this context the typical ranges in the areal extent of the various components in the land system hierarchy should be noted: land systems 10 $km^2$ – > 1000 $km^2$, land units 1 $km^2$ – > 100 $km^2$, land elements 100 $m^2$ – > 1 $km^2$.

| FACET No. | FORM | SOILS, MATERIALS AND HYDROLOGY | LAND COVER |
|---|---|---|---|
| 1 | **Lava plain.** Level to moderate slopes; variable extent; usually with abrupt margins. | Lava stone mantle over dark brown plastic loam to clay loam with $CaCO_3$ concretions to about 30 cm over weathered lava. | Desert scrub — very sparse *Aristida hertigtumis*. |
| 2 | **Steep dissection slopes.** Straight or irregular with numerous V-shaped gullies up to 100 m high. | Mainly lava boulders and lava outcrop. | As 1. |
| 3 | **Valley floors.** Narrow (50–200 m wide), flat bottoms, including stream channel which is frequently braided. | Loose sands and lava boulders. Channel flooded for short periods. | Sparse *Acacia misera*. |
| 4 | **Hills.** Conical or elongated; usually with steep straight sides but may be very irregular; up to 100 m high. Standing on plateau surface. | As facet 2. | As facet 1. |
| 5 | **Volcanic cones.** 500–1,500 m across with: (a) Crater floor. (b) Very deep inner slopes. (c) Steep to moderate outer slopes. | Volcanic ash. | |
| 6 | **Ash plains.** Level to moderately sloping, very variable in extent — irregular abrupt margins. Crossed by diffuse stream channels. | Volcanic ash and pumice. | |

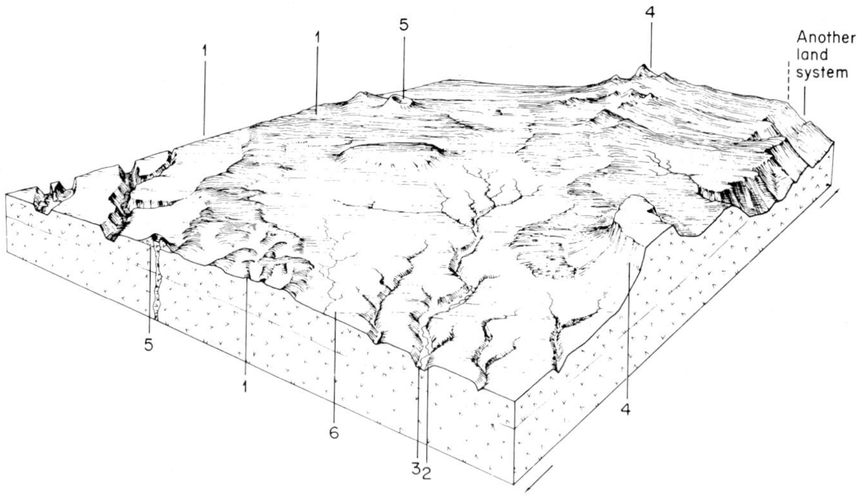


**LOPET KE 19**

| | |
|---|---|
| **Climate:** | Rainfall 150–250 mm. April. |
| **Geology:** | Tertiary to Recent volcanics, mainly basalt but including some trachyte, phonolite, etc., and ashes, pumice and agglomerates. |
| **Landscape:** | Extensive plateau dissected by frequent deep, narrow valleys with a dendritic pattern, locally with volcanic cones and irregular lava flow features. |
| **Soil:** | Lava boulders (Rock and rock debris). |
| **Vegetation:** | Desert scrub. |
| **Relief:** | about 100 m. |
| **Altitude:** | 370–2, 130 m. |

Fig. 6 The Lopet land system. An example of one of the sixteen land systems in the Maralal transect adopted from the *Land Systems Atlas of Western Kenya* (Scott *et al.*, 1971). (*Reproduced by permission of Royal Armament Research and Development Establishment, Christchurch.*)

Thus, land systems are clearly above the resolution limits of the Landsat system, whereas land elements are close to or below the resolution threshold.

The Maralal transect occupies approximately one-quarter of the image area of Landsat scene 181–059 which covers the greater part of Samburu District in north central Kenya. In the period 1972–79 twelve colour composite images (bands 4, 5, 7) are available for the transect. Of these, only seven have less than 30% cloud cover and are of sufficiently good quality to warrant detailed analysis (Table 2).

## MAPPING LAND SYSTEMS BOUNDARIES FROM LANDSAT

Mapping of the land system boundaries in the Maralal transect was carried out using a Zoom Transferscope (Bausch and Lomb ZT3). One of the better quality images (2062–07052, 25 March 1975) with minimum cloud cover (< 10%) was set up on the transillumination surface of the instrument (Fig. 7). A reduction of the Maralal topographic sheet (NA-37-A 1:250 000) of the Survey of Kenya was used as a base map at an effective scale of 1:500 000. Delineation of the

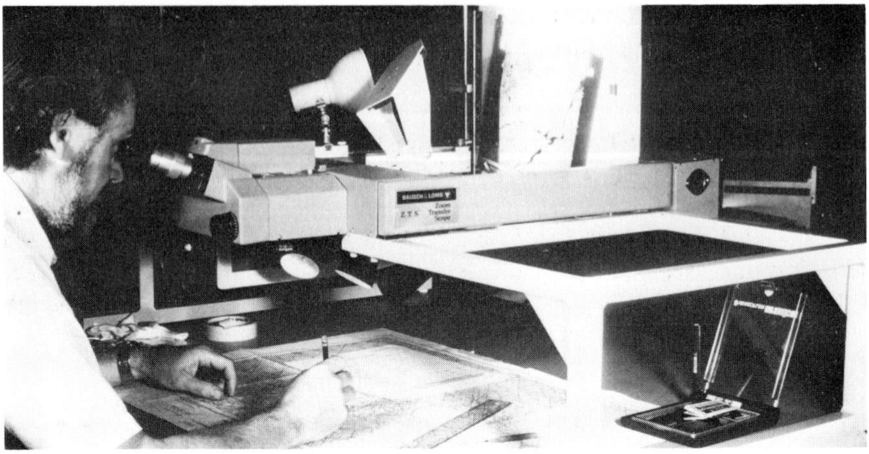

Fig. 7 Zoom Transferscope (Bausch and Lomb ZT3).

land system boundaries was carried out on a transparent overlay without reference to the land systems map of Scott *et al.* (1971), although the land system names are retained in this discussion for ease of reference.

The procedures followed in identifying land system boundaries were essentially subjective. The visible image characteristics of colour, pattern, texture, size, shape, and shadow were assessed and related deductively to the landscape elements—landform, surface material and soil, drainage, and vegetation—following the scheme outlined by Verstappen (1977) and Hilwig (1980). In situations where image characteristics indicated a change in landscape elements a boundary was drawn following the line of demarcation or transition.

It was found that in areas of marked relief land system boundaries could be readily drawn on the basis of landform alone. In areas of more subdued relief it was necessary to give more careful consideration to landscape texture, the colour of surface materials, and vegetation. Significant colour differences were frequently found to be associated with changes in surface texture, and minor variations in landform pattern and vegetation were often found to be correlated. It should be noted that the areas delineated were not necessarily homogeneous with regard to texture and colour. For example, vegetation differences (with the associated colour changes) were accepted within a land system if there were no corresponding changes in surface texture or background (soil) colour. Similarly soil colour differences were accepted if there were no corresponding changes in landscape texture or landforms.

The Maralal land system of the upland massifs (Karisia Hills and Mathews Ranges) was readily delineated on the basis of landform and texture. The gneisses of the Basement Series provide a unique texture which is discernible even on ultra-small scale imagery. The Lopet land system, consisting of a series of lava-filled valley remnants, now occurring as inverted relief forms, is easily recognizable by its steep-sided relief and sinuous outlines. The step faults, fault splinters, and deep valleys of the Tikamur land system permit easy recognition and delimitation. Mapping the Barsaloi, Baragoi, and Laijok land systems proved to be more difficult, and was based on the identification of subtle indicators of surface texture. Differences in spectral reflection between surficial deposits, such as volcanic ash, lava, and residual soils, were found to be quite well marked and provided a valid base for delimiting the Lodeke, Silali, Suk, and Loima land systems. It is mainly in those areas where major relief differences are absent that difficulties in using Landsat imagery for land system analysis are likely to be encountered.

## Validating the land system boundaries obtained from Landsat: the ABD test

In order to obtain a measure of the validity of the land system boundaries obtained according to the subjective procedures described above an objective test of boundary strength was devised. The across boundary difference (ABD) test was elaborated to determine the extent to which the land system boundaries, delineated using Landsat imagery, represented actual differences between adjacent land systems. Two Landsat images (25 March and 21 September 1975) were prepared as prints at a scale of 1:500 000 and overlaid with a transparent acetate containing the land system boundaries obtained in the analyses described above. The procedure for carrying out the ABD test is illustrated in Fig. 8 and complete details are given in Williams (1983). The boundary of each land system

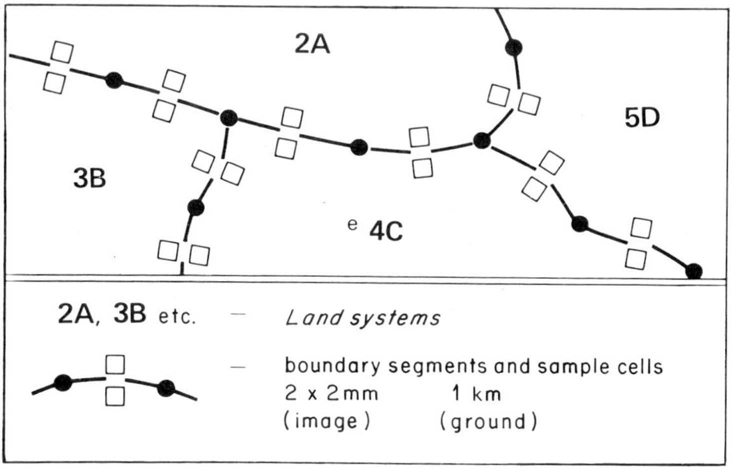


Fig. 8 Sampling procedure used in carrying out the across boundary difference (ABD) test.

was divided into equi-spaced segments and terrain conditions were examined using an acetate overlay containing two $2 \times 2$ mm grid cells arranged as a sampling pair and positioned in each segment so that the sampling cells straddled the boundary. Terrain conditions within the sample cells were examined and the level of difference from one side of the boundary to the other was established based on consideration of four parameters: (1) a measure of surface albedo, (2) a measure of surface texture, (3) a measure of drainage pattern, (4) an assessment of unit landform.

Surface albedo was measured using a Macbeth R.D. 400 densitometer at the amber filter position with a sampling aperture of 1 mm. The instrument measures diffuse reflection density as defined by the American Standards Institute. A threshold level was established in order to determine whether the reflectance densities obtained from the sample cells on either side of the land system boundary were to be considered similar or dissimilar. The threshold was set at one-third the range value for all sample points along the boundary between the two contiguous land systems. Thus, if the range value (difference between extreme minimum and extreme maximum values for all sample points) was 0.54, the threshold was set at $0.54 \div 3 = 0.18$. If the difference in reflection density between sample grid cells exceeded this value the terrain on either side of the land system boundary was considered dissimilar. However, if the values were less than the threshold value the terrain was considered similar.

Surface texture within the sample cells was examined visually using a zoom macroscope with a magnification up to $30\times$. Three criteria were used in the assessment—spacing, arrangement, and orientation. The sampling cells were considered dissimilar when the surface texture on either side of the boundary differed with respect to one or more of the criteria. The terrain was considered similar when no effective difference in surface texture was found.

Drainage pattern was examined in a similar fashion. The decision as to whether or not the terrain within the sampling cells was similar or dissimilar with regard to drainage pattern was based on three criteria: drainage type or form, e.g. trellis, dendritic, etc., orientation, and degree of integration (Lueder, 1959).

Unit landform is defined by Lueder (1959, p. 20) as 'a terrain feature, usually of the third order . . . that can be described and recognised in terms of typical features wherever it may occur, and which, when identified provides dependable information concerning its own structure and either composition and texture or uniformity.' Land systems are essentially geomorphological units and so a change in unit landform across a land system boundary is to be expected. Where this occurred the terrain at the sampling point was considered dissimilar.

The ABD test rating for each boundary segment was arrived at by totalling the number of dissimilarity points that had been scored in assessing each of the four parameters. Major contrasts across a boundary resulted in ABD scores of 3.0, 3.5 or 4.0, indicating definite or strong boundaries. Minor contrasts produced ABD scores of $\leqslant 2$, indicating indefinite or weak boundaries.

A continuous 'hard' boundary can be taken as evidence of a definite terrain discontinuity, whereas 'soft' boundaries indicate more gradual transitions in the landscape.

## Seasonal variation in the detectability of land system boundaries

Land systems are an expression of the morphology of the landscape, and as such should be recognizable in both wet and dry seasons. However, several of the parameters used in the identification and delimitation of land systems using aircraft or spacecraft imagery are affected to some degree by seasonally dynamic factors. Landscape colours, textures, and patterns respond to phenological changes, and so some discrepancies between land system boundaries interpreted at different seasons may be expected. In order to establish the optimum season for mapping land system boundaries and to identify the seasonal changes in boundary definition that occur, the ABD test was applied to two Landsat scenes, one imaged in very dry conditions (25 March 1975), and one obtained after a period of rains (21 September 1975). The state of the ground and the rainfall in the period immediately prior to the Landsat overpass are shown in Table 2 and Fig. 4. The rating scores for boundary detectability in the two images are very similar (Fig. 5(b)) with just over half the boundary segments rated as definite (an ABD test score of $\geq 3$): 57.1% on the scene imaged after the rains (September 1975) compared with 55.3% for the dry season image (March 1975).

The somewhat better scores obtained from the September image do not reflect an overall superiority of wet season images for detecting land system boundaries. The improved boundary ratings occur only in the eastern part of the transect as a result of the response of the dynamic landscape elements to the onset of the rains. The very high albedo of areas such as the Sware Plains, which makes the detection of land system boundaries so difficult in the dry season, is significantly reduced by the green flush of vegetation after rain. There is an improvement in the ABD test scores for several contiguous land systems: Baragoi-Barsaloi, Laijok-Baragoi, and Barsaloi-Maralal. This comes about because of sharpened gradients in the landscape texture and easier differentiation of drainage patterns, changes which are directly related to vegetation growth along the drainage lines.

In contrast, in the western section of the transect, the ABD test scores obtained on the September image were somewhat worse than those for the dry season image. There is a decrease in the test scores for several contiguous land systems: Samburu-Ngapoi, Lodeke-Suk, Hannington-Silali, Samburu-Suk, etc. Land systems developed on old land surfaces with somewhat featureless relief display decreased boundary detectability on the wet season image as a result of masking by active vegetation growth stimulated by the abundant rainfall of the preceding eight weeks.

On the basis of these results the optimum period for land system boundary detection would appear to occur at the beginning of a period of rains when the first flush of vegetation provides an enhancement of landscape texture. Boundary detectability deteriorates in wet season conditions because of vegetation growth which masks both surface texture and colour differences related to soil or rock type. Reduction in boundary detectability scores also occurs in long dry periods when the vegetation is dormant, soil moisture minimal, and surface albedo very high.

## TESTING THE ACCURACY OF LAND SYSTEM DELIMITATION ON LANDSAT

The accuracy that can be attained in delimiting land systems is clearly the critical factor in determining whether or not Landsat imagery can be effectively applied in land system analysis. In order to assess the accuracy of the land system maps obtained by Landsat analyses a test of areal correspondence was carried out using the map in the *Land Systems Atlas of Western Kenya* (Scott *et al.*, 1971) as the reference base. The Atlas map was assumed to be correct, and the land system outlines derived from Landsat images were compared with this using a statistic proposed by Minnick (1964)—the coefficient of areal correspondence, which is essentially an extension of traditional map overlay techniques expressed in numerical terms.

The coefficient of areal correspondence ($C_a$) is defined as the area of intersection of two mapped phenomena ($AB$) divided by the total area covered by both phenomena ($A + B$).

$$C_a = \frac{AB}{A + B}$$

If the two distributions are completely distinct, then $AB = 0$ and $C_a = 0$; conversely, if the two distributions correspond exactly $AB = A + B$ and $C_a = 1$. In effect, the coefficient relates the area of direct areal correspondence to that of possible correspondence.

In order to obtain the coefficient values a digitizer and integrated graphics computer were used (Hewlett Packard 9874A and 9845C with 16 megabyte disc storage), as shown in Fig. 9. Land system boundaries were digitized starting with the reference base from the Atlas and continuing with the overlays obtained from each of the seven Landsat images. Precise registration of the map and overlays was achieved using the corner control points of the transect.

Two samples of digitized maps obtained from Landsat imagery (25 March 1975 and 21 September 1975) are presented in Fig. 10. $C_a$ values and the extent of boundary mismatch are shown for each land system. Although there are numerous minor discrepancies in boundary matching, the majority of $C_a$ values

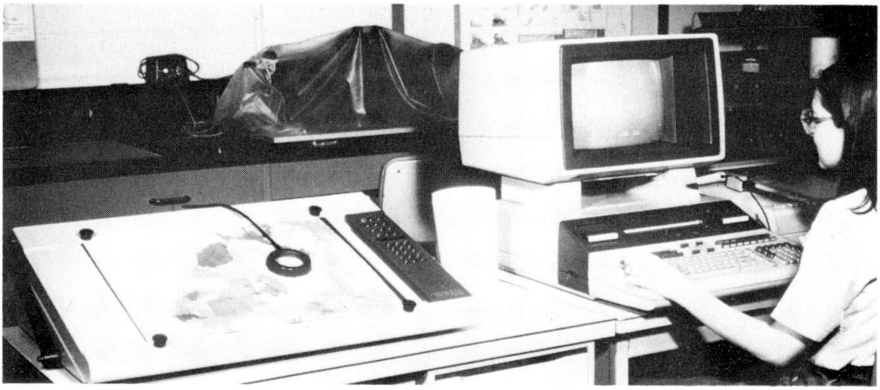

Fig. 9 Digitizer and computer (Hewlett Packard 9874A and 9845C).

fall in the range 0.4–0.9 with a mean of all values at 0.614 (Fig. 11). In general, higher $C_a$ values ($>0.75$) were obtained for large land systems (areas $>200\,km^2$) indicating a decreasing relative error with increasing size. This suggests that larger land systems can be mapped relatively accurately from Landsat. Mismatch of land system boundaries can be attributed primarily to the scale of the Landsat imagery. Most of the low $C_a$ values were obtained for land systems with small areal extent which makes them difficult to delimit. However, it should be noted that in several instances it was felt that the boundaries plotted by Scott *et al.* (1971) were at fault, and that the Landsat images provided a more accurate outline of the actual land system. Conventional land system analysis requires the examination of large numbers of air photos obtained under different illumination conditions and printed at varying contrast levels. It is often difficult in these circumstances to be consistent in delineating boundaries. Landsat with its synoptic view and uniform illumination may well provide a more appropriate land systems data base.

There are only minor variations in the mean $C_a$ values for Landsat scenes imaged at different seasons (Table 2). The best overall results were obtained with dry season images—9 September 1972, 31 January 1983, and 25 March 1975 with mean $C_a$ values of 0.652, 0.657, and 0.664, respectively. When the range in coefficient values for individual land systems is examined for the seven different image dates it is clear that in the majority of cases the highest $C_a$ value, indicating the best match of the boundary obtained from Landsat with the reference base map, was achieved with a dry season image. The Landsat scene with the largest number of best scores for individual land systems was that of 9 September 1975 with 16 out of a possible total of 40 land system occurrences.

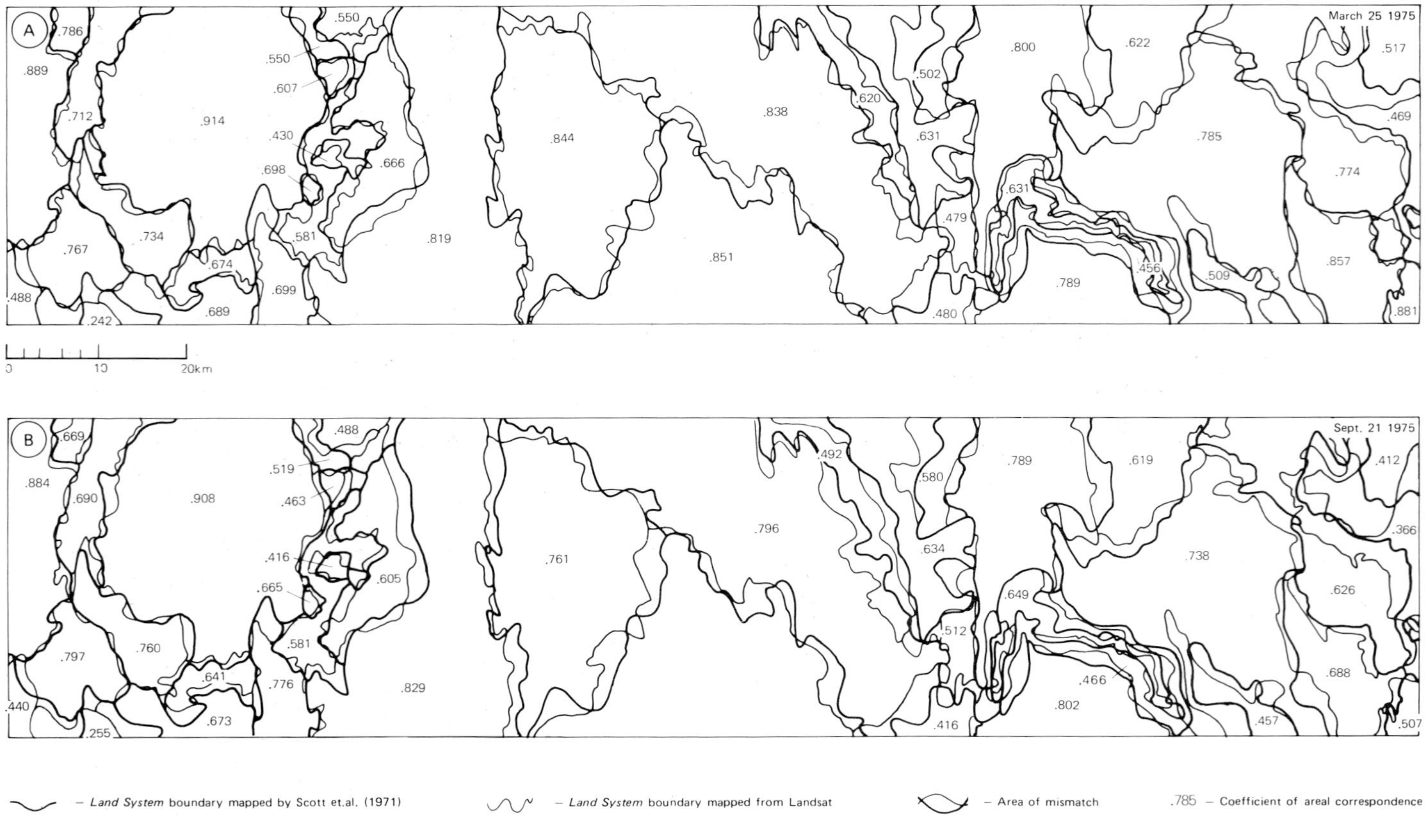


Fig. 10 Areal correspondence and mismatch of the land system boundaries of the Maralal transect interpreted from Landsat compared with those given in the *Land Systems Atlas of Western Kenya* (Scott *et al.*, 1971).

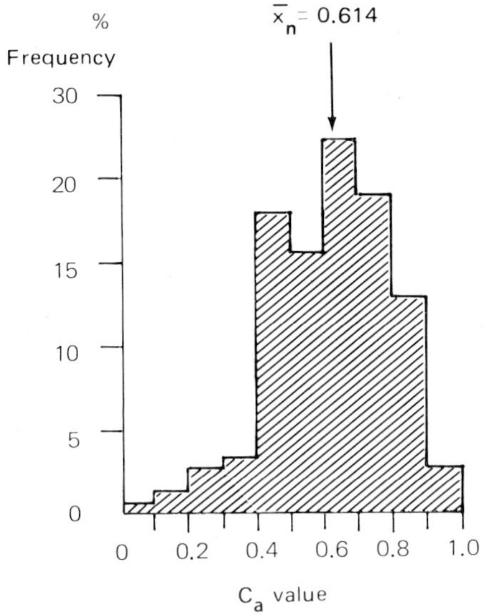


Fig. 11 Histogram showing the frequency of occurrence of $C_a$ values (coefficient of areal correspondence) for all land systems in the Maralal transect. Data based on the analysis of seven Landsat scenes.

A more detailed examination of the $C_a$ values set in their regional context and viewed against the vegetation growth stage associated with the antecedent rainfall is attempted in Table 4. The results are instructive because they illustrate some of the difficulties in arriving at valid generalizations for an area with such marked ecoclimate gradients. For the 22 land system occurrences in the western and central part of the transect the best coefficient values were obtained in dry season conditions with dormant vegetation. In the eastern section, $C_a$ values were significantly lower at all seasons indicating a weaker correspondence between land systems delimited in the Atlas and those obtained from Landsat. This may be the result of extraneous effects such as the greater amount of dust in the atmosphere in the eastern plains compared with other parts of the transect, or it may reflect weaker contrast in the actual land system boundaries. The best coefficient values in the eastern section were obtained at the beginning of a period of rain when the vegetation flush enhanced the landscape texture, thereby improving land system boundary definition. For example, the landscape of the Lodeke land system consists of extensive plains with a veneer of sand dunes and alluvial fans with narrow drainage lines. These features are not readily discernible in dry season conditions; however, with the onset of the rains differential vegetation growth enhances the textural detail thus permitting more precise delineation of the land system boundaries.

Table 4. Regional differences in coefficients of areal correspondence ($C_a$ values) stratified according to state of the ground and vegetation growth stage.

| Section of the transect | Nos. of land system occurrences | Mean $C_a$ value | Date* | Antecedent rain/stage of vegetation growth |
|---|---|---|---|---|
| | | 0.623 | Sep 21 1975 | H/A |
| | | 0.630 | May 24 1979 | E/A |
| West | 19 | 0.632 | Sep 9 1972 | E/B |
| | | 0.637 | Nov 20 1972 | L/D |
| | | 0.649 | Mar 25 1975 | Dy/D |
| | | 0.653 | Jan 31 1973 | Dy/D |
| Central | 3 | 0.794 | Sep 21 1975 | H/A |
| | | 0.819 | Sep 9 1972 | Dy/D |
| | | 0.849 | Jan 31 1973 | Dy/B |
| | | 0.856 | Mar 25 1975 | Dy/D |
| | | 0.527 | May 24 1979 | E/B |
| East | 18 | 0.535 | Jan 13 1973 | L/D |
| | | 0.549 | Nov 20 1972 | H/A |
| | | 0.574 | Sep 9 1972 | Dy/D |
| | | 0.579 | Sep 21 1975 | S/F |
| | | 0.591 | Jan 31 1973 | H/F |

*State of the ground/vegetation*
A: active growth, B: brown off, D: dormant, Dy: dry, E: end of rain period,
F: vegetation flush, H: heavy rain, L: light rain, S: start of rain period
*The number of Landsat scenes available for study varies because of differences in the percentage cloud cover

## EFFECTIVENESS OF LANDSAT IMAGERY AT DIFFERENT LEVELS IN THE LAND SYSTEM HIERARCHY

One of the significant features of land system theory is its hierarchical structure, that is, the concept that large tracts of land are composed of constituent elements. It follows that one can deal with land at any of several levels of abstraction, but before embarking on a land systems analysis using small-scale imagery, such as Landsat, it is essential to determine the effective limit set by the resolution of the system.

The relationship between the scale of analysis and the appropriate level or levels of operation in a land system hierarchy has been examined by Mitchell (1973). He has indicated that land systems are best studied at scales ranging from 1:250 000 to 1:1 000 000, land units at scales between 1:10 000 and 1:80 000, and land elements or sites at scales larger than 1:10 000. The standard Landsat products are available at scales between 1:1 000 000 and 1:250 000. Thus, in terms of scale, Landsat images would seem appropriate for the examination of land systems, but less well suited to the analysis of land elements or sites. The resolution limits of the Landsat MSS which sets a pixel size of $79\,m \times 79\,m$ ($6241\,m^2$) clearly preclude the study of sites, which exhibit a typical size range from $100\,m^2$ to $1\,km^2$. Land units, which typically have areas of $1\,km^2$ to $100\,km^2$, are above the resolution limits of Landsat. It follows that land units should be detectable on Landsat imagery if there is sufficient contrast in the spectral response of the unit compared with its neighbours.

In order to test the effectiveness of Landsat in identifying terrain features at the mid-level of the land system hierarchy a systematic assessment of the detectability of land units in the Maralal transect was carried out (Plate 8). The same two images were used as in the ABD test described above (a dry season image—25 March 1975, and a wet season image—21 September 1975). After careful consideration of the land unit descriptions in the *Land Systems Atlas of Western Kenya* (Scott *et al.*, 1971) an attempt was made to detect individual land units on the imagery. The ease of detection was scored using a simple detectability rating scale from 0 to 3, in which 0 indicates a failure to detect the land unit and 3 indicates immediate and easy recognition.

In the establishment of detectability scores a purely subjective procedure was followed involving direct visual examination of the Landsat image using the Zoom Transferscope (Fig. 7). A score of 3 was allocated to all land units that were immediately identifiable. For features that displayed good contrast with their backgrounds, where edging effects were minor, and the character of the feature could be determined from its contextual surroundings a score of 2 was assigned. If the identification was only achieved with difficulty, or certain reservations remained about its validity, a score of 1 was given.

The results of the analysis are summarized in histogram form in Fig. 12. Of the 86 land units present in the transect only five remained undetected giving

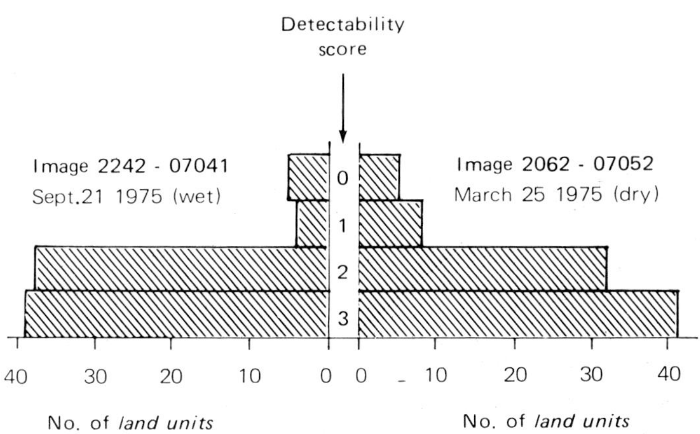


Fig. 12 Histogram showing the detectability of land units on Landsat images of the Maralal transect. Data based on two Landsat scenes.

a detectability score of 94.2%. Approximately half of the land units were immediately recognizable in both wet and dry conditions with a marginally superior score (47.6% compared with 45.3%) achieved using the dry season image. However, there is no significant seasonal advantage because the same five land units were undetectable at either season. For some land units, such as foot slopes and drainage lines, the wet season image was superior; for others, the dry season image was preferable. Simple tectonic forms and features with spectral signatures in marked contrast to their backgrounds were readily detected at all seasons. The less distinct land units, particularly those on old land surfaces of low relief, were most easily identified after light rains when landscape texture is significantly enhanced by the vegetation flush.

## CONCLUSION

The results of this research on the applicability of Landsat imagery in the analysis of land systems in semi-arid Kenya can be summarized in five ways.

1. Within the Maralal transect the sixteen land systems defined by Scott *et al.* (1971) in the *Land Systems Atlas of Western Kenya* were readily identifiable on Landsat imagery. The distinctive character of each land system was recognizable on the imagery and the mapping of land system boundaries was accomplished with relative ease using the appropriate equipment and interpretation procedures.
2. Detailed examination of the land system boundaries demonstrated that they differed quite markedly in their inherent visibility as a function of the level of contrast in such terrain parameters as albedo, surface texture, drainage pattern, and constituent unit landforms from one side of the boundary to the other. The relative 'strength' or definition of land system boundaries can be assessed at the regional level using Landsat imagery. The ability to distinguish between 'hard' boundaries indicating definite terrain discontinuities and 'soft' boundaries reflecting a gradual transition in landscape conditions has special significance for rangeland management and land use planning (Gwynne, 1977).
3. No marked differences in the detectability of the land systems of the Maralal transect from season to season were noted. The results obtained using an across boundary difference (ABD) test indicate that the optimum period for mapping land system boundaries in semi-arid areas occurs at the beginning of a period of rains when the green flush of vegetation enhances the surface texture and drainage pattern.
4. The accuracy achieved in defining the areas of land systems as measured by a coefficient of areal correspondence indicates that Landsat imagery would provide acceptable results for reconnaissance level resource surveys. The areas of the larger land systems ($>200\ km^2$) of the Maralal transect were

measured with consistent accuracy at all seasons of the year. The increasing relative error in area measurement with decreasing size of the land system is considered to be a direct function of the scale of the imagery and could probably be improved with digital procedures.

5. The scale factor limits the usefulness of Landsat imagery in the examination of smaller components in the land system hierarchy. The land units of the Maralal transect were readily recognizable at all seasons. A detectability score of 94.2% (81 out of 86 land units) was achieved; however, the small scale of the imagery precludes any possibility of mapping land units. Land elements or sites are generally not detectable because of the resolution limits of the Landsat system. No marked difference in the detectability of land units at different seasons was found. In some districts the dry season image provided more useful details because the obscuring effect of a full vegetation cover was absent, whereas, in other areas, the enhancement effect of the first flush of vegetation at the onset of the rains was beneficial in increasing textural contrast in the landscape.

## REFERENCES

Beckett, P. H. T. and Webster, R., 1965. *A Classification System for Terrain*. Rept. 872, Military Engineering Experimental Establishment, Christchurch, England.

Bourne, R., 1931. Regional survey and its relation to stocktaking of the agricultural resources of the British Empire. *Oxf. For. Mem.*, **13**, 16–18.

Christian, C. S. and Stewart, G. A., 1968. Methodology of integrated surveys. In *Aerial Surveys and Integrated Studies. Proceedings of the Toulouse Conference*. Paris: United Nations Educational, Scientific and Cultural Organization, pp. 233–280.

Griffiths, G. H. and Collins, W. G., 1981. Landsat imagery for mapping and monitoring vegetation in northern Kenya. *Proc. 9th Annual Conf. on Matching Remote Sensing Techniques and their Applications*, London, pp. 53–64.

Gwynne, M. D., 1977. *Landsat Image Studies of Green Flush Events in Kajiado District, Kenya*. Working Document no. 15, KEN/71/526, UNDP/FAO Wildlife Management Project.

Hellden, U., 1980. *A Test of Landsat 2 Imagery and Digital Data for Thematic Mapping Illustrated by an Environmental Study in Northern Kenya*. Rept. 47, Dep. Phys. Geography, University of Lund.

Hills, R. C., 1979. The structure of the Inter-Tropical Convergence Zone in equatorial Africa and its relationship to East African rainfall. *Trans. Inst. Br. Geogr.*, **54**, 329–352.

Hilwig, F. W., 1980. Visual interpretation of multi-temporal Landsat data for inventories of natural resources. *ITC Jl.* 1980, **2**, 297–321.

Kenworthy, J. M. and Glover, J., 1958. The reliability of the main rains in Kenya. *E. Afr. agric. J.*, **23**(4), 267–272.

Lind, E. M. and Morrison, M. E. S., 1974. *East African Vegetation*. London: Longmans.

Lueder, D. R., 1959. *Aerial Photographic Interpretation—Principles and Applications*. New York: McGraw Hill.

Mabbutt, J. A., 1968. Review of the concepts of land classification. In Stewart, G. A. (Ed.), *Land Evaluation*. Melbourne: Macmillan, pp. 11–28.

Minnick, R. F., 1964. A method for the measurement of areal correspondence. *Pap. Mich. Acad. Sci.*, **49**, 333–344.

Mitchell, C. W., 1973. *Terrain Evaluation*. London: Longmans.
Nieuwolt, S., 1978. Seasonal droughts in East Africa. *E. Afr. agric. For. J.*, **43**(3), 202–222.
Odenyo, V., 1979. *Application of Landsat Data to the Study of Land and Range Resources in the Narok Area, Kenya*. Research Rept. 79–4, Canada Centre for Remote Sensing, Ottawa.
Parry, J. T., 1983. Remote sensing and the cardinal points of the semiarid land compass—water, rangeland, animals and man. *Kenyan Geogr.*, **5**(1 and 2), 102–111.
Pratt, D. J., Greenway, P. J., and Gwynne, M. D., 1966. A classification of East African rangeland with an appendix on terminology. *J. appl. Ecol.*, **3**, 369–378.
Pratt, D. J. and Gwynne, M. D., 1977. *Rangeland Management and Ecology in East Africa*. London: Hodder and Stoughton.
Pugh, J. C., 1955. Isostatic readjustment and the theory of pediplanation. *Q. Jl. geol. Soc. Lond.*, **111**, 361–374.
Rango, A., 1980. Landsat MSS mapping of drainage features and vegetation in the African Rift zone, Kenya. *5th U.N. International Training Course in Remote Sensing Applications for Water Resources and Hydrology*, Rome, 87–93.
Scott, R. M., Webster, R., and Lawrence, C. J., 1971. *A Land Systems Atlas of Western Kenya*. M.E.X.E. Rept., Soil Science Laboratory, Department of Agriculture, Oxford.
Shackleton, R. M., 1946. *Geology of the Country between Nanyuki and Maralal*. Rept. 11, Geol. Survey of Kenya.
Trapnell, C. G. and Griffiths, J. F., 1960. The rainfall altitude relation and its ecological significance in Kenya. *E. Afr. agric. J.*, **25**, 207–213.
Veatch, J. O., 1933. *Agricultural Land Classification and Land Types of Michigan*. Special Bull. 231, Michigan Agric. Exp. Station.
Verstappen, H. T. H., 1977. *Remote Sensing in Geomorphology*. Amsterdam: Elsevier.
Williams, M., 1983. *Temporal Variations in the Detectability of Land Systems as Seen on Landsat Imagery, Maralal, Kenya*. Unpublished dissertation, Diploma in Environmental Remote Sensing, Department of Geography, McGill University.

Remote Sensing and Tropical Land Management
Edited by M. J. Eden and J. T. Parry


# CHAPTER 7

# *Landsat, Land Systems, and Livestock Disease Control in Northern Australia*

PETER LAUT
*Division of Water and Land Resources, CSIRO, Australia*

Visual interpretation of Landsat imagery has proved to be a cost-beneficial tool for broad-scale land resource inventories in Australia, where costs, including field sampling, must be kept to a minimum (Laut *et al.*, 1977; Graetz *et al.*, 1982). Landsat imagery is especially useful as a compilation tool for very large areas where land information is incomplete. This paper reports on the use of Landsat colour composite images to determine local vegetative cover for a project which will provide the Australian Government with information on the feasibility and costs of eradicating bovine tuberculosis in Northern Australia. Landsat imagery offers a number of advantages for the task since it is readily available for all parts of the continent, is relatively cheap per unit area, and can be used in conjunction with other sources of information.

## THE LIVESTOCK MUSTERING COSTS PROJECT

In 1981, a project involving officers from the Australian Bureau of Animal Health, the Bureau of Agricultural Economics and the then Division of Land Use Research CSIRO was initiated to determine the costs of eradicating bovine tuberculosis from the remainder of Northern Australia. Eradication was to be based on traditional mustering and testing methods, and by shooting livestock which could not be mustered. It was the author's responsibility in this task to evaluate the degrees of difficulty for livestock mustering presented by the landscape of Northern Australia. This required an assessment of the major landscape types where livestock are known to graze across the northern section of the continent and an evaluation of the range of difficulties that would be encountered in mustering. The project involved the following stages:

1. Definition of initial aims, methods, and study area.
2. A pilot survey of 42 cattle stations in Northern and Central Australia.

Main project:

3. Re-definition of aims, method, and study area of the project.
4. Field survey of an additional 38 cattle stations in Northern Australia.
5. Stratified random sampling of 1835 sites.
6. Data capture for sample sites.
7. Data analysis.

This study is based on the use of Landsat imagery in part of Stage 6 above: the method of data capture for vegetative cover for sample localities. A short summary of the earlier stages of the project will be presented together with a brief introduction to the land resources of Northern Australia and the methods of livestock mustering. These two sections are intended solely to provide a context for the study.

## Summary of Stages 1 to 5

The initial aim of the project, as defined by the Bureau of Animal Health, was to estimate costs of cattle mustering using a relatively small sample of 42 cattle stations or pastoral holdings selected by the Bureau of Agricultural Economics for economic studies. These cattle stations are located in northern South Australia, throughout the Northern Territory, in the lower portions of the Ord and Fitzroy River basins in Western Australia, and the Gulf of Carpentaria district of Queensland. A desk inventory of the land resources of these holdings and a brief field examination by land and air indicated that the sample was not representative of the land resources of Northern Australia and, more especially, of holdings experiencing the greatest degrees of difficulty in mustering cattle. Discussions with owners and managers emphasized the limitations of the sample and the need to redefine the study area. These discussions also provided additional information on mustering techniques and the restrictions on mustering imposed by the landscape, as well as approximate costs of mustering livestock in different types of landscapes.

At the conclusion of the pilot survey, the study area was redefined (Fig. 1), an additional sample of 38 properties was selected for inspection in the following dry season, and cattle stations in northern South Australia and southern Northern Territory were removed from the sample. It was also decided to obtain a new sample of landscape types in the newly defined study area. This was intended to provide a more satisfactory basis for estimating the range of landscape types, their extent and approximate distribution, and finally the costs involved in a clean muster of livestock, i.e. one in which all livestock are mustered.

The sample of landscape sites was selected using a stratified random base. The population from which the sample was selected consisted of all intersection

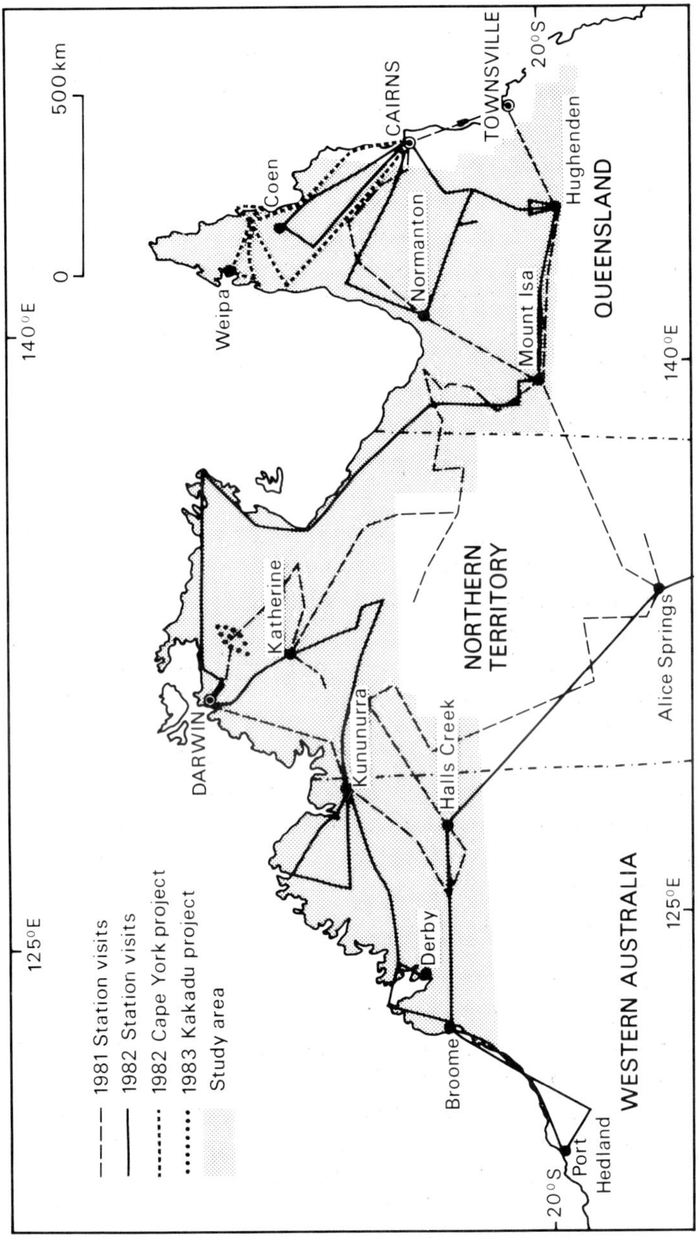


Fig. 1 The study area showing routes of low level reconnaissance and verification flights.

points of the 10 km × 10 km grid lines of the Australian Map Grid. The stratification was based on the relative areal significance of those biophysical regions which occurred within the study area (Laut *et al.*, 1975), and it was carried out by numbering the total population of intersections, allotting each to its biophysical region, and using computer-generated random numbers to fill the quota for each biophysical region. In all, 1835 grid intersections were selected to provide a 12.5% sample.

## Land resources and livestock in Northern Australia

The study area is entirely within the zone of very marked summer rainfall (Köppen climate types AW and BSwh). Precipitation declines from over 1500 mm per year in the vicinity of Darwin and northern Cape York Peninsula to around 400 mm along the southern boundary of the area. The extreme seasonality of the rainfall has a marked impact on timing for mustering. During the wet season, cattle disperse as watering points and pastures become more widely available. At the conclusion of the wet season the water level in creeks is still high, limiting vehicle access, and the general dispersion of cattle makes mustering impossible. Consequently, mustering can only proceed towards the mid-dry season, especially in the more northerly high rainfall areas, or even later, where watering points are numerous.

The landforms of the study area include extensive floodplains which are inundated throughout the wet season, areas of low relief above flood levels, dissected plateaux, and rugged ranges of high hills and mountains. Large floodplains occur along all major streams throughout the study area. The floodplains, especially those with permanent waters, make mustering difficult well into the dry season, and exacerbate livestock management problems when fences are damaged by floodwaters. The more rugged landscapes hinder mustering by providing natural barriers which limit the ease of local movement. The Kimberleys of northern Western Australia, the dissected plateaux and ranges of the Victoria River basin and western Arnhem Land, and the northern portions of the Eastern Highlands in north Queensland pose enormous difficulties for livestock mustering and hence for the eradication of bovine tuberculosis.

The structure of the vegetation cover may also impose considerable difficulties in livestock mustering. Tall shrubs and trees hinder and restrict the movements of horsemen. Thickets and forests provide hiding places for livestock and limit the effectiveness of aircraft both as spotting platforms and as mustering vehicles. Almost all the study area has a vegetative cover of open or closed woodland and forest. In addition, many streams are lined by narrow but dense gallery forest and even areas of woodland contain a variety of thorny thickets which provide security for wary stock. The isolated areas of short grassland which occur mainly on low relief areas of residual soils in the southern portion of the study area offer few difficulties to mustering. The taller grasses of the northern floodplains are more restrictive. In the Northern Territory the grasslands are the habitat of the feral water buffalo (*Bubalus bubalis*) which is readily infected with bovine tuberculosis and can be a difficult and dangerous animal to muster in many circumstances.

Cattle stations in the study area are Crown leases measured in hundreds of square kilometers, most without continuous boundary fences. Cattle densities range from less than 10 to 25 per square kilometre, providing low returns per unit area which limit capital investment to a bare minimum of stockyards and holding paddocks. There is a generally low level of management in the area which militates against disease eradication. This is exacerbated by the movement of cattle into non-leased areas.

### The nature of livestock mustering in Northern Australia

A variety of methods are employed for mustering cattle in the study area. The traditional mode of mustering is by stockmen on horseback who gather cattle together in a mob and move it to the stockyards. Skilled stockmen are no longer readily available for seasonal work and, wherever possible, mechanical aids supplement the stockmen. Where permanent water is scarce, traps or large yards may be used to hold stock around watering points. Where the terrain is suitable, motor vehicles may be used both to complement horsemen and to catch recalcitrant stock. Fixed-wing aircraft are commonly used in hilly or partly forested terrain for spotting groups of cattle and, if the country is sufficiently open, for driving cattle towards the main herd. Over the past decade small helicopters have been used as a partial replacement for stockmen. Pilots locate and drive cattle to either central holding areas, or towards small herds or 'coaches' of relatively quiet cattle.

Within the study area, a 'clean' muster involving all the livestock only occurs along the margins of the Barkly Tableland and on some of the smaller properties in the southern part of the Gulf of Carpentaria. Elsewhere, only a proportion are mustered because the cost of a 100% or 'clean' muster is not warranted. Generally, as the degree of difficulty imposed by the landscape increases, so the level of cattle management and the mustering rate decreases.

## LIVESTOCK MUSTERING AND THE DEGREE OF DIFFICULTY OF THE LANDSCAPE

Levels of management, stockmen's skills, and helicopter pilot capabilities vary too greatly to allow the development of a deterministic model of the relationship between mustering costs, livestock management, and the degree of difficulty imposed by the landscape at the individual property level. However, in order to establish a basis for comparison across a large number of properties and to assess the total cost for which each State or Territory would be liable, it is justifiable to assume a common level of livestock management including mustering skills. The degree of difficulty imposed by landscape factors in livestock mustering may be expressed by the following relationship:

$$DM = f(L, R, V, W)$$

$DM$ is the degree of difficulty,
$L$ is the type of landform or morphology,
$R$ is the type of lithology and associated microrelief including surface rockiness,
$V$ is the type of vegetative cover,
$W$ is the availability of watering points for livestock.

The great variety of combinations of landform, microrelief, vegetative cover, and stock watering situations in the study area precluded examination of the difficulties imposed by each combination. Discussions with property owners and managers suggested that the function was additive and subsequent analysis of mustering costs in relation to landscape difficulties for the sample stations demonstrated an effective correlation between summation of the factors and actual mustering costs. The description and evaluation procedure is described below, and the scores used for each component are given in Appendix 1.

**Data capture for sample sites (Stage 6)**

At the outset of the study it was clear that there would be no opportunity to undertake site surveys in the field and so most data had to be derived from previously published material. Low level flights between the properties visited and brief on-site field checks at individual properties provided supplementary data but this was not considered as field site data gathering or checking in the usual sense. Fortunately, there were a number of potential sources of land information, albeit at disparate scales (Beadle, 1981; Beard, 1979; Carnahan, 1976; CSIRO 1952–1976, Northcote *et al.*, 1967–1968).

Details of landforms and water features were obtained from the 1:100 000 topographic map series; lithology was taken from the 1:250 000 geological map series and vegetative cover was derived from a variety of sources. However, no single source of information provided adequate vegetative cover data: the problem was compounded by the absence of vegetation maps at a suitable scale, a lack of consistency between reports in the same series and classification schemes that were inappropriate.

Data were captured for two grid cells surrounding each sample intersection. An inner 4 $km^2$ cell was used to provide information about the particular locality and an outer 100 $km^2$ cell was used to provide a context for the inner cell. Context is frequently very important in assessing the degree of difficulty for livestock mustering imposed by the landscape as perceived by station owners and managers. For example, a small grassy plain which dominates the 4 $km^2$ grid cell may be ignored in normal livestock mustering activities when it is surrounded by steep, forested hills in an isolated part of the station. The areal extent of the different landform types, lithology, and vegetative cover classes was recorded as rounded tenths of grid cell areas for both the inner and outer cells.

Since the emphasis of this study is on the use of Landsat imagery in the synthesis of vegetative cover data, other aspects of data capture and evaluation are ignored here. A fuller account of these aspects will be published later (Andrews *et al.*, in preparation; Laut and Nahinga, 1985).

## Synthesis of vegetative cover data for sample cells

The most relevant sources of vegetation information for the project were: the 1:100 000 topographic map series; the land systems reports available for parts of Northern Australia; the 'National Vegetation' map from the Atlas of Australian Resources—scale 1:6 m (Carnahan, 1976); and Landsat images. The 1:100 000 topographic map series includes some vegetative cover information but there are serious inconsistencies. The two agencies which produce this series, the Division of National Mapping and the Australian Survey Corps, do not appear to have a common standard for mapping vegetation. The Division of National Mapping has adopted a classification which includes three density classes (sparse, medium, and dense) and two height classes (forest and scrub) for tree vegetation. Survey Corps maps show only the density classes. In addition, there are frequent discontinuities of density classes from map sheet to map sheet. Although both agencies map grasslands, swamplands, mangroves, and rain forests consistently, neither indicates the extent of low shrubs such as spinifex (*Triodia* spp.) or saltbush (*Atriplex* spp.) although these names may be printed on the map. Thus, although maps of the 1:100 000 series indicate the distribution of vegetation types, such as mangrove, rain forest, open forest, and woodland which are important to the present study, additional vegetation information is required to determine the height and density of trees and the presence of understory or understories.

Between 1952 and 1976, twelve land systems surveys and associated reports were published by the Division of Land Research, which covered all but a small portion of the study area (Fig. 2). Each report included information on vegetative cover. In some cases (e.g. CSIRO, 1960), this was limited to pasture types; in others, vegetation was fundamental in the delimitation of the land systems and was described in great detail. However, in most cases, only a general chapter on the vegetation of the survey area was provided together with a set of land system descriptions in which vegetation was an important component and, in later reports, a large scale vegetation map of the area. Land systems were generally mapped at a scale of 1:250 000, and at this scale the distribution of vegetative cover was frequently obscured. The vegetation maps were usually concerned only with the broad distribution of dominant vegetation communities and small pockets of minor communities were generally ignored.

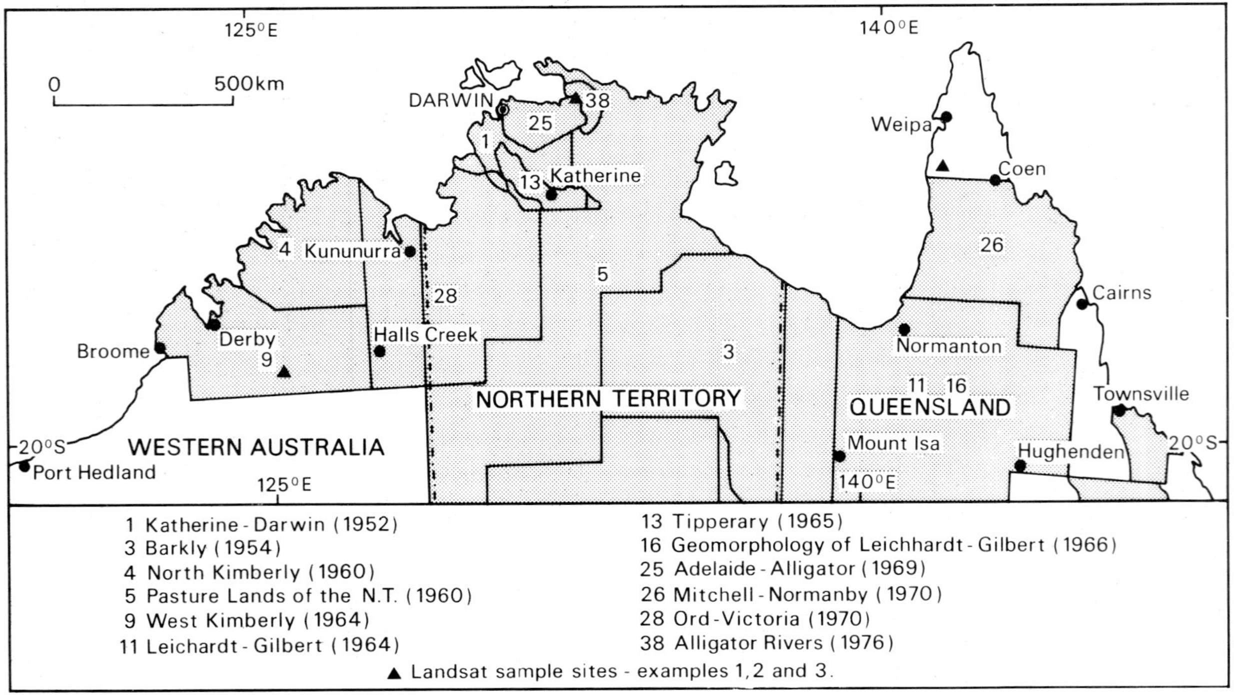


Fig. 2 Northern Australia showing selected Landsat sample sites and areas covered by the Land Research Series, CSIRO, Canberra.

The Natural Vegetation map of the Atlas of Australian Resources (Carnahan, 1976) provides a source of relatively uniform vegetative cover data for the continent. Vegetation information and land systems boundaries from the Land Research Series noted above provide the data base for most of the northern portion of this map and the framework is provided by a modified form of Specht's classification of Australian vegetation (Specht, 1972). This is based on the physiognomic structure of plant communities as they are assumed to have been before European settlement, with reference to the dominant families in each community. Fortunately, only very small areas of Northern Australia have been affected by intensive forms of land use. Although the species composition of some vegetation communities may have been modified, the physiognomic structure is thought to be similar to that of the original vegetation. The most serious difficulty is the very small scale of the published map (1:6 000 000) which limits its usefulness for the task at hand, since the smallest areas mapped are seldom less than several hundred square kilometres.

The third source, Landsat, provides the most useful data source in the interpretation of the vegetative cover. Where Landsat data are provided in the form of computer compatible tapes (CCTs), a digital classification of spectral data may be obtained if the necessary system software is available. While the level of spectral and spatial accuracy obtained by computer classification is very high, successful interpretation of the results requires detailed field checking, which for the present project would be both costly and time-consuming. CCTs are expensive, and the costs of developing a supervised classification system and providing suitable hard copy output are relatively high.

The alternative is visual interpretation of Landsat, either in the form of single-band black and white or colour composite images. This approach has a decided cost advantage, an important factor when 73 Landsat scenes are required to cover the study area. However, the process of visual interpretation is subjective and lacks the spatial and spectral precision associated with digital classification.

For the type of vegetative cover classification required and the level of spatial accuracy needed, it is obvious that the cost of digital classification could not be justified for this project. Therefore, it was decided to use visual interpretation of Landsat colour composite images at a scale of 1:500 000 to provide additional vegetative cover data for the 1835 sample locations.

As with other types of image interpretation, the more detailed the local knowledge available to the interpreter, the more useful his analysis. The interpretation process involves three steps: recognition of patterns; association of patterns with phenomena; and classification of the phenomena. Most of the images required were available before the first visit to sample cattle stations. Flight paths between stations were marked, patterns identified before departure, and notes made on the forms of vegetative cover and other phenomena associated with each pattern (Fig. 1). During the brief visits to the stations, local vegetation forms were noted and described. The same procedure was adopted during the visit to the second sample of cattle stations. In this way, the author was able to 'calibrate' his interpretation base by noting the effects of seasonal and short-term weather conditions and fire on adjacent scenes.

Each of the sources of data identified above offered some advantages and some disadvantages for vegetative cover mapping. Landsat images offer particular benefits in that they provide unclassified data which can be evaluated using information from the other sources, as well as the personal experience of the interpreter. It was decided to draw upon all four data sources using additional material to corroborate and expand the description obtained from the initial sources.

## VISUAL INTERPRETATION OF LANDSAT IMAGERY

Landsat colour composite images (bands 4, 5, and 7) at a scale of 1:500 000 were used in the study. The interpretation of colour composite imagery involves the evaluation of the three elements of the colour signature—hue, value, and chroma—and their spatial arrangement. Six basic types of vegetative cover in the study area could be identified by colour alone on the Landsat imagery: (1) closed forest, (2) eucalypts, (3) non-eucalypt open forest (predominantly *Acacia* spp.), (4) grasses, (5) lands previously burnt, and (6) bare lands. Allowing for those variations which normally occur from scene to scene during image preparation, and seasonal variations in the condition of vegetation, the closed forest areas appear on the Landsat image as deep red, eucalypts as orange-red, and non-eucalypt open forest areas as blue-black or purple-black. Grasslands vary significantly according to their density, seasonal condition, and the underlying soil, and range from blue-white to yellow. Burnt areas are black. Open and closed woodlands, according to the relative proportion of grasses and trees, are represented by intermediate hues.

In general, patterns as displayed by the arrangements of colours play a more limited role in Landsat image interpretation of vegetative cover than in air photo interpretation. However, patterns are frequently very important in recognizing the land systems and land units of an area which in turn provide a basis for identifying the types of vegetation that are associated. For example, longitudinal sand dunes are readily recognized on Landsat images and their presence is a significant clue to determining the vegetative cover type of the area.

Although 39 of the vegetation structure classes identified by Carnahan (1976) were recognized in sample areas, many of these were not distinguishable from Landsat images alone. For example, it was not possible to identify positively and consistently the five major types of grasslands without reference to other information. The *Astrebla* spp. tended to appear on the Landsat image as blue-white; the *Dichanthium* spp. tended to be more yellow, and *Triodia* spp. which form only a sparse cover were most readily identified by their inability to conceal the underlying rock structures. However, in many cases, it was necessary to refer to the land systems reports and to the Natural Vegetation map (Carnahan, 1976) to positively identify the species.

Each sample site was located as precisely as possible on the image and an initial set of vegetative cover boundaries identified on the basis of colour signatures in the Landsat image. These boundaries were compared with information for the site on the 1:100 000 topographic map sheets, and where additional information was available this was incorporated in both boundary definition and the tentative classification of the vegetative cover type. Finally, land systems reports and the Natural Vegetation map were consulted, and the land systems description and the map classification used to determine the most likely vegetative classes for the area. In effect the data recorded for each sample site were synthesized from the four major sources; they were best estimates rather than actual measurements. This was considered satisfactory for the present study given the time and funds available for its completion.

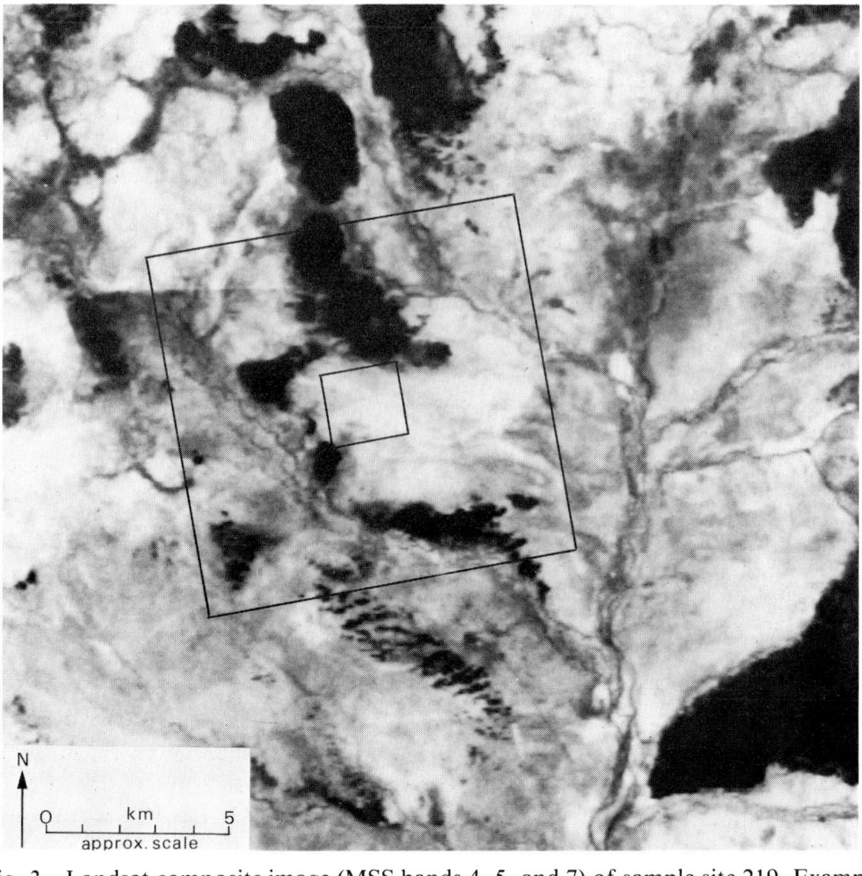


Fig. 3 Landsat composite image (MSS bands 4, 5, and 7) of sample site 219, Example 1. Image ID no. 114–073 181081. (*Reproduced by permission of NOAA*.)

The procedure was found to be quick and effective to use and in most cases involved direct interpretation of Landsat images with confirmation of the details from the other sources. However, interpretation for a number of sample localities proved to be more difficult. In some cases, it was a matter of estimating the relative area of particular types of vegetative cover; in others, it was a matter of assessing the vegetative cover types in the sample locality which were not mentioned by the other sources; and in more difficult cases it was a question of deciding whether or not vegetative cover types occurred according to the descriptions of complex land systems given in the literature. Three examples are provided below to illustrate these difficulties. In the examples the final synthesis of vegetative cover is classified according to the Natural Vegetation map (Carnahan, 1976).

**Example 1**

| | |
|---|---|
| Sample 219. Location: | 18° 25′S 125° 25′E |
| Topographic map sheet: | ANTRIM 4561 |
| Landsat scene: | 114-073 (Fig. 3) |
| Land Research Series: | Land of the Ord-Victoria Area, W. A. and N. T.<br>Map scale 1:1 m |

*Data for inner cell*

| | |
|---|---|
| 1. Topographic map sheet: | 100% no category |
| 2. Landsat interpretation: | 100% grassland, possibly with small areas of low shrubland |
| 3. Land systems: | 100% Inverway—mainly Barley Mitchell grass (*Astrebla pectinata*) |
| 4. Natural Vegetation map: | 100% aG3 (*Astrebla* tussock grassland with >70% foliage cover) |
| 5. Synthesis: | 100% *Astrebla* tussock grassland |

*Data for outer cell*

| | |
|---|---|
| 1. Topographic map sheet: | 80–90% no category<br>5–10% scattered scrub<br>5–10% medium scrub |
| 2. Landsat: | 5–10% woodland<br>5–10% scattered woodland<br>70–80% grassland<br>5–10% shrubland |

3. Land systems: —Inverway—mainly Barley Mitchell grass (*Astrebla pectinata*), some blue bush shrubland (*Chenopodium avricomum*)
—Geebee—mainly low woodland (*Eucalyptus brevifolia*) or woodland (*E. dichromopholoa*) with soft spinifex (*Triodia pungens*), with some sparse low woodland (*E. pruniosa* or *E. argillacea*) with mid-height grass (*Aristida pruniosa*)

4. Natural Vegetation map: 80% aG3 (*Astrebla* tussock grassland with >70% foliage cover)
20% ehlyG (low eucalypt trees of <1% foliage cover with an understory of tussock grasses)

5. Synthesis: 80% *Astrebla* tussock grassland
5–10% low eucalypt woodland with hummock grasses (e12th)
5–10% low eucalypt open woodland with tussock grasses (ehlsG)
5–10% mixed chenopod shrubland/tussock grassland

Interpretation of vegetation for the inner cell presents no particular problems as all sources generally agree. However, a bluish tinge on the Landsat image suggests there may be some sparse areas of low chenopod shrubland adjacent to the lower water course areas. This impression is confirmed over the larger area of the outer cell and chenopod shrubland is regarded as an identifiable class of cover in this sample. During the second series of visits to cattle stations, this area was examined from a low-flying aircraft and the existence of low shrubs, probably chenopods, was confirmed.

**Example 2**

Sample 964. Location: 12° 15′S 133° 0′E

Topographic map sheet: CAHILL 5472

Landsat scene: 112-069 (Fig. 4)

Land Research Series: Lands of the Alligator Rivers Area, N.T.
Map scale 1:250 000

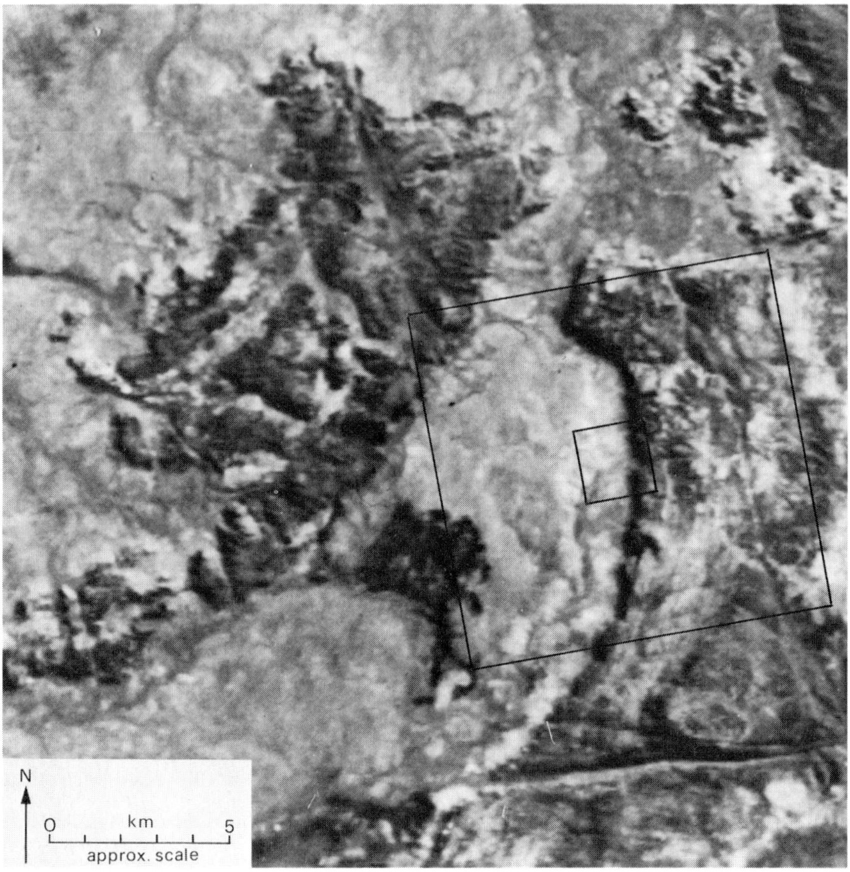


Fig.4 Landsat composite image (MSS bands 4, 5, and 7) of sample site 964, Example 2. Image ID no. 112-069 180781. (*Reproduced by permission of NOAA*.)

*Data for inner cell*

1. Topographic map sheet: 30% no category
   70% medium forest
2. Landsat interpretation: 10% bare rock surface
   10% low shrubs and heath
   10% *Allosyncarpia* forest
   20% open forest
   50% woodland
3. Land systems: 20% Baldiva: mainly bare rock with scattered sandstone scrub and sandstone scrub with small areas of woodland and occasional patches of *Allosyncarpia* forest

| | |
|---|---|
| 3. Land systems (*cont.*) | 10% Amhurst—tall open eucalypt forest with mixed shrubby woodland and an understory of perennial grasses, and scattered patches of *Allosyncarpia* forest<br>70% Kysto—eucalypt woodland with an understory of annual grasses, and tall open eucalypt forest with a mixed grasses and shrub understory |
| 4. Natural Vegetation map: | 80% eM2G—medium eucalypt trees with a foliage cover of 10–30%<br>20% eLl—low eucalypt trees with a foliage cover of <10% |
| 5. Synthesis: | 10% bare sandstone with some low shrubs and heath with occasional low eucalypts (Nil, Zl, eLl)<br>10% *Allosyncarpia* forest (xM4)<br>40% medium and tall open eucalypt forest with an understory of perennial grasses (eM, T3yG)<br>40% eucalypt woodland with an understory of mixed grasses and shrubs (eM2yG, xZ) |

*Data for outer cell*

| | |
|---|---|
| 1. Topographic map sheet: | 47% no category<br>3% dense forest<br>50% medium forest |
| 2. Landsat interpretation | 10% bare rock surface<br>20% low shrubs and heath<br>15% *Allosyncarpia* forest<br>15% open forest<br>40% woodland |
| 3. Land systems: | 30% Baldiva—mainly bare rock with scattered sandstone scrub and sandstone scrub with small areas of woodland and occasional patches of *Allosyncarpia* forest<br>10% Amhurst—tall open eucalypt forest with mixed shrubby woodland and an understory of perennial grasses and scattered patches of *Allosyncarpia* forest |

| | |
|---|---|
| 3. Land systems (*cont.*) | 60% Kysto—eucalypt woodland with an understory of annual grasses, and tall open eucalypt forest with a mixed grass and shrub understory |
| 4. Natural Vegetation map: | 60% eM2G—medium eucalypt trees with a foliage cover of 10–30%<br>40% eLl—low eucalypt trees with a foliage cover of <10% |
| 5. Synthesis: | 30% bare sandstone with some low shrubs; and heath with occasional low eucalypts (Nil, Zl, eLl)<br>15% *Allosyncarpia* forest (xM4)<br>15% medium and tall open eucalypt forest with an understory of perennial grasses (eM, T3yG)<br>40% eucalypt woodland with an understory of mixed grasses and shrubs (eM2yg, xZ) |

The topographic and the Natural Vegetation maps provide quite misleading information for both the inner and outer cells of this sample. The vegetation description provided by the land system report is better and includes details of the various components of the vegetative cover. Unfortunately, the level of generalization used to describe the overall cover of each land system introduces a somewhat false impression of the proportional distributions of each type of cover. However, when combined with the Landsat interpretation, it is possible to overcome this weakness. This site was visited by helicopter and the synthesis appeared to provide a satisfactory description of the sample area.

**Example 3**

| | |
|---|---|
| Sample 1691. Location: | 13° 34′S 141° 45′E |
| Topographic Map Sheet: | ARCHER RIVER 7270 |
| Landsat scene: | 105-069 (Fig. 5) |
| Land Research Series: | Nil |
| Replacement: | Plant Communities of Cape York Peninsula<br>Map scale 1:2 m |

*Data for inner cell*

| | |
|---|---|
| 1. Topographic map sheet: | 80% medium vegetation<br>20% dense vegetation |

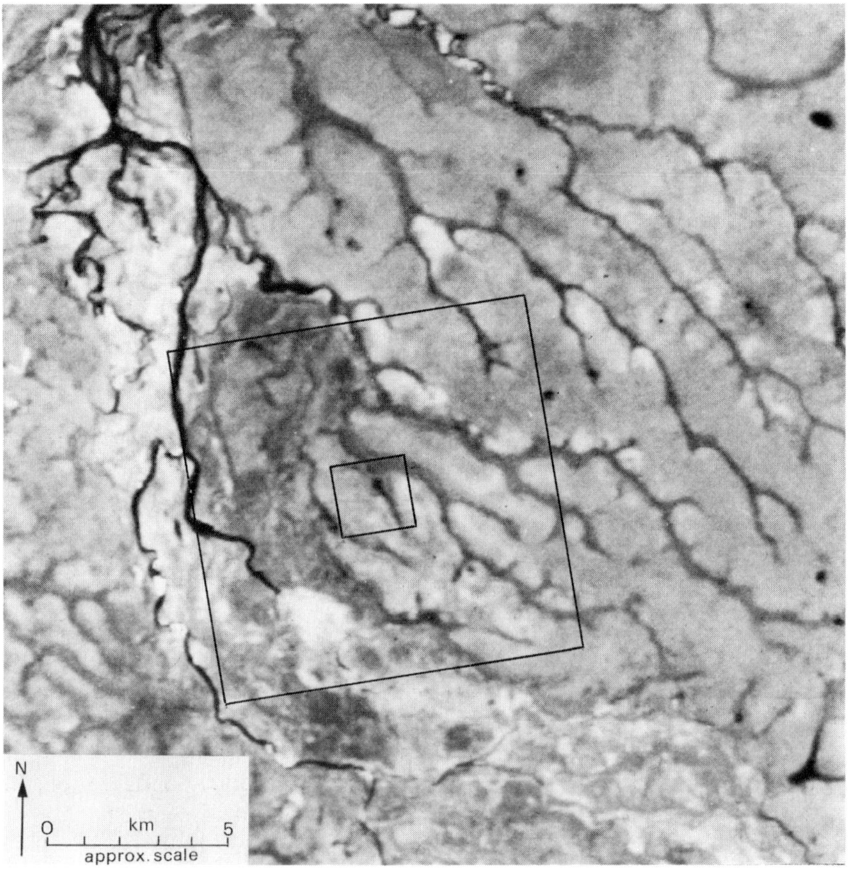


Fig. 5 Landsat composite image (MSS bands, 4, 5, and 7) of sample site 1691, Example 3. Image ID no. 105-069 261081. (*Reproduced by permission of NOAA*.)

| | |
|---|---|
| 2. Landsat interpretation | 15% gallery forest<br>30% grassland<br>55% open woodland |
| 3. Replacement for Land Research Series: | 100% open forest |
| 4. Natural Vegetation map: | 100% eM3L—medium eucalypt forest with a foliage cover of 10–30% |
| 5. Synthesis: | 15% dense mixed gallery forest (x, eM3, 4)<br>30% tall grassland (yG3)<br>55% open eucalypt woodland with a tall grass understory (eM2yG) |

*Data for outer cell*

| | |
|---|---|
| 1. Topographic map sheet: | 80% medium vegetation<br>10% dense vegetation<br>10% sparse vegetation |
| 2. Landsat interpretation: | 10% gallery forest<br>20% grassland<br>40% open woodland<br>30% forest |
| 3. Replacement for Land Research Series: | 100% open forest |
| 4. Natural Vegetation map: | 100% eM3L—medium eucalypt forest with a foliage cover of 10–30% |
| 5. Synthesis: | 10% dense mixed gallery forest (x, eM3, 4)<br>20% tall grassland (yG3)<br>30% eucalypt forest with low tree understory (eM3L)<br>40% open eucalypt woodland with a tall grass understory (eM2yG) |

In this sample, Landsat interpretation suggested two additional classes of vegetative cover not indicated by other sources. These were confirmed by light aircraft reconnaissance while working on another project.

## THE VALUE OF LANDSAT IMAGERY IN THE VEGETATION SYNTHESIS

It is difficult to judge the overall contribution of Landsat information to the synthesis of vegetative cover in this project. A subjective assessment was made by evaluating one-sixth of the sample locations according to whether the Landsat imagery simply confirmed the boundaries and classes identified from other sources; indicated a change in the proportions of cells allotted to classes; or

Table 1. Contribution of Landsat interpretation to the synthesis of vegetative cover type.

| | Inner cells (%) | Outer cells (%) |
|---|---|---|
| Confirmation only | 61 | 39 |
| Change of class proportions | 24 | 34 |
| Addition of new class or classes | 15 | 27 |
| | 100 | 100 |

provided new classes to be included in cell descriptions. A summary of the results is provided in Table 1.

There are two important limitations on these findings which must be taken into account. Fire is a common land management tool in Northern Australia and severe fires immediately prior to satellite overpass frequently obliterate details of the vegetative cover. Thus, wherever fire had occurred in the sample areas it was not possible to interpret local vegetative cover using Landsat imagery. Approximately one-third of sample locations in the 'Confirmation only' category showed some evidence of fire. It must be recognized that the assessment criteria are only crude indicators of the usefulness of Landsat since they provide identification of a change in class proportions, or the need to add new classes. They do not indicate how great a change or how important the changes were.

Despite these limitations, it is apparent that visual interpretation of Landsat images did make an important contribution to the final description and classification of local vegetative cover, adding as it did, both spatial and structural indicators to data derived from other sources.

## EVALUATION OF THE SIGNIFICANCE OF THE LANDSCAPE COMPONENTS

Discussions with cattle station owners and managers and independent mustering contractors indicated that it would be possible to rank the landscape components (water features, landforms, vegetative cover, and microrelief) according to their relative importance in livestock mustering. For example, the frequency, permanence, and degree of control of water resources are more important elements in determining mustering costs and the covariant, the proportion of livestock mustered for a given cost, than landforms or vegetative cover, which in their turn are more important than microrelief.

The same discussions indicated that it would also be possible to rank the relative importance of the elements which comprise the water features, landforms, vegetative cover, and microrelief. For the purposes of this project it appeared more sensible to rank individual elements occurring within the sample separately, and then evaluate the overall degree of difficulty the landscape imposed on livestock mustering.

The elements of each of the landscape components were ranked according to the following criteria, with the degree of difficulty for livestock mustering increasing from 1 to 5:

1. Elements that exert positive effects on mustering; such elements offer no difficulties and may even attract livestock to them, e.g. grassland or edible shrublands.

2. Elements that impose a minor degree of difficulty on mustering either by partially limiting the movements of musterers or by obscuring the presence of individual cattle, e.g. rolling topography.
3. Elements that impose a definite degree of difficulty on mustering, either by seriously limiting the movement of musterers or by allowing individual cattle to hide, e.g. open forests.
4. Elements that impose a severe degree of difficulty on mustering either by making the movement of musterers impossibly slow or by allowing groups of cattle to remain undetected.
5. Elements that prevent livestock mustering altogether and offer considerable difficulty for livestock shooting.

In the case of vegetative cover, the 35 classes used to describe the cell landscapes were reduced to 13 vegetation structure classes, and ranked according to the above criteria (Appendix 1). The ranking was applied to each sample cell by multiplying the proportion of the cell under each type of cover by its ranking and totalling the values.

Similar calculations were undertaken for landform, water features, and lithology. As noted above, the preliminary analysis of mustering costs in relation to landscape for sample cattle stations suggested that the relationship between the degree of difficulty imposed on mustering and the landscape components was additive. Thus the undefined function:

$$DM = f(L, R, V, W)$$

can be rewritten as:

$$DM = aL + bR + cV + \mathrm{d}W,$$

where the weights *a*, *b*, *c*, and *d* are yet to be determined.

At the time of writing, no decision has been made as to the form of the function that will be adopted in the final analysis, but the additive relationship is being used in current analysis.

## CONCLUSION

In the evaluation of landscapes required by the Livestock Mustering Cost project, Landsat has contributed to the description of one of the four main components: vegetation cover type. It is an important contribution, providing additional information and accuracy beyond the capability of other data sources, and well below the cost of traditional forms of field examination. With regard to land management problems, such as the eradication of bovine tuberculosis in Northern Australia, Landsat must be seen as one tool among many, each of which makes its own unique contribution to the solution of the problem at hand.

## REFERENCES


Andrews, L., Johnston, J. and Laut, P. (in preparation). Landscape and locational factors and their influence on herd disease eradication strategies in Northern Australia.

Beadle, N. C., 1981. *The Vegetation of Australia*. Cambridge: Cambridge University Press.

Beard, J. S., 1979. *Vegetation Survey of Western Australia—Kimberley*. Perth: University of Western Australia.

Carnahan, J. A., 1976. Natural Vegetation. In *Atlas of Australian Resources*, 2nd Series. Canberra: Division of National Mapping.

CSIRO, 1952–1976. *Land Research Series*, vols 1, 3, 4, 5, 9, 11, 13, 16, 25, 26, 28, 38. Canberra: Commonwealth Scientific and Industrial Research Organization.

CSIRO, 1960. Pasture lands of Northern Territory, Australia. *Land Research Rept. 5*, CSIRO, Melbourne.

Graetz, R. D., Tongway, D. J. and Pech, R. P., 1982. An ecological classification of the lands comprising the southern Simpson Desert and its margins. *Tech. Memo 82/2*, CSIRO, Deniliquin.

Laut, P., Margules, C. and Nix, H. A., 1975. *Australian Biophysical Regions*. Canberra: Department of Urban and Regional Development.

Laut, P. *et al.*, 1977. *Environments of South Australia. Provinces 1–8 and Handbook*. Canberra: Commonwealth Scientific and Industrial Research Organization.

Laut, P. and Nahinga, P., 1985. Landscape data for cattle disease eradication in Northern Australia. Tech. Paper 47, Division of Water and Land Resources, CSIRO, Australia.

Northcote, K. H. *et al.*, 1967, 1968, 1968. *Atlas of Australian Soils*. Sheets 7, 8, and 9. Melbourne: CSIRO–Melbourne University Press.

Pedley, L. and Isbell, R. F., 1971. Plant communities of Cape York peninsula. *Proc. R. Soc. Qd.*, **82**(5), 51–74.

Specht, R. L., 1972. *The Vegetation of South Australia*, 2nd edn. Adelaide: Government Printer.


## APPENDIX 1. RANKINGS FOR LANDSCAPE ELEMENTS INDICATING INCREASING DEGREE OF DIFFICULTY FOR LIVESTOCK MUSTERING.

| LANDSCAPE ELEMENTS | Ranking |
|---|---|
| I. WATERING SOURCES | |
| Bore | 1 |
| Well | 1 |
| Dam | 1 |
| Small water hole or spring | 1 |
| Dry stream-bed | 1 |
| Land seasonally inundated | 2 |
| Intermittent stream | 3 |
| Numerous springs or small water holes | 5 |
| Long permanent water holes | 5 |
| Permanent river | 5 |

*continued*

Appendix *continued*

II. VEGETATIVE COVER (after Carnahan, 1976)

| | | |
|---|---|---|
| 1. Grasslands only | aG3 | 1 |
| | dG3 | 1 |
| | yG3 | 1 |
| | yG4 | 1 |
| H2 | | 1 |
| 2. Grasslands with occasional tall shrubs (i.e. top stratum provides <10% of cover) | wSltH | 2 |
| | wS2tJ | 2 |
| 3. Predominantly grasslands with low shrubs | kZ2G | 1 |
| 4. Grasslands with occasional trees (i.e. top stratum provides <10% of cover) | eL1yG | 2 |
| | eL1dG | 2 |
| | xL1yG | 2 |
| | eL1tH | 2 |
| | oL1tH | 2 |
| | eM1G | 2 |
| 5. Low shrublands | xZ3 | 1 |
| 6. Low shrublands with occasional trees (i.e. top stratum provides <10% of cover) | eL1xZ | 2 |
| | eL1Z | 2 |
| 7. Tall shrublands with occasional trees (i.e. top stratum provides <10% of cover) | eL1wS | 5 |
| | eL2sS | 5 |
| 8. Tall shrublands | sS2H | 3 |
| | wS2G | 3 |
| | wS3H | 3 |
| 9. Low woodlands | eL2H | 3 |
| | eL2G | 3 |
| | xL2G | 3 |
| | mL2G | 4 |
| | mL2H | 4 |
| 10. Medium height woodlands | eM2G | 3 |
| | eM2L | 4 |
| 11. Low open forest | wL3 | 5 |
| | eL3 | 5 |
| 12. Medium height open forest | eM3Z | 4 |
| | eM3L | 5 |
| 13. Medium height forest | xM4 | 5 |
| | eM4L | 5 |

*continued*

Appendix *continued*

III. LANDFORMS

| Relief/distance category | Landform | Rank |
|---|---|---|
| < 20 m/km | Plains—tidal flat | 1 |
| | beach | 1 |
| | floodplain | 4 |
| | pediment plain | 1 |
| | inclined plain | 1 |
| | may have—occasional dunes | 1 |
| | occasional parallel dunes | 1 |
| | dunefield | 2 |
| | parallel dunefield | 2 |
| 20–40 m/km | Undulating plain—pediment plain | 1 |
| | inclined plain | 1 |
| | may have—occasional dunes | 1 |
| | occasional parallel dunes | 1 |
| | dunefield | 2 |
| | parallel dunefield | 2 |
| | Very low escarpment | 2 |
| | Breakaway country | 2 |
| 40–60 m/km | Rolling plain | 2 |
| | Inclined rolling plain | 2 |
| | may have—occasional dunes | 2 |
| | occasional parallel dunes | 2 |
| | dunefield | 3 |
| | parallel dunefield | 3 |
| | Low escarpment | 3 |
| 60–100 m/km | Hills | 3 |
| | Ridges | 3 |
| | Isolated hillocks | 3 |
| | Escarpment | 4 |
| 100–300 m/km | High hills | 4 |
| | High ridges | 4 |
| | Isolated high hills | 4 |
| | High escarpment | 5 |
| > 300 m/km | Mountains | 5 |
| | Mountain ridges | 5 |
| | Isolated mountains | 5 |

N.B. A plateau is a planar surface, usually edged by escarpments or rugged hills, and may be flat, undulating or rolling. It is ranked as a plain.

*continued*

Appendix *continued*

| | | |
|---|---|---|
| IV. LITHOLOGY | | |
| | Beach sands | 1 |
| | River alluvials | 1 |
| | Aeolian sands | 1 |
| | Residual soils | 1 |
| | Deep sandy soils | 1 |
| | Deep volcanic soils | 1 |
| | Coarse sediments without limestones | 4 |
| | Fine sediments without limestones | 3 |
| | Limestones | 4 |
| | Metasediments | 3 |
| | Acid coarse-grained volcanics | 3 |
| | Acid fine-grained volcanics | 5 |
| | Non-acid coarse-grained volcanics | 5 |
| | Non-acid fine-grained volcanics | 5 |
| | Duricrusts | 2 |

# PART III

# Land Use Survey

Remote Sensing and Tropical Land Management
Edited by M. J. Eden and J. T. Parry


# CHAPTER 8

# *Land Use and Forest Resources of Kauai and New Caledonia: the Role of Landsat*

G. Ross Cochrane
*Geography Department, University of Auckland, New Zealand*

The requirement to inventory, classify, and monitor large areas of forest, bush, and crop land in a timely and cost-effective manner is a major concern of resource managers in both the private and public domain. Landsat classification has proven valuable as a convenient source for such data in large-scale surveys, but a major limitation arises when the classification procedures are unable accurately to separate categories that may require quite different management policies.

This study is part of a long-term project for mapping tropical island environments using Landsat data. A range of analytical procedures is being tested and evaluated for the mapping of geology, landforms, and land cover in the broadest sense, including vegetation, forest type, and agricultural land use. Reported here are the results of a variety of image enhancement and information extraction techniques that have been applied in the analysis of land cover and forestry for the islands of Kauai in the Hawaiian chain and New Caledonia in the southwest Pacific.

The accuracy and versatility of land cover classification can be improved by the application of three general types of information:

1. ancillary information, such as digital elevation data;
2. synthetic texture data;
3. a range of derived products resulting from digital image processing and data compression.

## ANCILLARY INFORMATION—DIGITAL ELEVATION DATA

Studies by Strahler *et al.* (1978, 1979) have shown that a significant improvement in natural vegetation classification accuracy (from 51% to 85%) is achieved by incorporating data on digital terrain (elevation) and aspect. Unfortunately, suitable terrain data are not available for either Kauai or New Caledonia to allow further testing of this technique.

## SYNTHETIC TEXTURE DATA

In a study of coniferous forests in northern California, Strahler (1981) demonstrated that it was possible to obtain a measure of vegetation texture (the spatial arrangement of tonal elements) on Landsat images by measuring the standard deviation of brightness values for small groups of pixels. The data are then used as a separate channel in an unsupervised clustering procedure to provide an additional dimension in the feature space. This direct use of the distinctive texture of a vegetation unit led to a considerable improvement in forest class mapping in the area. The forest communities of northern California are relatively simple in structure, single-species dominant, and largely even-aged, and they display strong ecological site preferences. In contrast, the tropical forests of Kauai and New Caledonia are structurally complex, floristically diverse, and confined to a much lower elevation range. Additionally, the forests have been substantially modified by cultural practices, notably fire and the introduction of aggressive exotic species. In these circumstances, the generation of a standard deviation texture channel, as an input for classification, is likely to be less useful than in the northern Californian softwoods.

## DIGITAL IMAGE PROCESSING AND DATA COMPRESSION

The two categories or suites of data compression that have been widely used for Landsat studies are principal components analysis (PCA) and band ratioing. Principal components analysis 1 (PCA 1) basically gives a 'brightness' value and PCA 2 provides a 'greenness' or biomass measure. Both are useful in that data from several Landsat bands are compressed into a single channel.

The 'tasselled cap' linear transform of Kauth and Thomas (1976) is effective for remote sensing studies of agricultural crops in temperate environments. It is dependent on the temporal change of radiance of a crop pixel through measurement space as the crop calendar sequence progresses through germination, crop development/increased ground cover, maturation, senescence, and finally harvest. These temporal variations of crop pixels are less important in the case of tropical plantation crops, such as sugar cane and pineapple. The temporal component is not seasonal, but spread over several years. For tropical tree crops, distinctive spectral signatures are independent of season. A time-trajectory of crop pixel change through measurement space is absent. Only at the early stages of development, if there is a strong soil component in the pixel, is there a noticeable contrast to the signature of the mature crop.

Similarly, with tropical forest vegetation, temporal spectral variations are of minimal importance in remote sensing applications. Spectral signatures are governed primarily by life-form characteristics, i.e. the relative permanence of plant geometry and canopy cover. They are also influenced by topographic location because of the effect of aspect on illumination and the complex reflectance distribution function. Two important questions arise: firstly, whether various data compression and enhancement techniques used elsewhere are applicable to tropical island environments and, secondly, whether any of these provide improved classification accuracy over raw data. In an attempt to answer these questions, investigation of the following techniques—principal components analysis (PCA), linear transform (band ratioing), raw data enhancement, supervised classification, and unsupervised classification—has been carried out on Landsat data of Kauai in the Hawaii group and Grand Terre, the main island of New Caledonia.

## KAUAI: GENERAL CHARACTERISTICS

Kauai is a nearly circular island of approximately 40 km diameter at the western end of the Hawaiian chain. It is a volcanic island, rising to over 1500 m, with deep, radiating ravines descending to the flanking lowlands on the north, east, and south (Fig. 1). The precipitous cliffs of the Na Pali form the coast on the west and northwest.

Situated at longitude 159° 30′ W and latitude 22° N, Kauai lies in the northeast trade wind belt. Marked precipitation differences exist between the windward (northeastern) side and the leeward (southwestern) slopes. There is a dramatic precipitation gradient from the central peaks of Wai'ale'ale (1570 m) and Kawaikini (1600 m), which are reputedly the wettest areas on earth with a mean annual rainfall in excess of 11 800 mm, to truly desert conditions on the southern coastal lowland where the annual total averages less than 250 mm.

The higher slopes of the island are extremely inaccessible, and extensive areas of relatively unmodified forest remain. Modified forest and shrubland occupy the moderately steep to steep slopes at intermediate elevations. On the lower peripheral areas, slopes are more gentle. Variable quality pastures support grazing (ranching) on the northern lowlands. Patches of open bush and shrub are also present. Large scale commercial sugar cane growing is widespread on the gentle slopes and coastal lowlands of the east and south. Sugar cane fields also extend up to elevations of 500 m along the broad interfluves of old, deeply weathered lava flows in the southern district of Hanapepe (Waimea to Mana), wherever irrigation water is available and soils are stable. Erosion is prevalent on steeper slopes with sparser vegetation at all elevations. It is especially severe on the southern and southwestern slopes at intermediate elevations (300–700 m). Culturally accelerated erosion is widespread at lower and intermediate elevations.

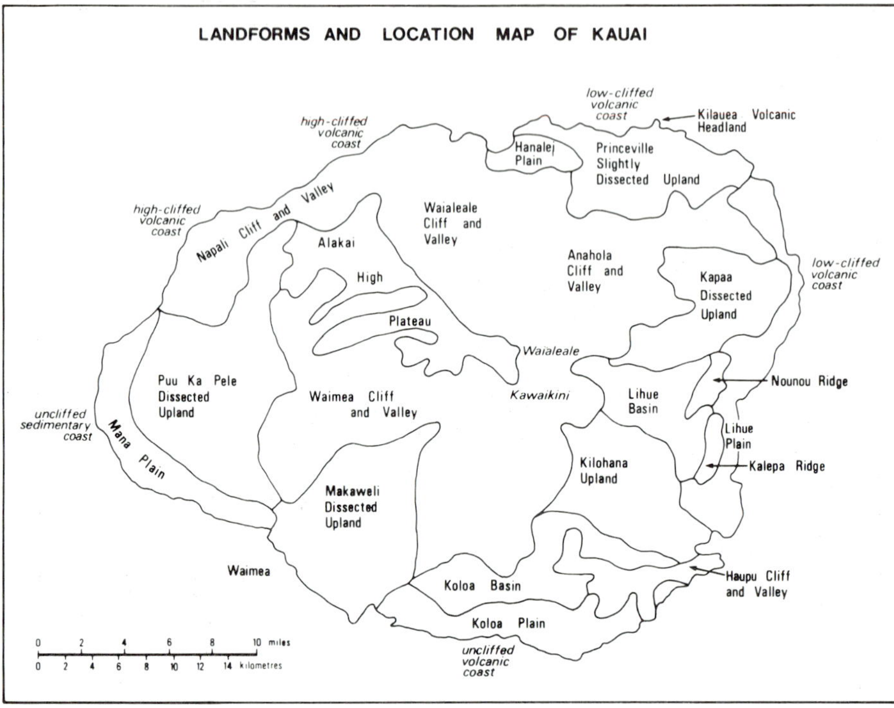


Fig. 1 Generalized physiographic units of Kauai (after Armstrong, 1973).

## NEW CALEDONIA: GENERAL CHARACTERISTICS

The French island of New Caledonia is situated at longitude 185° E and latitude 21° S, within the southeast trade wind belt. The island is narrow and mountainous, extending in a northwest-southeast direction for some 400 km, with a maximum width of only 80 km (Fig. 2). There is a marked contrast between the windward (eastern) slopes with an annual precipitation of more than 4000 mm and the drier, western coastal lowland (< 1000 mm), although the gradient is not so dramatic as in Kauai.

Mountains rise abruptly from the coast along most of the eastern side of the island to form a massive central core. The southern third of the mountain chain is composed of the ultrabasic rock, peridotite. A series of steep, isolated peridotite massifs occur along the southern side of the central mountains, becoming progressively smaller and lower to the west. The central area is a complex of metamorphic and volcanic rocks. The northern third of the island is dominated by the ancient schists and gneisses of the Massif du Panié, with

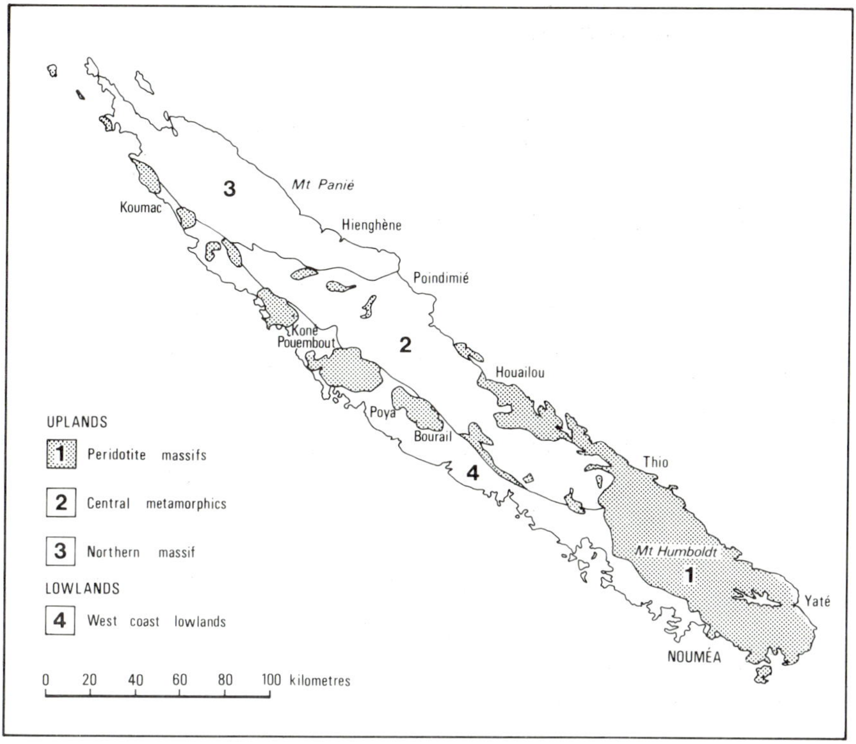


Fig. 2 Generalized physiographic units of New Caledonia (after Paris, 1981).

flanking hills of Mesozoic and Tertiary sediments. Along the west coast there is an irregular lowland from 3 to 20 km in width, which in the north is developed over Tertiary basalts.

A combination of natural and human influences has produced strongly site-specific plant communities in New Caledonia. Modification of the indigenous forest cover by anthropic burning, possibly over a period of centuries, has frequently accentuated the natural controls by the removal of buffering ecotones. Many plant communities display strong geologic, topographic, edaphic, ecologic, and pyric interrelationships which result in sharply defined community boundaries. The communities can be mapped as vegetation or land cover units in their own right, but also serve as surrogates for soils and landforms.

Some areas of dense evergreen forest remain in the mountains, but most of the vegetation has been extensively modified by fire. All stages of degradation from the original forest through secondary forest, savanna, shrubland, depauperate herb, and fern (*Dicranopteris*) communities to bare areas are present. Fire-induced savanna woodlands and more open savanna are dominated by the myrtaceous sclerophyllous fire-resistant *niaouli* (*Melaleuca quinquenervia*). These communities are particularly prominent on the western side of the island, occurring from sea level to the summits of the Massif du Panié at about 1600 m. As a result of frequent burning and shifting cropping, especially around Nouméa, the savannas often develop dense thickets of introduced weed species, including mimosa, guava, and lantana.

Low shrubby maquis communities appear to be an edaphic climax normally confined to black soils derived from ultrabasic rocks. Grassy formations, found typically on the thin soils overlying basalts and shales, are dominated by *Heteropogon*, *Imperata*, and *Aristida* species.

Mangrove communities are present along much of the western coast, but are highly localized within a few major estuaries on the steep eastern coast.

The major economic activity in New Caledonia is mining nickel and chromium from the ultrabasic rocks. Erosion scars from mining activities are prominent features of the landscape. There is little agricultural activity, apart from subsistence cropping interspersed with coconut and coffee growing in small, dispersed pockets of alluvial land along the east coast. Some grazing occurs in the north of the island and in alluvial valleys of the central zone. A localized tourist industry exists at Nouméa.

## FIELD WORK IN THE STUDY AREA

Following discussions with scientists from the *Bureau de Recherches Géologiques et Minières* (BRGM) and the *Office de la Recherche Scientifique d'Outre Mer* (ORSTOM) in Nouméa, a period of approximately one month was devoted to field survey in New Caledonia. This involved two phases: an initial, broad reconnaissance survey of the whole island, followed by more detailed studies, using Landsat 1 images, of the northern half of the island and of selected areas in the central and eastern parts. Measurements, field sketches, and notes were recorded. Colour photographs were taken of plant community characteristics, cover classes, soils, geomorphic units, and bedrock outcrops.

A shorter period of fieldwork was required for the much smaller island of Kauai. Problems of location in the field and exact orientation experienced when using Landsat images in New Caledonia were largely overcome in Kauai through the use of a 'photomap'. This was prepared by superimposing an enlarged Landsat colour composite, at a scale of 1:125 000, on a topographic base containing details of roads, settlements, place names, spot heights, and contours at 50 ft (16.5 m) vertical interval. The photomap proved to be very useful in the field for location purposes and for the selection of field sites to serve as training sets for the computer analysis of land cover types. In addition to the collection of general field information, colour photographs and field notes were recorded for the selected test sites.

## LANDSAT IMAGE ANALYSIS

Two types of analysis were employed in the treatment of Landsat data, namely, visual analysis of both single-band Landsat images and multiband colour composites, and digital analysis involving various data compression and enhancement techniques.

### Single-band visual analysis

Positive photographic prints of the Kauai scene (070-045) of 12 October 1977 (ID E 20994-19533) were produced for each of Landsat MSS bands 4, 5, 6, and 7, at the following scales: 1:1 000 000, 1:500 000, and 1:150 000. Thirteen Landsat images of New Caledonia were available, and photographic prints of band 5 were prepared for all of these at 1:1 000 000 scale. Subsequently, four scenes (ID E 10692-22222 and E 10692-22225, 15 June 1974; E 30224-22135, 15 October 1978; and E 30277-22080, 7 December 1978) were selected as the best cloud-free images. Photographic prints at scales of 1:1 000 000, 1:500 000, 1:200 000 and 1:166 000 were made for each of the four MSS bands of these scenes. These were used in preliminary mapping of the land cover and of the landforms and geology. Landsat 1 images of the northern half of the island were selected for final study. These proved the most useful for detailed surveys of land cover. Density slicing of band 5 was carried out on several subscenes.

### Multiband colour composites

An $I_2S$ Mini Addcol Additive Viewer was used to produce simulated colour infrared composites, as positive transparencies, using 70 mm film chips of bands 4, 5, and 7 for both Kauai and New Caledonia. These were prepared at a scale of 1:1 000 000 for New Caledonia and at 1:250 000 for Kauai. Subsequently, colour print enlargements were made at scales of 1:1 000 000 and 1:500 000 for New Caledonia and 1:150 000 and 1:100 000 for Kauai. Some subscene areas of the northern half of New Caledonia were later enlarged to scales of 1:500 000, 1:200 000, and 1:166 000. These enlarged positive colour composite prints were used for initial mapping of land cover classes and geologic-geomorphic units and the results were compared with those obtained from the single-band, black and white, positive print enlargements.

One feature that became obvious in the study of the New Caledonia scenes was the significance of season. The two June (winter) scenes were greatly superior to the spring and summer scenes for vegetation differentiation. Precipitation has a spring-summer peak and the herbaceous forbs (ferns) and grasses grow profusely to provide a relatively uniform cover over most of the area. In contrast, because of drier conditions in the winter months, highly variable soil moisture availability causes dramatic contrasts in plant growth. Differences in moisture stress and reduced plant vigour, as well as variable plant canopy/soil gap patterns within and between plant communities, result in a more complex pattern being recorded. These contrasts in spectral signature facilitate the mapping of land cover classes. It was this factor that lead to the selection of the two northern New Caledonia scenes for detailed computer analysis.

**Digital image processing**

Various types of digital image processing were carried out in France at the *Laboratoire Analyse Numerique Image* (LANI) of the *Office de la Recherche Scientifique d'Outre Mer* (ORSTOM) and at the excellent facilities of the *Groupement pour le Développement de la Télédétection Aerospatiale* (GDTA) and the *Centre National d'Etudes Spatiales* (CNES) at Toulouse. At a later stage, batch and interactive digital processing was carried out at the Jet Propulsion Laboratory (JPL) in Pasadena, California, using the Video Image Communication and Retrieval system (VICAR) .

A detailed discussion of all the procedures employed is not possible. All that can be attempted here is a review of the results of selected data analysis, focusing on the two basic questions posed in the initial discussion of digital image processing: are the available data compression techniques suitable for the task of mapping land cover types in tropical island environments, and are the resulting classifications superior to those obtained using only the raw data?

*Raw data processing*

The 15 June 1974 Landsat scene of New Caledonia (image ID E 10692-22222) was selected for raw data analyses to compare with the results of the data compression processing. MSS bands 4, 5, and 7 were depicted on the colour monitor of the TRIM interactive system at CNES as blue, green, and red, respectively. A linear contrast stretch was applied and the full scene and subscene enhancements were recorded and analysed. Differentiation of the major plant communities became increasingly more accurate with subscene enlargements at 2, 4, and 6 times. The results were subsequently compared with enlargements from ratio images and with the products of principal components analysis.

A second New Caledonia scene (image ID E 10692-22225) was processed in the same manner using bands 4, 5, and 6 data on the VICAR system. Full scene and subscene images were produced and recorded in hardcopy using a camera to photograph the visual display.

For the Kauai scene (image ID E 20994-19533), four types of processing were undertaken on the VICAR system, as follows:

1. linear contrast stretch of bands 4, 5 and 6, with colour enhancement and preparation of full and half-scene images and of subscenes at 3, 5 and 6 times enlargement;
2. supervised classification;

Table 1. Land cover types of Kauai.

| Group or category | Nos of classes | Description of classes |
|---|---|---|
| Forest | 5 | *Ohia* forest, candlenut (*kukui*) forest, closed forest (types 1 and 2), open forest and scrub |
| Sugar cane | 3 | Mature cane, bare or thin cane, bare burned areas |
| Pasture | 1 | – |
| Eroded land | 3 | Classes 1, 2, 3 (in order of decreasing severity) |
| Reservoirs | 1 | – |
| Urban areas, beach | 1 | – |

3. unsupervised classification using the same subscenes as for linear contrast stretch in order to obtain exact area comparisons;
4. D-stretch, combining raw data processing, and data compression.

In order to illustrate the techniques outlined above, the processing of the Kauai image is considered in more detail, especially in respect of supervised classification. The following procedure was adopted:

1. Location of the test sites using the photomap and a linear contrast stretched image.
2. Preparation of histograms of spectral radiances for MSS bands 4, 5, and 6 of the test sites. These were inspected, and refined where necessary, to establish the 14 training sets for the land cover types listed in Table 1.
3. Preparation of single theme maps for each of the cover classes.
4. Preparation of a composite land cover map of Kauai using maximum likelihood classification.
5. Generation of composite maps of subcategories, including five forest classes and three erosion classes.
6. Preparation of fully classified subscenes at three times and five times magnification covering most of the island.
7. Production of hardcopy records (colour photographs) of all maps produced in steps 3–6 above.
8. Field checking of the results. The ravages of Hurricane Iwa on 23 November 1982 severely limited the usefulness of the field check.

The land cover maps of Kauai derived from the supervised classification (steps 4–6 above) provided a spatial display of the land type categories defined in the training sets. The land cover classes can be presented individually as thematic maps, progressively as partly classified maps, or collectively as fully classified land cover maps (Plate 1).

*Forest classes.* Five land cover classes of forest are identified in the supervised classification. Four of these are closed canopy forests, whereas the fifth is more open. *Ohia* communities are dominated by dark-toned, sclerophyllous small-leaved *ohia lehua* (*Metrosideros collina* subsp. *polymorpha*). They appear as a distinctive class on all photographic products and, on colour composites, exist as dark brown areas in contrast to the red-browns of other forest types (Plate 2). The *ohia* distribution is closely related to a specific elevation range, climatic regime, and soil environment. Two subcategories of *ohia* forest have been recognized, namely, relatively 'pure' *ohia* communities, with *ohia* dominance in the upper canopy, and a mixed or ecotonal community representing the transition to more mesophytic forest classes.

Candlenut (*kukui*) forest communities have a broadly similar and partially overlapping spectral response to mature sugar cane, but differentiation is relatively easy because of location. *Kukui* forest appears as thin, distinctively light-toned bands at high elevations (Plate 1) in contrast to the broad geometric patterns of sugar cane at lower elevations and on gentler slopes. Use of digital terrain data as an additional channel would enable an operator to classify these *kukui* communities directly by digital means.

Other closed canopy forest communities (closed forest types 1 and 2) have an overlapping areal distribution throughout the higher, central area of the island. The species making up the different communities are present in a complex pattern related to local habitat differences (Plate 1).

Open forest communities are widespread. They consist of scrub species such as guava, *Acacia*, and other xeric shrubs. Because of the limited range of strongly contrasted colours that can be used on the visual display when a full classification is undertaken, the various scrub communities are merged.

*Agricultural land classes: sugar cane*. Three categories of land cover associated with commercial sugar cane production are recognized (Plate 1), as follows: firstly, dense mature cane characterized by 100% plant cover; secondly, less dense cane, either youthful thin growth or poor older growth in neither of which full ground cover is achieved so that bare-ground reflectance contributes to the spectral signature; thirdly, bare ground prepared for replanting cane after harvesting of the previous crop and burning of the stubble: this class was readily separated from most bare eroded areas with the exception of some of the very extensive upper slopes of Waimea Canyon.

*Agricultural land classes: pasture*. Relatively large areas of pasture were accurately classified, but there were small areas, especially within the Waimea and Kahana gorges, where herbaceous vegetation with similar spectral reflectance values to pasture was included in the category.

*Eroded land classes*. The supervised classification used in the study provided effective discrimination of the widespread and complex pattern of eroded land in Kauai (Plate 3). Three separate eroded land categories were identified, as follows:

erosion class 1, comprising severely eroded areas with much bare ground on steep slopes; erosion class 2, comprising bare ground, usually less steep, with scattered low shrub vegetation generally providing less than 20% plant cover; and erosion class 3, which is a complex category including bare ground with sparse vegetation, partly burned or ploughed ground with weeds, and areas of recently planted sugar cane.

This classification probably provides the best record to date of the extent and types of erosion in Kauai. If available, additional terrain data used as corollary information and providing elevation and slope classes would improve the separation of newly planted cane fields on gentler slopes and the steeper, bare eroded areas in erosion class 3.

In addition to the supervised classification of Kauai, fully classified subscenes enlarged to a suitable scale were used for detailed land cover mapping of selected areas of interest. Such a subscene, the Koloa area of Kauai, is shown in Plate 4 at a three times enlargement. A direct comparison can be made with the same area enlarged from a simple linear contrast stretch with colour enhancement, reproduced as a simulated colour infrared image (Plate 5). The general pattern of the land cover classes can be identified in Plate 5, but the classified image (Plate 4) provides superior detail.

### *Data compression techniques*

The purpose of data compression techniques, such as principal components analysis and band ratioing, is to retain, and often enhance, essential information while removing the redundancy inherent in the Landsat system which records in four separate bands (Haralick *et al.*, 1972; Ready and Wintz, 1972; Billingsley, 1973; Deering *et al.*, 1975; Fontanel *et al.*, 1975; Kauth and Thomas, 1976; Richardson and Weigand, 1977; Soha and Schwartz, 1978; Moik, 1980; Logan and Strahler, 1983). Data compression transforms have the potential for improved classification accuracy with a reduction in computer processing time and costs, although it must be remembered that most transforms result in some information loss during the compression process. The degree of improvement in the classification is strongly dependent upon both the purpose and the methods used. Some success has been demonstrated in forestry studies (Deering *et al.*, 1975; Logan and Strahler, 1983), in agricultural applications (Kauth and Thomas, 1976), in water studies (Alföldi and Munday, 1977; Lindell, 1981), and in geology (Soha and Schwartz, 1978; Moik, 1980). However, methods that are successful in one environment may not necessarily be as useful elsewhere.

*Complex transforms*. Logan and Strahler (1983) tested eleven different Landsat transforms and compared these with four bands of raw data in an attempt to find an optimal transform for the classification of coniferous forest types. Only one of the eleven transforms tested gave more information than the original

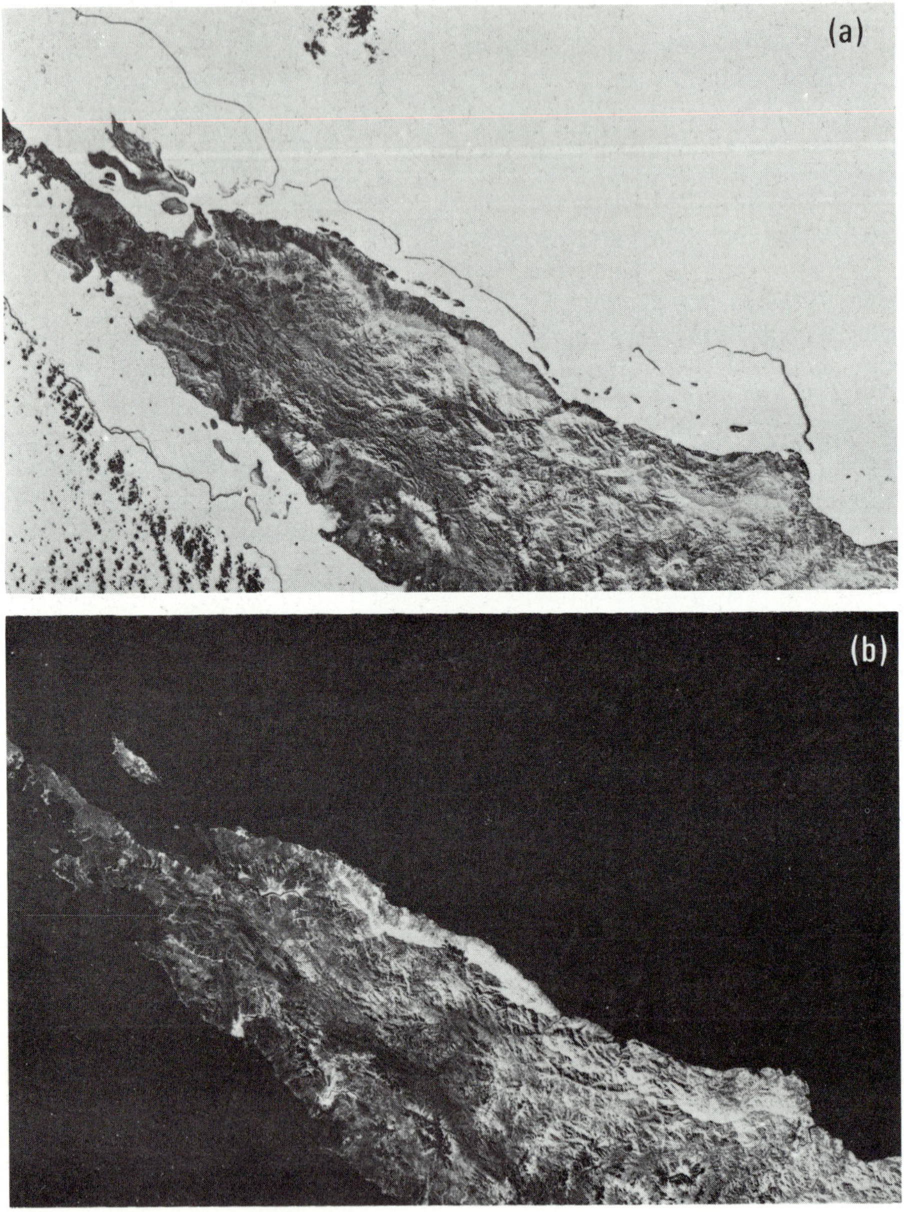


Fig. 3 Comparison of different transform products, New Caledonia. Landsat scene 088-074, image ID E 10692-22222, 15 June 1974 (*Reproduced by permission of NOAA*). (a) Band ratio 4:5, (b) Band ratio 7:5, (c) TVI 7 transform, $\sqrt{\frac{\text{Bands } 7-5}{\text{Bands } 7+5}}$

(d) Principal components analysis (PCA 2).

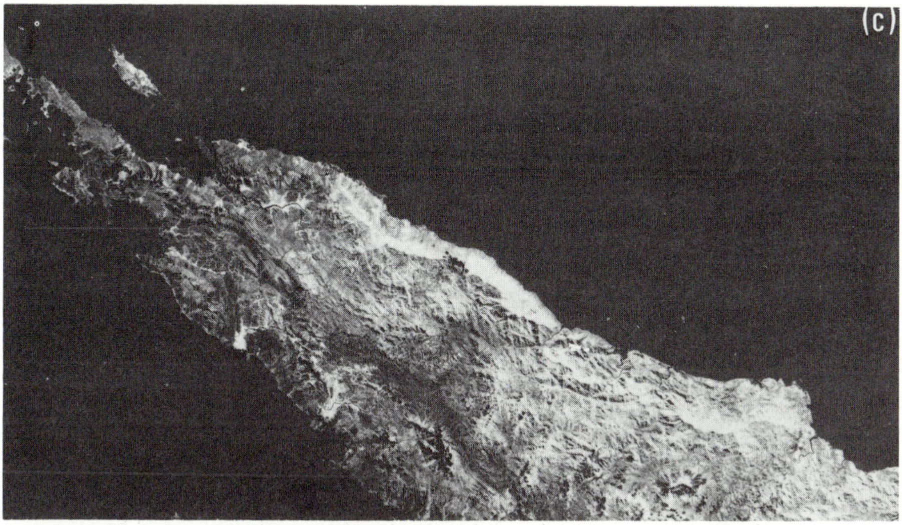
(c)

(d)

individual inputs. This transform, RAVE/E2, compares the ratio of averages to the ratio of Eigenvector 2 and requires both band ratioing and principal components analysis. Although successful in Logan and Strahler's study of the coniferous forest of northern California, it proved to be of little benefit in the analysis of Pacific island environments.

D-stretch, a complex transform developed by the Jet Propulsion Laboratory using the VICAR system was employed on the Kauai data. In this process, raw Landsat radiance data from MSS bands 4, 5, and 6 are contrast enhanced using a gaussian stretch of the principal components prior to inverse rotation (Soha and Schwartz, 1978).

This technique had been previously employed with success on a Landsat subscene (ID E 1556-21331) in New Zealand (Cochrane and Logan, 1979; Cochrane, 1980). In the Auckland region, it provided a substantial improvement over linear contrast stretching for recognizing agricultural and horticultural patterns, identifying myrtaceous forest of *Metrosideros excelsa* (similar to Kauai's *ohia* forest), and refining the definition of urban functional areas. In the Kauai study, the results were disappointing (Plate 6). No substantial improvement in classification accuracy was achieved compared with the simple linear contrast stretch. D-stretch techniques were not applied to the New Caledonia scenes.

*Band ratioing*. Band ratioing techniques are useful because they can be carried out both digitally and photographically. In the study of natural vegetation, the most commonly used two-band data compression ratios are MSS bands 5:6 and 5:7 (Billingsley, 1973). The band 4:5 ratio has also been employed. None of these provided any improvement over single band data for scenes of New Caledonia (Figs 3(a) and 3(b)) or Kauai.

Two other transforms use four MSS channels rather than two. These are RAVE (the ratio of averages) and RASD (the ratio of sums divided by the differences). Neither of these provided any improvement in classification accuracy compared with the single bands.

Deering (1973) and Richardson and Weigand (1977) found the Transform Vegetation Index (TVI 6) a useful ratio for distinguishing the relative 'greenness' of vegetation:

$$\text{TVI } 6 = \sqrt{\frac{\text{MSS } 6 - \text{MSS } 5}{\text{MSS } 6 + \text{MSS } 5} + 0.5}$$

In the present study it was found that superior results were obtained with bands 7 and 5 data (Fig. 3(c)):

$$\text{TVI } 7 = \sqrt{\frac{\text{MSS } 7 - \text{MSS } 5}{\text{MSS } 7 + \text{MSS } 5} + 0.5}$$

This transform compressed more vegetation information into a single channel than any of the original, individual channels of raw data. In New Caledonia, it proved particularly valuable in enhancing two important land cover categories, namely, sparsely vegetated areas on steeply eroded hillslopes, and dense forests on highland areas and river floodplains. Other categories of 'greenness' were depicted as satisfactorily on this transform as on the original individual channels. The transform was also found to be slightly better than a principal components analysis (PCA 2) for the definition of 'greenness'; most interband wavelength contrasts were removed and subtle variations within image features were enhanced (Fig. 3(c)). The TVI 7 transform has not yet been tested for Kauai, but good results are anticipated.

*Principal components analysis*. Fontanel *et al.* (1975), Jenson and Waltz (1978) and Lefevre (1982) have demonstrated that principal components analysis (PCA) processing is a useful tool for the compression of Landsat data. The first principal component (PCA 1) has all four channels positively weighted according to the magnitude of the standard deviations. PCA 1 provides a measure of overall albedo, illumination and brightness. It is a good index of the effects of topographic influence on illumination. Thus, while no better than any of the individual channels for vegetation studies, it is a valuable tool for terrain analysis.

The second principal component (PCA 2) emphasizes the contrasts between the visible bands (4 and 5) and the near infrared bands (6 and 7). As such, it provides an image that is similar to the colour composite and has the advantage of enhancing vegetation features (Fig. 3(d)). Other principal components analysis (PCA 3 and PCA 4) provided minimal information and only vaguely approximated the original data.

*Linear transforms*. The steep slopes and narrow gorges of much of upland Kauai are always in shadow at the time of Landsat overpass. In any analysis using raw Landsat data, shadows appear as a distinct class. They cannot be separated or merged with other categories, and so the whole range of land cover themes could be concealed within the shadowed areas. Several shadow subclasses were present in the unsupervised dendrogram of radiances, although these represent differences in degree (i.e. of relative darkness) rather than variation of kind. Although band ratioing tends to reduce the effects of shadow, traditional ratioing combinations do not fully answer the shadow problem.

Linear transform procedures tested on the Kauai scene appear very successful in eliminating the effects of shadow. To date, these compressed data products have only been examined at the photo interpretation level. If supervised digital classification techniques prove as accurate, they could provide a major benefit in the analysis of Landsat data for the rugged terrain of tropical volcanic islands, even though initial inspection of the photographic product suggests that there may be some information loss in the forest classes.

## CONCLUSIONS

Much of the research described above is still in progress and so the conclusions are preliminary. However, there is sufficient evidence to justify a report on six specific aspects.

1. The photomap produced by superimposing the Landsat colour composite on a topographic map base proved to be an invaluable field tool in Kauai. A base map of this kind should be a prerequisite to any fieldwork involving Landsat imagery. It is an extremely effective tool that can be produced simply and cheaply by photographic superimposition. For organizations that have more sophisticated computer processing capabilities and the appropriate digital data base, registration and superimposition can be achieved by digital processing.
2. Enlarged subscenes of linear contrast stretched, colour enhanced data using three or four Landsat bands were found to provide acceptable results for land cover analysis in both Kauai and New Caledonia. Such processing should be adequate for most types of classification, unless absolute accuracy is of prime concern.
3. Both unsupervised and supervised classifications of raw data were found to provide superior detail in land cover mapping compared with simple linear contrast stretch.
4. Very subtle differences that show up in unsupervised classifications can be incorporated into a detailed vegetation classification, provided the operator or analyst can contribute precise, local field knowledge. Conversely, considerable clustering of related categories is required, if the classification does not benefit from local information.
5. Of the numerous data compression techniques utilized, the Transform Vegetation Index (TVI 7) and the principal components analysis (PCA 2) appear to provide more information than any of the simple processing techniques for purposes of land cover classification. These were the only multichannel transforms that produced better results than the procedures using raw data from MSS bands 5 and 7. It is clear that data compression techniques must be carefully selected for specific applications to particular vegetation types.
6. Evaluation of supervised and unsupervised classifications using a single channel input rather than four MSS channels of raw data is yet to be undertaken. Further improvement in classification accuracy is expected. The use of single rather than multiple channel data offers substantial savings in computer time, especially for unsupervised classification.

## ACKNOWLEDGEMENTS

Grateful acknowledgement is expressed to Mr M. Carbonatto, French Cultural Attaché in New Zealand, and to the Foreign Affairs Department of the French Government for providing a Visiting Fellowship to conduct research in France at GDTA, CNES, ORSTOM, and other government institutions. Very sincere thanks are due to Mr L. Laidet (GDTA) for arranging excellent facilities within GDTA and CNES at Toulouse, and to other French colleagues. Sincere thanks are given to JPL at Pasadena, California for technical and other assistance, notably from Dr N. Bryant and Dr T. L. Logan. Thanks are due to Ms H. Harvey of Koloa, Kauai, to Dr J. P. Paris, BRGM, and to the Director and Mr J. Iltis, ORSTOM, New Caledonia; also to the Geography Department, University of Hawaii, Honolulu. Financial assistance is acknowledged from the University of Auckland.

## REFERENCES

Alföldi, T. T. and Munday, Jr, J. C., 1977. Progress towards a Landsat water quality monitoring system. *Proc. 4th Canadian Symp. on Remote Sensing*, Quebec, 325–340.

Armstrong, R. W. (Ed.), 1973. *Atlas of Hawaii.* Honolulu: University Press of Hawaii.

Billingsley, F. C., 1973. Some digital techniques for enhancing ERTS imagery. In Anson, A. (Ed.), *Management and Utilization of Remote Sensing Data.* Sioux Falls: American Society of Photogrammetry, pp. 284–293.

Cochrane, G. R., 1980. *Rural Smallholdings in the Auckland Region.* Vol. 5, *Landcover Mapping from Landsat Satellite Data.* Auckland: Regional Authority Planning Division.

Cochrane, G. R. and Logan, T. L., 1979. Satellites reveal wealth of vital data. *Bldg Prog., Dunedin*, **2**, 9–15.

Deering, D. W., Rouse Jr, J., Haas, R. and Schell, J., 1975. Measuring 'forage production' of grazing units from Landsat MSS data. *Proc. 10th int. Symp. Remote Sensing Environ.*, **2**, 1169–1178.

Fontanel, A., Blanchet, C. and Lallemand, C., 1975. Enhancement of Landsat imagery by combination of multispectral classification and principal component analysis. *Proc. NASA Earth Resources Symposium*, vol. 1-B, NASA-TMX-58168, 991–1012.

Haralick, R. M., Young, J. D., Goel, D. K. and Shanmugam, K. S., 1972. A comparative study of data compression techniques for digital image transmission. *4th Ann. Earth Resources Program Review*, vol. II, NASA-TM-X-68397, 27–33.

Jenson, S. K. and Waltz, F. A., 1978. *Canonical Analysis and Principal Components Analysis in Remote Sensing.* Rept. under U.S. Geol. Surv. Contract No. 14-08-0001-16439, Technicolor Graphic Services Inc., Sioux Falls.

Kauth, R. J. and Thomas, G. S., 1976. The tasselled cap—a graphic description of the spectral-temporal development of agricultural crops as seen by Landsat. *Proc. Symp. on Machine Processing of Remotely Sensed Data*, West Lafeyette, Purdue University, 41–51.

Lefevre, M. J., 1982. *Method de Recherche par Télédétection: Anomalies Geobotaniques en Prospection Minière.* GDTA Rept., Toulouse.

Lindell, L. T., 1981. Mapping of water quality using Landsat imagery. *Proc. 15th int. Symp. Remote Sensing Environ.*, **3**, 1375–1385.

Logan, T. L. and Strahler, A. H., 1983. Optimal Landsat transforms for forest applications. *Proc. Symp. Machine Processing of Remotely Sensed Data*, Purdue University, West Lafeyette.

Moik, J. G., 1980. *Digital Processing of Remotely Sensed Images.* NASA SP-431, Washington.

Paris, J. P., 1981. *Géologie de la Nouvelle-Calédonie.* Memoire du Bureau de Recherches Géologiques et Minières 113, Orléans.

Ready, P. J. and Wintz, P. A., 1972. Information extraction, SNR improvement, and data compression in multispectral imagery. *IEEE Transactions on Communications*, vol. COM-21, No. 10, 1123–1130.

Richardson, A. J. and Weigand, C. L., 1977. Distinguishing vegetation from soil background information. *Photogramm. Engng Remote Sensing*, **43**, 1541–1552.

Soha, J. M. and Schwartz, A., 1978. Multispectral histogram normalization contrast. *Proc. 5th Canadian Symp. on Remote Sensing*, 86–93.

Strahler, A. H., 1981. Stratification of natural vegetation for forest and range land inventory using Landsat digital imagery and collateral data. *Int. J. Remote Sensing*, **2**, 15–41.

Strahler, A. H., Logan T. L. and Bryant, N. A., 1978. Improving forest cover classification accuracy from Landsat by incorporating topographic information. *Proc. 12th int. Symp. Remote Sensing Environ.*, **2**, 927–942.

Strahler, A. H., Logan, T. L. and Woodcock, C. E., 1979. Forest classification and inventory system using Landsat, digital terrain, and ground sample data. *Proc. 13th int. Symp. Remote Sensing Environ.*, **3**, 1541–1557.

Remote Sensing and Tropical Land Management
Edited by M. J. Eden and J. T. Parry


# CHAPTER 9

# *Agricultural Land Use Inventory and Mapping in Nigeria: the Application of Remote Sensing*

PETER O. ADENIYI
*Department of Geography, University of Lagos, Nigeria*

In Nigeria, as in many other tropical areas, it is often difficult to obtain reliable and up-to-date information on agriculture and land use. In a recent report to the Federal Ministry of National Planning it was noted that essential data for planning effective land management were often unavailable. Even basic information on the type of crops grown, crop yields, farming practices, and fertilizer and pesticide usage was difficult to obtain and often unreliable (Anon., 1981). The Agricultural Research Institutes of Nigeria, which are responsible for cash crops and staple produce, including cocoa, rubber, oil palm, cereals, and root and garden crops, have a very imprecise knowledge of the crop distributions, the size of holdings, and the land productivity. Before effective land use policies can be defined, it is essential that the details of existing land use are collected and accurately mapped at suitable scales. The urgency of a national land use mapping programme was stressed more than five years ago (Agboola, 1979).

The research project described in the following pages is an attempt to address this need. The main objectives of the project are to develop a multilevel, land use classification scheme for Nigeria and to test the effectiveness of different remote sensing techniques for the mapping of agricultural land use. In the paper, a brief review of the present status of agricultural land use data collection and mapping in Nigeria is followed by a discussion of image processing techniques for Landsat multispectral scanner data.

## AGRICULTURAL LAND USE DATA COLLECTION AND MAPPING IN NIGERIA FROM 1950s TO THE PRESENT

Although information on the current agricultural use of the land is a fundamental requirement for land use planning and management, no single agency in Nigeria has sole responsibility for this task; hence no comprehensive land use survey has been carried out. Several attempts have been made by individuals and institutions to collect some agricultural information and to prepare local or regional surveys, but these are commonly incompatible because different mapping systems have been employed.

Nigeria has had an annual agricultural census for the whole country since 1950–1951, and the data are published in the Annual Abstracts of Statistics. The items on which data are usually collected include household characteristics, area under cultivation, types of crops grown, yield estimates for the major crops, numbers of domestic livestock, and details of agricultural practice. Although generalized regional crop distribution maps have been derived from these statistics, they are inadequate for the mapping of agricultural resources in Nigeria. In this respect, there are fundamental problems with the census procedure, in that (1) incompatible and inconsistent data collection systems are used, not only from one time period to another, but also from one region to another, (2) the data cannot be properly geo-coded because, although the size of farm is known, the location is not, and (3) the number of villages from which data are collected is insufficient to be representative of the true situation in the country (Adeniyi, 1979; Agboola, 1979).

Apart from this source, there have been several special projects in which agricultural land use data have been collected directly in the field or interpreted from aerial photographs or radar mosaics. Using the field traverse method, Prothero (1952) employed undergraduate students to make an inventory of agricultural land use near Ibadan in western Nigeria. Progress was extremely slow. It required nearly a week to cover an area of 780 ha. Some of the difficulties in mapping land use with the field traverse method in Nigeria include: (1) the lack of base maps at suitable scales, (2) the inadequacy of the road network, (3) the difficulty of movement and visibility in forest areas, and (4) the high cost in terms of time and labour (Wikkamatileke, 1959).

The mapping of agricultural land using aerial photographs has been limited by the fact that the available medium-scale panchromatic coverage was produced for purposes of topographic mapping. Such photographs are not suitable for

Table 1. Variation in farm size in Nigeria (adapted from Gleave and White, 1969).

| Farm size (ha) | Percentage of farms by size for each province | | | | | |
|---|---|---|---|---|---|---|
| | Benin | Ibadan | Niger | Benue | Kano | Sokoto |
| <0.1 | 2 | 3 | – | 2 | 2 | 2 |
| 0.1–0.2 | 5 | 8 | 5 | 6 | 4 | 10 |
| 0.2–0.4 | 11 | 16 | 6 | 12 | 14 | 12 |
| 0.4–1.0 | 42 | 42 | 21 | 27 | 27 | 26 |
| 1.0–2.0 | 24 | 18 | 25 | 28 | 26 | 29 |
| 2.0–4.0 | 11 | 11 | 22 | 20 | 19 | 17 |
| >4.0 | 5 | 2 | 21 | 5 | 8 | 4 |

agricultural mapping purposes because of their scale and relatively poor resolution. In addition, the system of agriculture in Nigeria makes the identification and mapping of crops very difficult. Farm plots are of variable size (Table 1), they are widely scattered, their boundaries are irregular, and the plot size frequently changes from season to season and from year to year. These factors can often result in the incorrect classification of land use. For example, perennial crops, especially cocoa in the forest belt, have at times been incorrectly mapped as regenerated forest with isolated trees (Oyelese, 1968; Larin-Alabi, 1978; Anon., 1979).

Side-looking airborne radar (SLAR) has recently been used for resource mapping in Nigeria. The Nigerian Radar Project (NIRAD) produced a series of 69 vegetation and land use maps, at a scale of 1:250 000 (Anon., 1978). The primary purpose of the project was vegetation mapping, but the types of crops grown in each map sheet area were also identified.

These sources provide useful background data for land use inventory and mapping in Nigeria. However, the complex environmental conditions, the subsistence character of the farming system and the generally uncontrolled form of land utilization in the country make the goal of a national, agricultural land use survey very difficult to attain.

## APPLICATION OF LANDSAT DATA TO AGRICULTURAL LAND USE INVENTORY AND MAPPING

Data from the Landsat multispectral scanner (MSS) have been successfully employed for the acquisition of various types of agricultural information in many parts of the temperate zone. The use of Landsat, especially for crop inventory in developing countries, has been less successful. Crop identification is most straightforward for large fields in which only one type of crop is grown. As field size decreases and the variety of crops increases (as is the case in Nigeria), crop types are more difficult to identify on the Landsat imagery, although broader land use categories can be differentiated (Dey, 1978; UNDP/FAO, 1979; Odenyo and Negash, 1981).

Landsat has certain basic characteristics that make it suitable for carrying out an inventory of land use. Firstly, it provides a synoptic coverage: the whole of Nigeria can be viewed in only 50 scenes compared with more than 150 000 aerial photographs at 1:40 000 scale. Secondly, it is repetitive. Landsats 1, 2, and 3 provided an 18-day repeat cycle which is now reduced to nine days with Landsats 4 and 5. Thus, the researcher can monitor changes in surface reflectance and relate these to a crop calendar for the specific region of interest (Fig. 1). Thirdly, Landsat data are digital. Surface brightness values are sensed, integer-truncated (rounded), and recorded on computer compatible tapes (CCTs) as digital numbers (DN values), using an eight-bit scale (0–255). The major advantage of digital data is that they are amenable to various

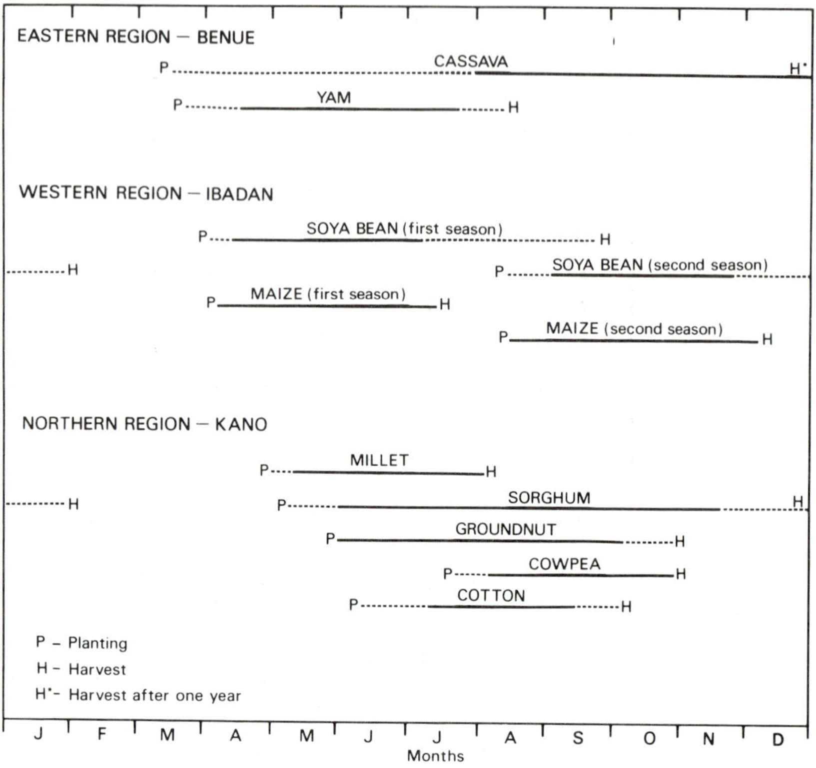


Fig. 1 Crop calendars for various regions of Nigeria.

types of image processing routines, some of which are examined in the following section.

## IMAGE PROCESSING: ENHANCEMENT AND INFORMATION EXTRACTION

The main objective of the present study was to examine selected image processing techniques that would increase the information content of Landsat imagery and improve the interpretation of agricultural land use in Nigeria. A test site was selected in west central Nigeria covering an area of approximately 1600 km$^2$, corresponding to a Landsat subscene of $500 \times 500$ pixels (Fig. 2). The subscene selected for study was imaged on 25 December 1973 during the dry season. Geometric correction, atmospheric correction, and image restoration procedures were applied prior to enhancement. The subscene characteristics are listed in Table 2.

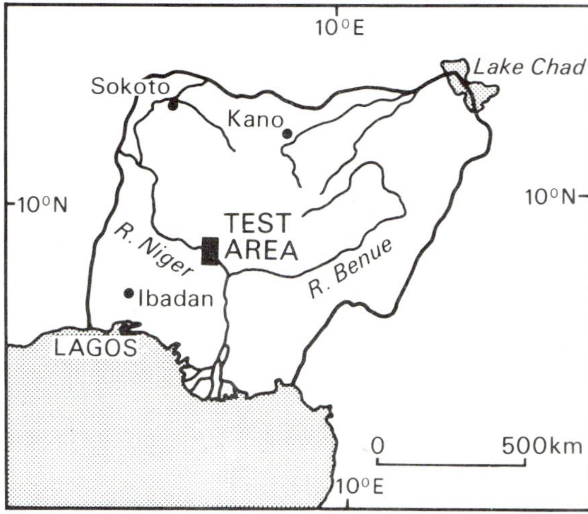


TEST AREA CHARACTERISTICS

| | |
|---|---|
| Location : | centred on latitude 8° 45′ N, longitude 6° 15′ E |
| Mean annual rainfall : | 1016 mm |
| Wet season duration : | 200 - 210 days (April - October) |
| Vegetation zone : | Southern Guinea Savanna |
| Soil types : | Alluvial / Ferralsol (reddish yellow) |
| Agricultural system : | rotational grass fallow |
| Types of crop grown : | rice, yam, sorghum, groundnut & cowpea |

Fig. 2 Location of the test area and summary of site characteristics.

Table 2. Landsat MSS scene characteristics for the test area.

| | |
|---|---|
| Date | 25 December 1973 |
| Cloud cover % | 0 |
| Sun angle ° | 44 |
| MSS bands | 4, 5, 7 |
| Quality | 8, 8, 8 |
| Frame ID path/row | 203/054 |
| Scene ID | 81520-091815 |
| Test site subscene | 500 × 500 pixels |

Table 3. Linear stretch parameters for MSS bands 4, 5, and 7 of the test area subscene.

| | Band 4 | | | | | Band 5 | | | | | Band 7 | | | | |
|---|---|---|---|---|---|---|---|---|---|---|---|---|---|---|---|
| Original DN value | 0 | 1 | 10 | 27 | 255 | 0 | 3 | 13 | 31 | 255 | 0 | 2 | 17 | 35 | 255 |
| Value allocated | 0 | 0 | 127 | 255 | 255 | 0 | 0 | 127 | 255 | 255 | 0 | 0 | 127 | 255 | 255 |

**Linear contrast stretch and high pass filtering**

Landsat images of the test area displayed very low contrast. The limited range of brightness values is apparent when one examines the frequency histograms of the raw data for each band. For example, the band 7 data are concentrated in a range of 43 DN brightness levels, utilizing only 16.8% of the available brightness range (Fig. 3). This limited range accounts for the low contrast ratio of the image. In order to enhance the image contrast a linear contrast stretch was carried out. A value in the low range of the original histogram was assigned to black and a value at the high end was assigned to white. The remaining pixel values were distributed linearly between these extremes, as indicated in Table 3.

In order to improve the dynamic range of the grey levels in the image still further, a high pass filter (HPF 3 × 3) with 100% add-back was applied. Filtering changes the DN value of a pixel as a function of the DN values of neighbouring pixels. High pass filtering amplifies the detail by enhancing the high frequency variations within the image. The decision as to the number of neighbouring pixels to select is determined by the variability of the original data and the purpose for which the final image product is to be used (Condit and Chavez Jr, 1979; Chavez Jr and Bauer, 1982). For the test area subscene, a 3 × 3 pixel size was used in the high pass filter.

After linear contrast stretch and high pass filtering were applied to each band, a colour composite image was generated with the usual false colour allocation: band 4 blue, band 5 green, and band 7 red (Plate 7(a)).

**Image ratioing**

Ratio images are generated by divided the DN value of each pixel in one MSS band by the DN value of the corresponding pixel in another band. The quotients

Plate 1 (Chapter 8) Fully classified map of 14 land cover themes for Kauai using a supervised classification. Colour key: brown – *ohia* forest; aqua – candlenut (*kukui*) forest; orange – closed forest (type 1); melon – closed forest (type 2); pink – open forest and scrub; white – urban areas, bare sand, cloud; light blue – reservoir; dark blue – unclassified (shadow); dark green – sugar cane; light green – old or thin cane; black – bare, burned areas; yellow – pasture; red – eroded land (class 1); magenta – eroded land (class 2); violet – eroded land (class 3). Data from Landsat MSS bands 4, 5, and 6, image ID E 20994-19533, 12 October 1977. (*Digital data courtesy of NOAA.*)

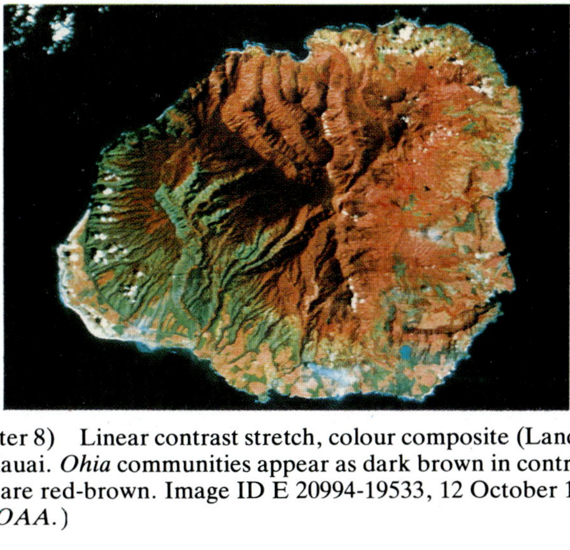

Plate 2 (Chapter 8) Linear contrast stretch, colour composite (Landsat MSS bands 4, 5, and 6) of Kauai. *Ohia* communities appear as dark brown in contrast to other forest classes which are red-brown. Image ID E 20994-19533, 12 October 1977. (*Digital data courtesy of NOAA.*)

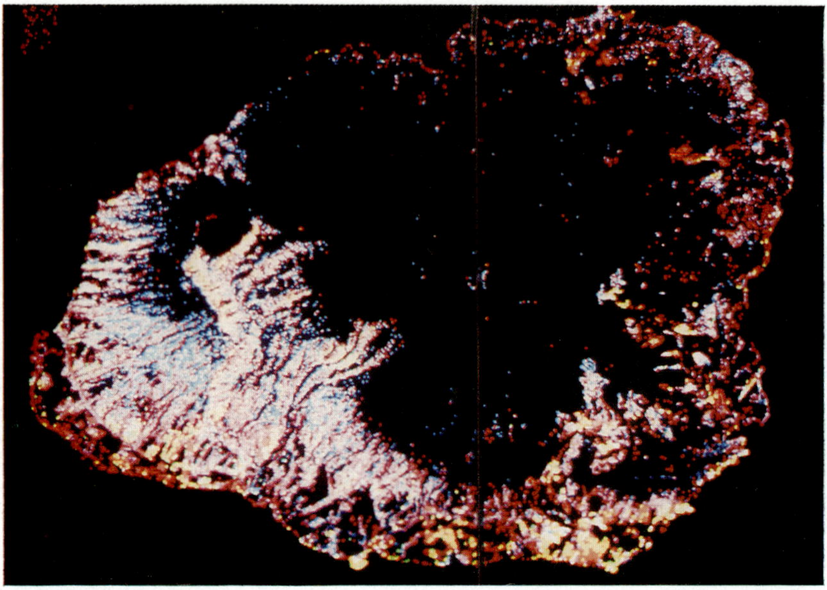

Plate 3(Chapter 8) Eroded land themes for Kauai using a supervised classification. Colour key: blue – class 1, severe erosion; yellow – class 2, moderate erosion; red – class 3, slight erosion (includes some areas of recently planted sugar cane). Data from Landsat MSS bands 4, 5, and 6, image ID E 20994-19533, 12 October 1977. (*Digital data courtesy of NOAA.*)

Plate 4 (Chapter 8) Fully classified map of 11 land cover themes for the Koloa subscene, southeastern Kauai. Supervised classification, 3 x enlargement. Colour key: aqua – candlenut (*kukui*) forest; orange – closed forest (type 1); melon – closed forest (type 2); green/cyan – sugar cane; light green/white – old or thin cane; black – bare or burned areas; violet – recently planted cane and eroded land (class 3); magenta – eroded land (class 1); yellow – pasture; white – urban areas and cloud. Data from Landsat MSS bands 4, 5, and 6, image ID E 20994-19533, 12 October 1977. (*Digital data courtesy of NOAA.*)

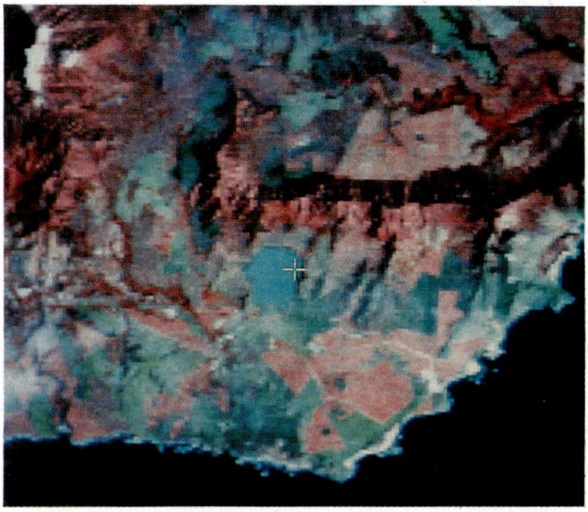

Plate 5 (Chapter 8) Koloa subscene, southeastern Kauai, 3 x enlargement. Raw data with linear contrast stretch and colour enhancement, reproduced as a simulated colour infrared composite. Colour key: dark red – forest; pink – sugar cane; white – urban areas, bare sand and cloud; cyan – bare or burned areas. The haze effect in the southwest is smoke from burning cane fields. Data from Landsat MSS bands 4, 5, and 7, image ID E 20994-19533, 12 October 1977. (*Digital data courtesy of NOAA.*)

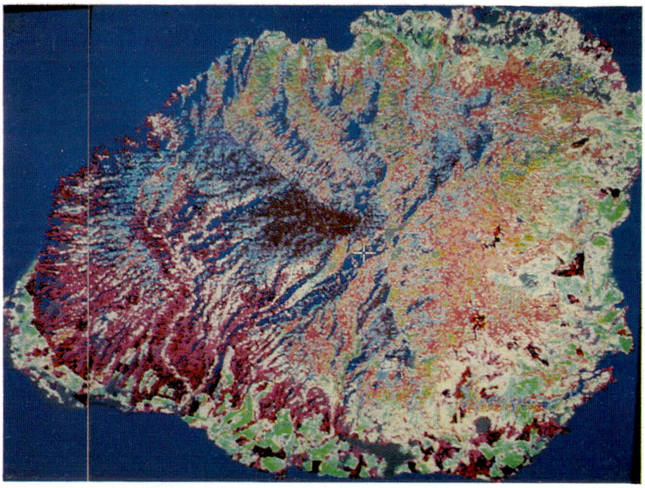

Plate 6 (Chapter 8) Fully classified map of 14 land cover themes for Kauai using an unsupervised classification and D-stretch transformation. Colour key: brown – dense *ohia* forest; violet – *ohia* forest; aqua – candlenut (*kukui*) forest; orange – closed forest (type 1); melon – closed forest (type 2); pink – open forest and scrub; grey – urban areas, reservoir, bare sand, coral reef; dark green – sugar cane; light green – old or thin cane; black – bare, burned areas; yellow – pasture; red – eroded land (class 1); magenta – eroded land (class 2); blue – sea and shadow. Data from Landsat MSS bands 4, 5, and 6, image ID E 20994-19533, 12 October 1977. (*Digital data courtesy of NOAA.*)

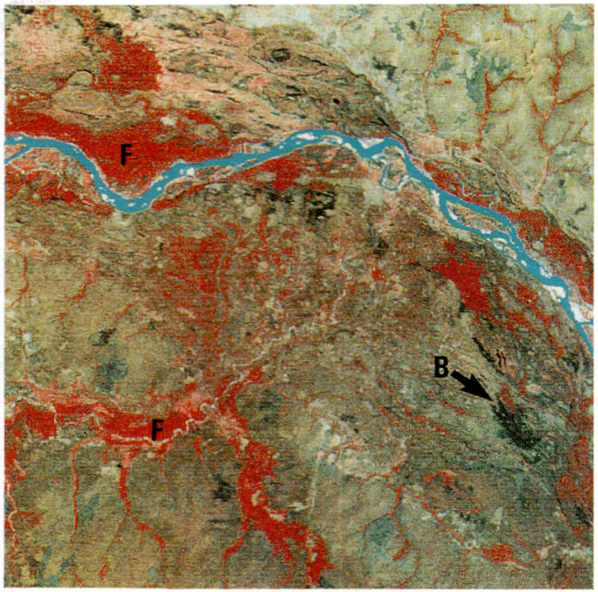


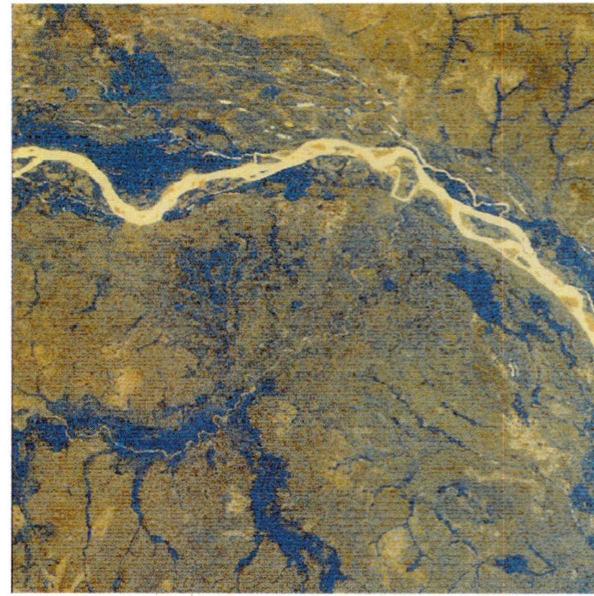

Plate 7 (Chapter 9) Katcha test area, west central Nigeria. Landsat image ID 81520-091815, 25 December 1973. (*Digital data courtesy of NOAA.*)

(a) Standard colour composite using Landsat MSS bands 4, 5, and 7 (blue, green, red) of the test area subscene. The riparian forest (F) and recently burnt areas (B) are well defined in this image

(b) (Chapter 9) Ratio colour composite using Landsat MSS bands 4:5, 4:7 and 5:7 (blue, green, red) of the test area subscene. Heavily vegetated areas (blue) are clearly visible in this ratio composite

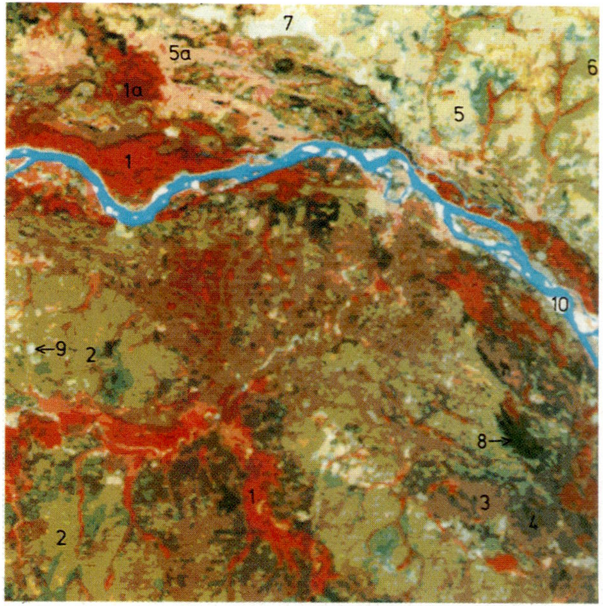


(c) (Chapter 9) Landsat digital classification of the land use of the test area subscene. Spectral class (classifier) labels 1-10 should be compared with the land type categories shown in (d)

1 Riparian forest
1a Dense riparian forest
2 Cropland (yam, rice) and wooded shrub
3 Wooded shrub, grass and cropland (rice, sorghum)
4 Aquatic grassland
5 Mosaic farmland and wet grassland
6 Dry upland grassland
7 Bare or cleared ground
8 Burnt area
9 Built-up area
10 Bare sand

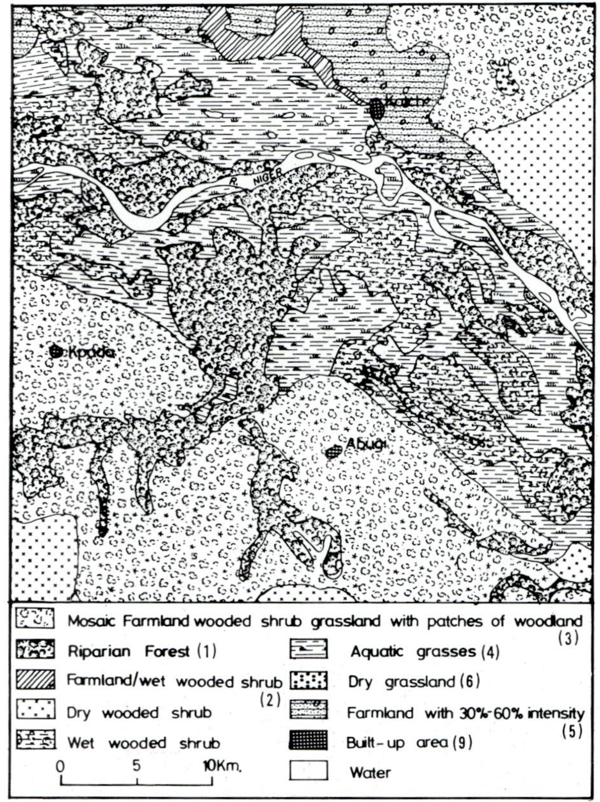


(d) (Chapter 9) Vegetation and land use of the test area, Sheet NC 32-13 Nigerian Radar Project (NIRAD) 1976-77; 2x reduction from the original scale. (*Reproduced by permission of the Federal Department of Forestry, Lagos.*)

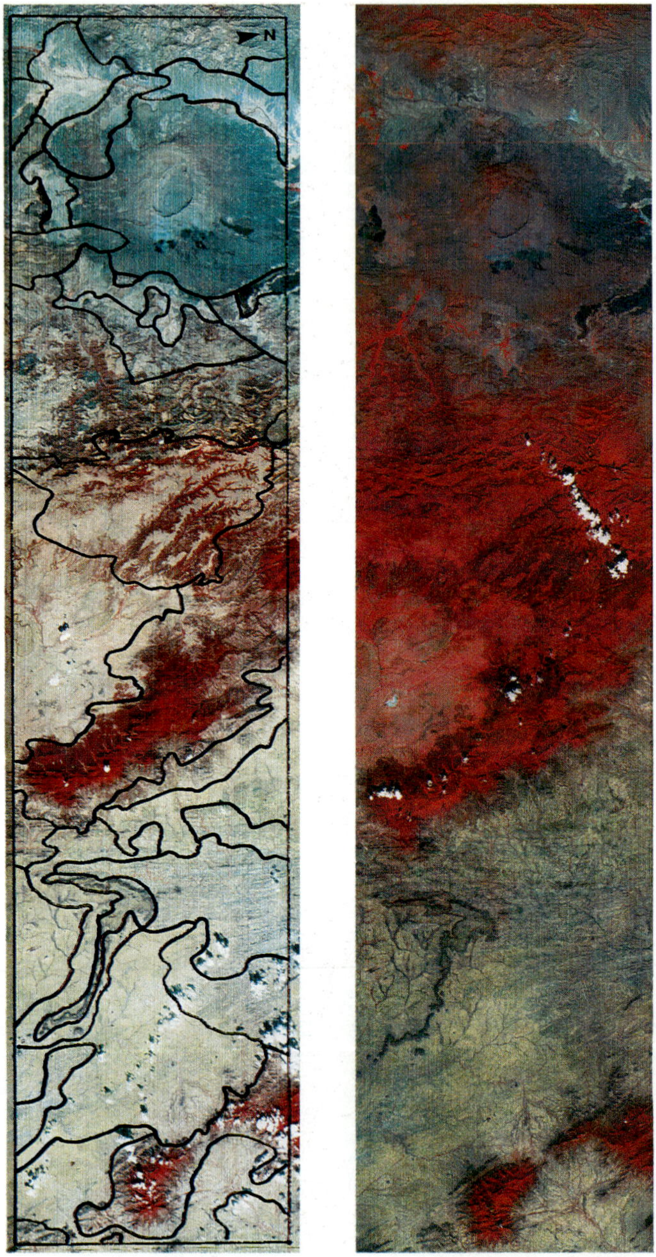


(a) (b)

Plate 8 (Chapter 6) Landsat colour composite images of the Maralal transect, MSS bands 4, 5, and 7. (*Reproduced by permission of NOAA.*) (a) March 25 1975 (dry) image ID 2062-07052, with land systems outline from the *Land Systems Atlas of Western Kenya* (Scott *et al.*, 1971) (b) September 21 1975 (wet) image ID 2242-07041.

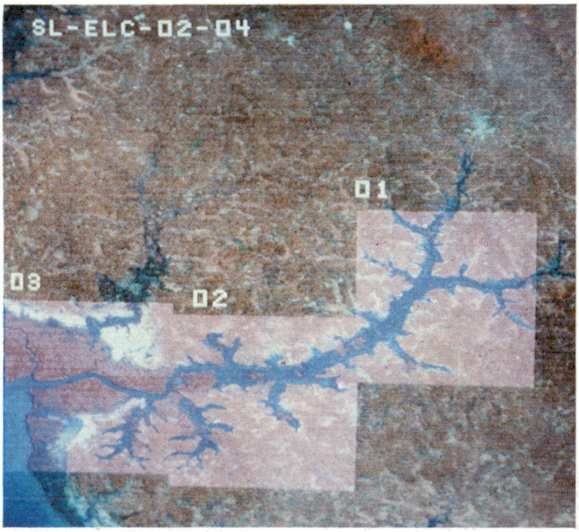


Plate 9 (Chapter 10) Colour composite (MSS bands 4, 5, and 7) of Bumpe estuary subscene 0204, with subdivisions showing upstream, midstream, and downstream areas, Sierra Leone. Landsat image ID 2403-10204, 29 February 1976. (*Reproduced by permission of NOAA and Canada Centre for Remote Sensing.*)

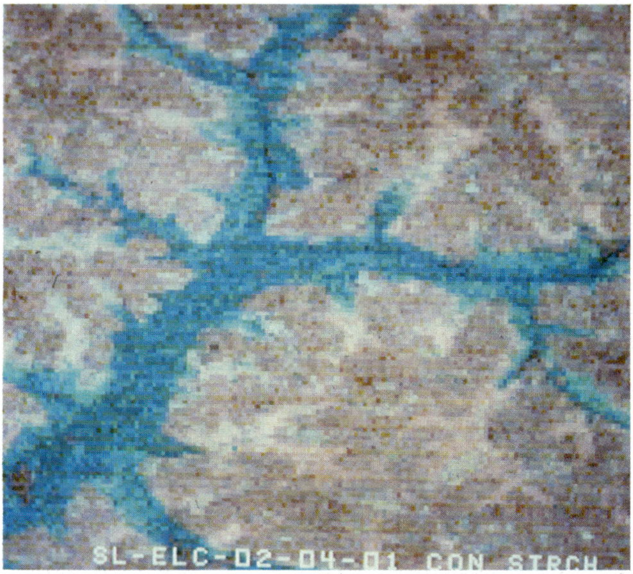


Plate 10 (Chapter 10) Linear contrast stretch of the Bumpe estuary subscene 020401, Sierra Leone. Blue – main channel and rice-growing lands; white – inland valley swamps. Landsat image ID 2403-10204, 29 February 1976. (*Reproduced by permission of NOAA and Canada Centre for Remote Sensing.*)

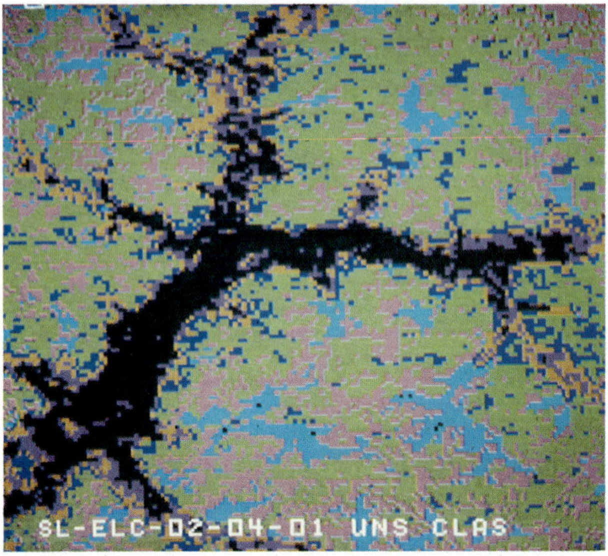


Plate 11 (Chapter 10) Unsupervised classification of the Bumpe estuary subscene 020401, Sierra Leone. Purple and orange – rice areas and remnant mangrove; blue – estuarine mangrove and cleared areas; green – forest regrowth and upland rice; pink – partly cleared land; light blue – inland swamp; black – river channel and rice areas. Landsat image ID 2403-10204, 29 February 1976. (*Reproduced by permission of NOAA and Canada Centre for Remote Sensing.*)

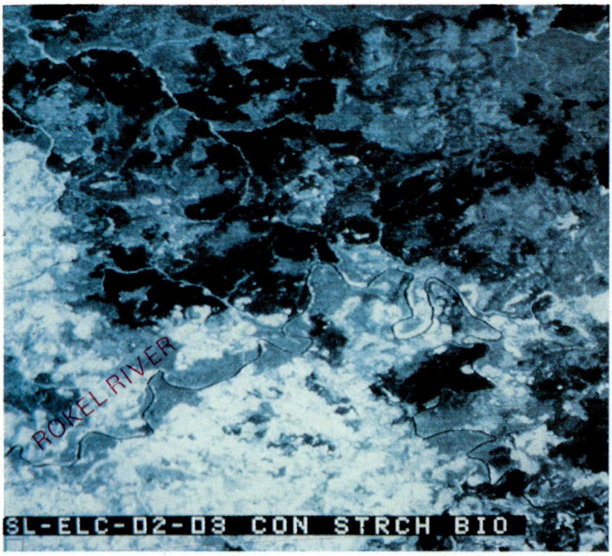


Plate 12 (Chapter 10) Bands 5:7 ratio stretch of the *bolilands,* subscene 0203, Sierra Leone. Black – inland *bolis*; blue grey – riverain *bolis* and inter-*boli* flanking slopes. Landsat image ID 2403-10204, 29 February 1976. (*Reproduced by permission of NOAA and Canada Centre for Remote Sensing.*)

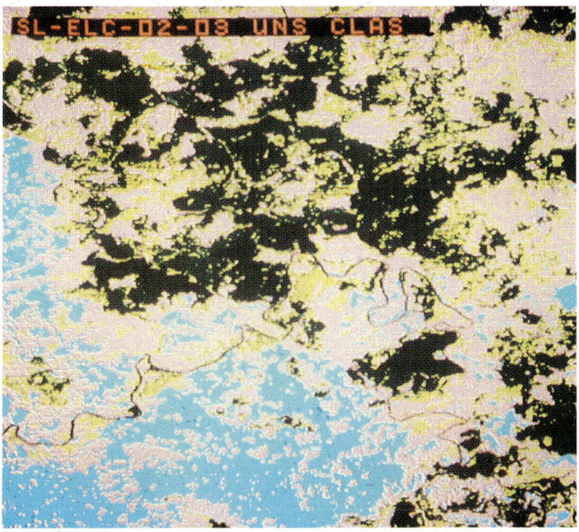


Plate 13 (Chapter 10) Unsupervised classification of the *bolilands*, subscene 0203, Sierra Leone. Black – inland *bolis*; yellow – riverain *bolis* and inter-*boli* flanking slopes. Landsat image ID 2403-10204, 29 February 1976. (*Reproduced by permission of NOAA and Canada Centre for Remote Sensing.*)

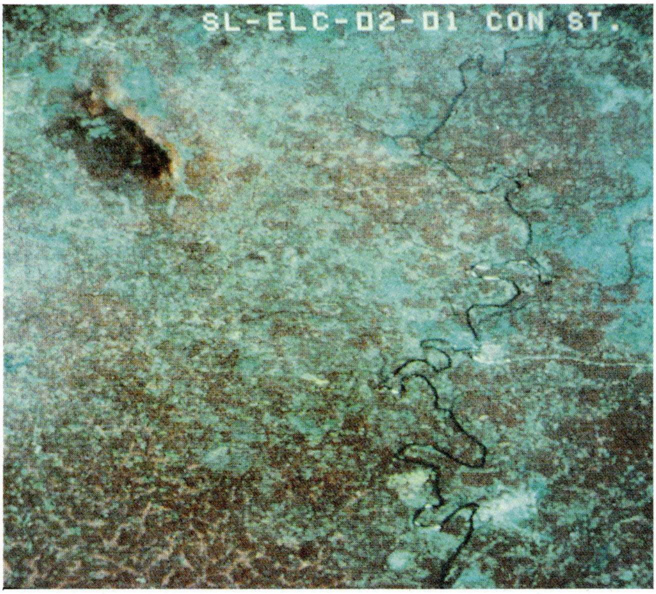


Plate 14 (Chapter 10) Bands 5:7 ratio stretch of the Njala area, subscene 0201. Light, linear features (southwest of subscene) – inland valley swamps; very dark grey and black – inland *bolis*. Landsat image ID 2403-10204, 29 February 1976. (*Reproduced by permission of NOAA and Canada Centre for Remote Sensing.*)

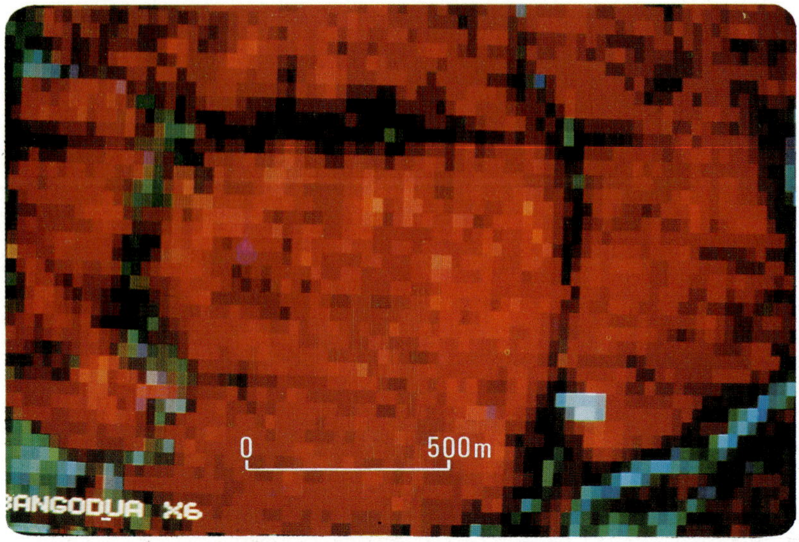


Plate 15 (Chapter 15) Landsat colour composite (MSS bands 4, 5, and 7) of a rice-growing area on the coastal plain of Java. The crop is at the late-vegetative stage, and some fields are beginning to ripen. Landsat subscene, 20 June 1976 (ID E 2515-02085). (*Reproduced by permission of NOAA.*)

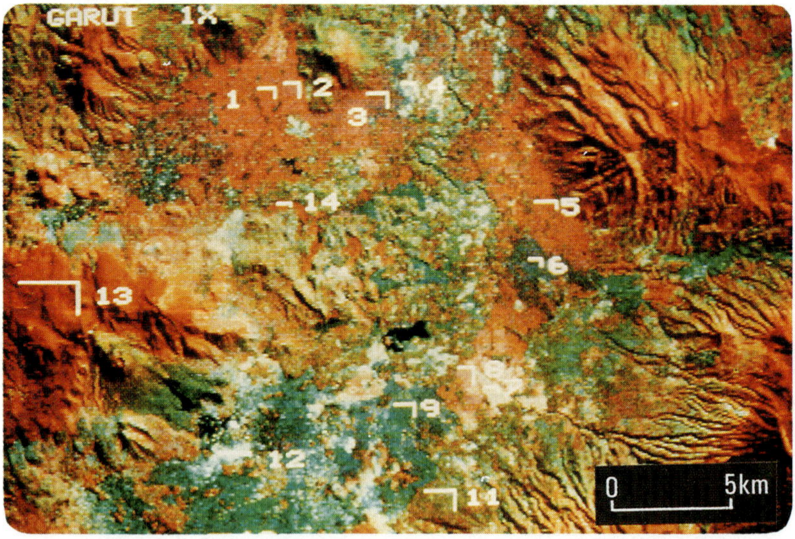


Plate 16 (Chapter 15) Landsat colour composite (MSS bands 4, 5, and 7) of a rice-growing area in a mountain plain of Java: fallow fields (12), flooding/field preparation (6, 9), young rice crop (3), full rice cover (1, 2, 5, 14), maturing crop (8); also dry farming area (11), and forest cover (13). Landsat subscene, 20 June 1976 (ID E 2515-02085). (*Reproduced by permission of NOAA.*)

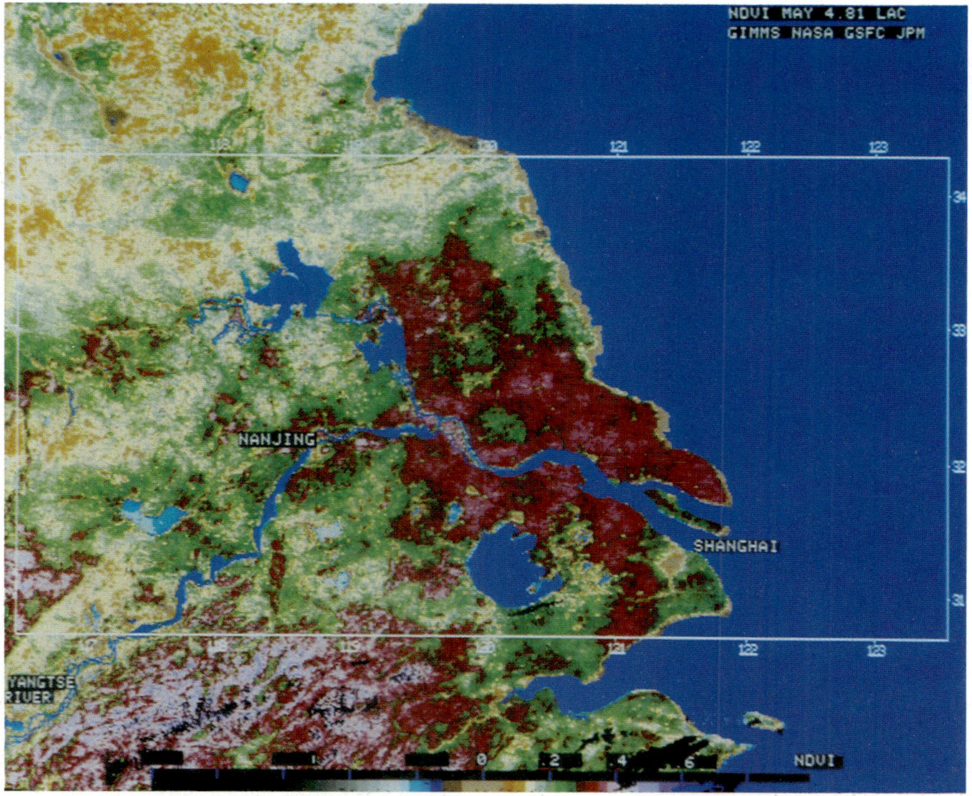


Plate 17 (Chapter 15) AVHRR vegetation index (NDVI) of the Shanghai region, Jiangsu Province, China for 4 May 1981, based on LAC data. Colours range from brown to orange, green, red, and blue as the index increases: red (near Shanghai) – winter wheat; red-blue (south) – forest cover. (*Reproduced by permission of GIMMS-NASA GSFC.*)

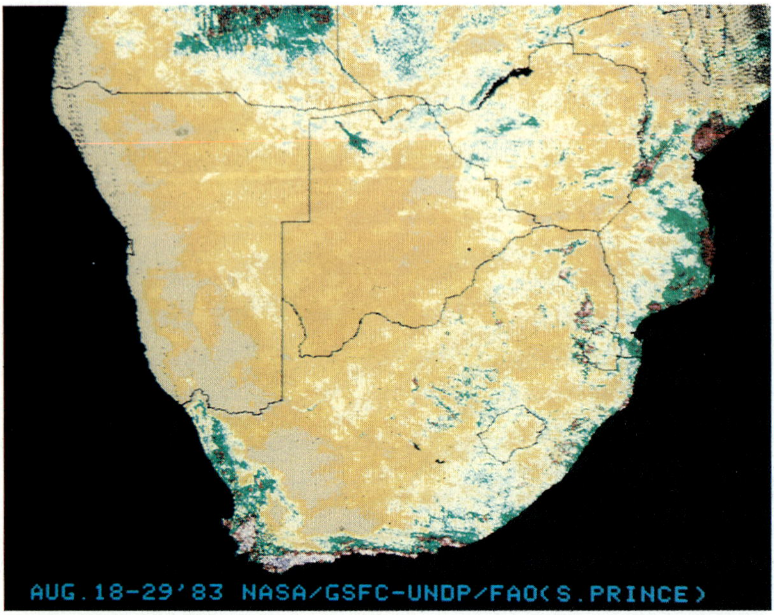


Plate 18 (Chapter 16)

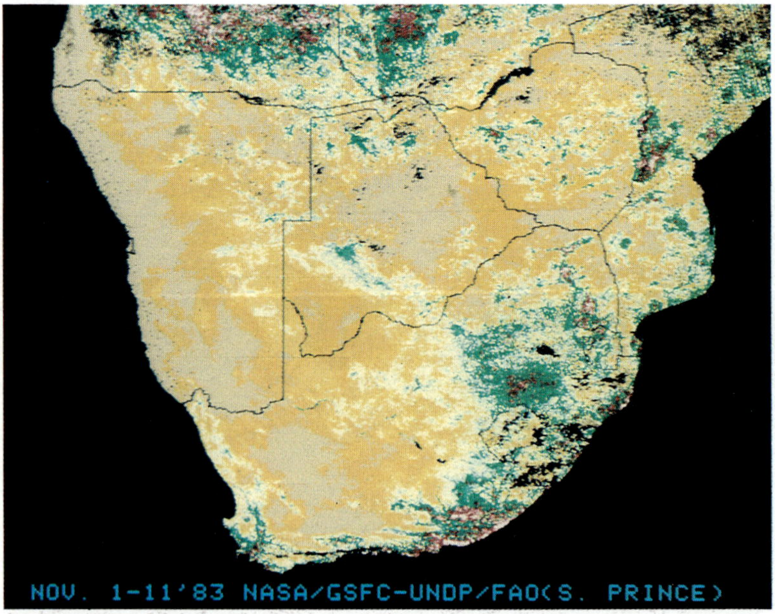


Plate 19 (Chapter 16)

NOAA-7 AVHRR GAC normalized difference vegetation index (NDVI) composites for southern Africa. Compositing period given on each image. Colour scale: brown (minimum), green, red, blue (maximum). (*Reproduced by permission of FOA.*)

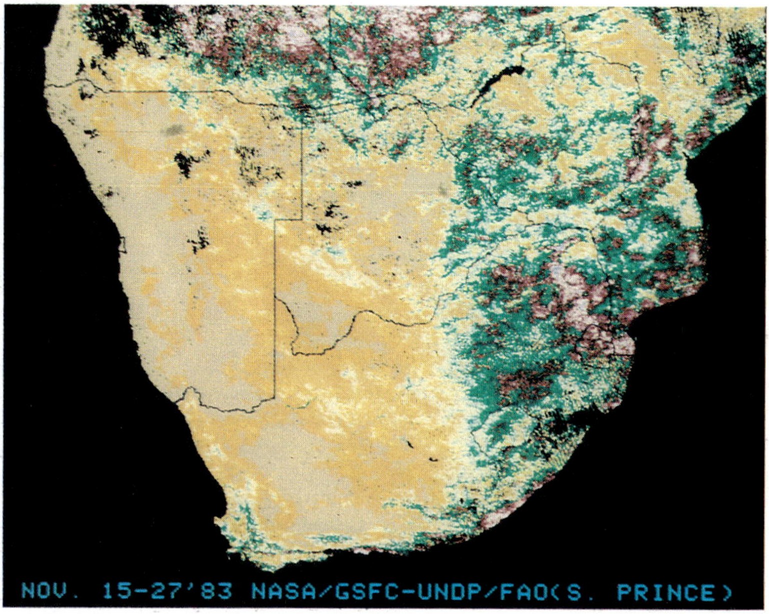


Plate 20 (Chapter 16)

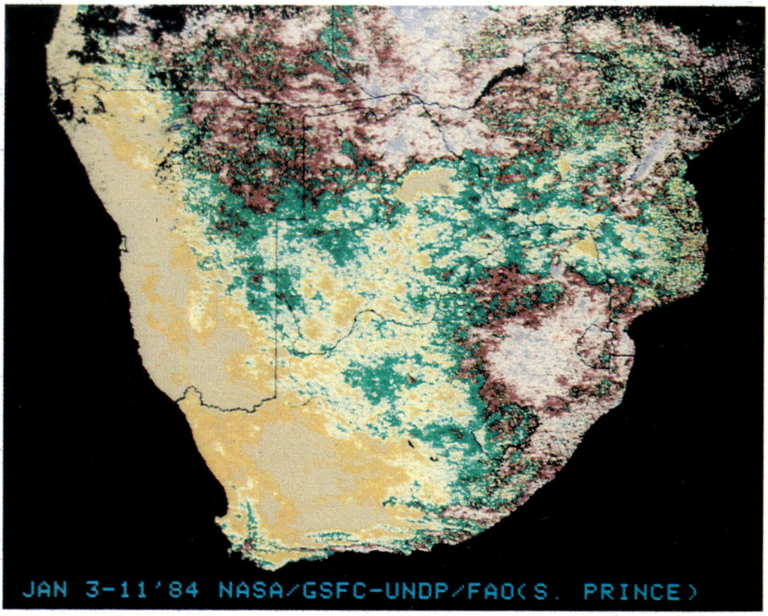


Plate 21 (Chapter 16)

NOAA-7 AVHRR GAC normalized difference vegetation index (NDVI) composites for southern Africa. Compositing period given on each image. Colour scale: brown (minimum), green, red, blue (maximum). (*Reproduced by permission of FOA.*)

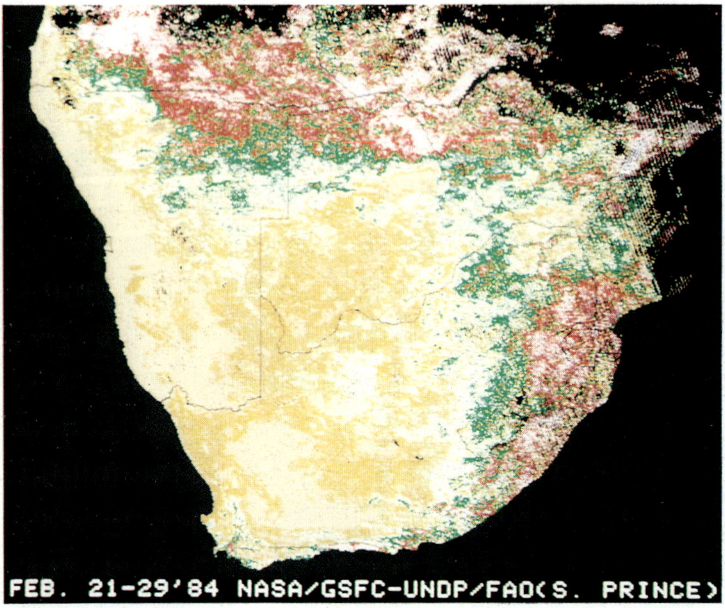


Plate 22 (Chapter 16)

Plate 23 (Chapter 16)

NOAA-7 AVHRR GAC normalized difference vegetation index (NDVI) composites for southern Africa. Compositing period given on each image. Colour scale: brown (minimum), green, red, blue (maximum). (*Reproduced by permission of FOA.*)

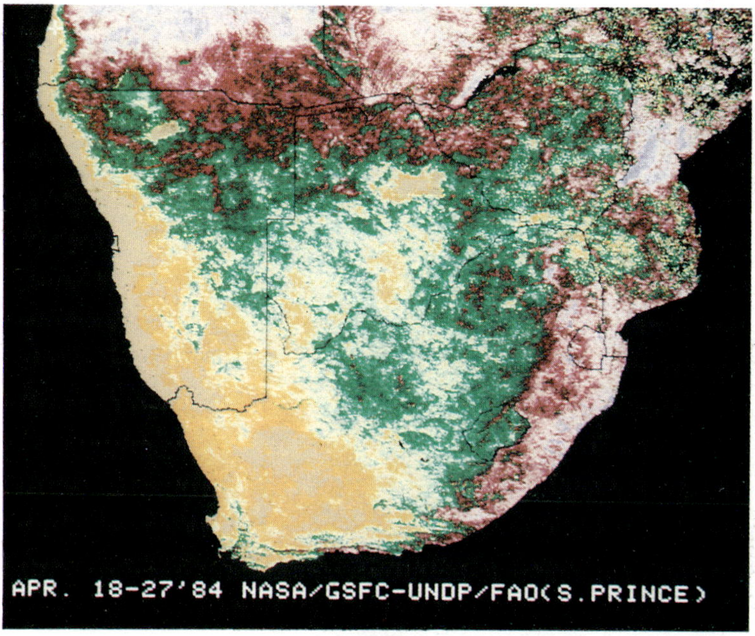


Plate 24

Plate 24 (Chapter 16) NOAA-7 AVHRR GAC normalized difference vegetation index (NDVI) composites for southern Africa. Compositing period given on each image. Colour scale: brown (minimum), green, red, blue (maximum). (*Reproduced by permission of FAO.*)

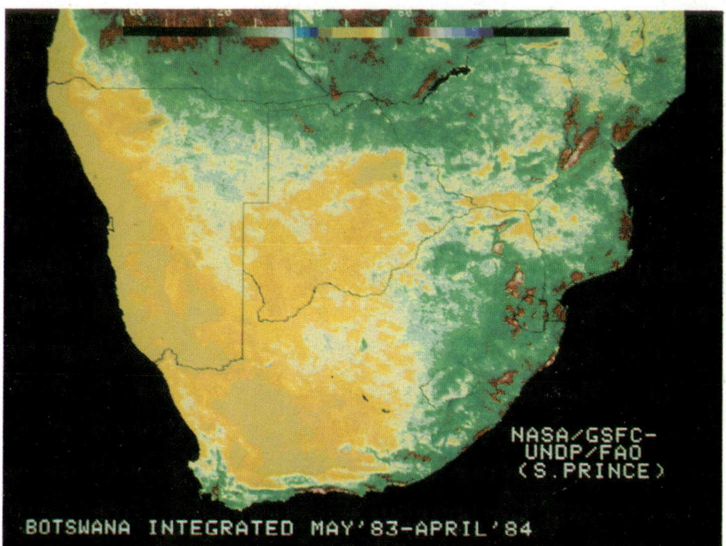


Plate 25 (Chapter 16) Integrated NDVI image for southern Africa for May 1983 to April 1984. Colour scale: brown (minimum), green, red, blue (maximum). (*Reproduced by permission of FAO.*)

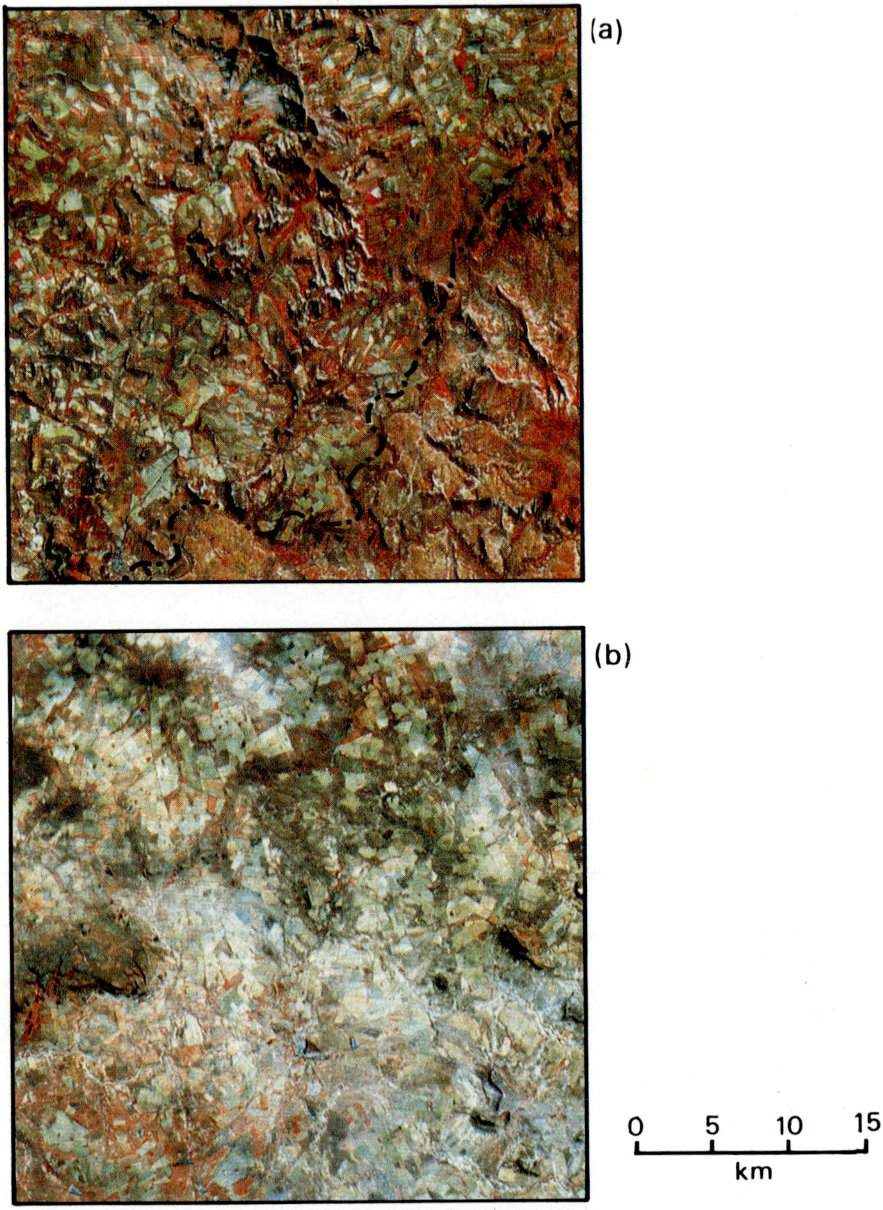


Plate 26 (Chapter 11) (a) Landsat colour composite (subscene) showing part of northwest Lesotho and adjacent areas of the Republic of South Africa

(b) Landsat colour composite (subscene) of the Senekal district, Republic of South Africa

Landsat MSS 4, 5, and 7, ID 21521-07094, 23 March 1979. (*Digital data courtesty of NOAA.*)

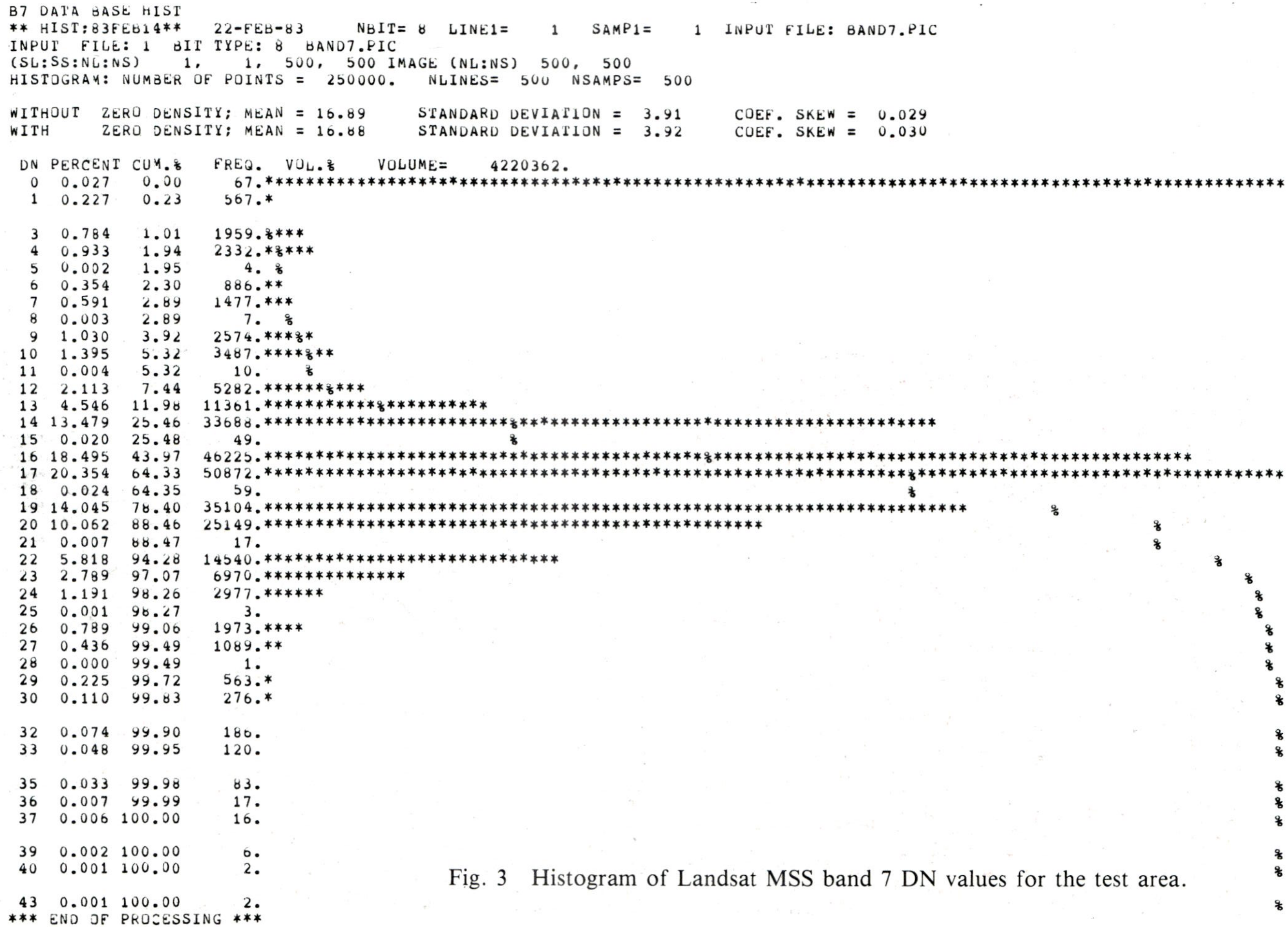


Fig. 3 Histogram of Landsat MSS band 7 DN values for the test area.

are multiplied by 100 to yield derived DN values ($DN^1$) in the range 0–255 and the data are then subject to a linear contrast stretch, for example:

$$DN^1 \left(\frac{4}{5}\right) = \frac{DN4}{DN5} \times 100 + \text{stretch}$$

Using the four MSS bands of the test area subscene, six sets of ratio data were derived, as indicated in Table 4.

On a ratio image, the extremes of the grey scale (black and white) represent the maximum differences in spectral reflectivity between the two MSS bands. The lightest tones correspond to surface types where the numerator is greater than the denominator. Thus, vegetation and crops have a bright signature on a band 4:5 ratio image because reflectance is greater in band 4 (green) than in band 5 (red).

In the present study, the major concern was to discriminate between vegetation, crops, and other surface types. An important characteristic of a healthy, growing crop is its strong infrared reflectance. This can be assessed using ratio techniques as the 'relative infraredness'—the relative level of the near-infrared response compared to the response in the visible wavelengths (Hay, 1982). After a careful evaluation of each of the six ratio products (Table 4), three were selected as providing the most useful information for land use mapping purposes: band ratios 4:5, 4:7, and 5:7. These were made into a colour composite by projecting the registered black and white transparencies with blue, green, and red light sources (Plate 7(b)).

## Unsupervised classification

Contrast stretching and ratioing techniques generally improved the interpretability of land use in the subscene test area. However, the identification and delineation of cropland remained difficult because of the small size of fields and their dispersion. In order to increase the efficiency of the analysis an automated multispectral classification was attempted. With unsupervised classification, or clustering procedures, statistically based classification rules are applied to digital multispectral data. During the classification, the computer retrieves the spectral values for each pixel and calculates the position the pixel occupies in a multidimensional feature space. Statistical descriptions or definitions of pixel clusters are then determined automatically.

In the analysis, data from three MSS bands (4, 5, and 7) were examined. In order to subdue the high frequency variations, a low pass filter was applied to each set of data. Decision boundaries defining each cluster were established with the computer in an interactive manual mode. In this procedure, the analyst compares the visual display with the information available from other sources,

Table 4. Ratio stretch parameters for the test area subscene.

| Band ratio | 4:5 | 4:6 | 4:7 | 5:6 | 5:7 | 6:7 |
|---|---|---|---|---|---|---|
| Original DN value | 0 21 71 255 | 0 13 40 255 | 0 17 55 255 | 0 10 48 255 | 0 12 79 255 | 0 96 135 255 |
| Value allocated | 0 0 127 255 | 0 0 127 255 | 0 0 127 255 | 0 0 127 255 | 0 0 127 255 | 0 0 127 255 |

and decides on the optimum number of classes in relation to the needs of the project and the statistical attributes of the data. In order to complete the classification, classifier labels (land use or crop type identifiers) are allocated to each of the spectrally defined classes, based on available information from field and other sources. An example of digital classification output—land use and land cover themes for the subscene test area—is shown in Plate 7(c).

## EVALUATION OF THE RESULTS

It is difficult to assess the usefulness of the different products of the digital imaging processing. The major difficulty lies in the lack of synchronous field data to serve as ground truth for the Landsat scene of 25 December 1973. Clearly, many details will have changed since then, but it is not unreasonable to assume that the basic categories of land use will have remained constant, and that the vegetation and land use boundaries outlined by the Nigerian Radar project in 1976–1977 (Plate 7(d)) can be taken as broadly indicative of the ground cover in 1973. Each of the subscene images generated in the digital processing was thus compared with the NIRAD vegetation and land use map (sheet NC 32-13), and a subjective assessment of the detail was made by a group of interpreters, eleven members of a final year undergraduate class in remote sensing at the Department of Geography, University of Lagos. The results of this evaluation are summarized below for each of the digital image types.

### Colour composite image (bands 4, 5, and 7) with linear contrast stretch and high pass filtering

This image (Plate 7(a)) was found to be superior to any single band image in terms of the interpretability of the land use. Major physiographic divisions, such as the floodplain of the Niger river, the Kpada plains on the Basement Complex in the south and southwest, and the rolling sandstone terrain of the Cretaceous deposits in the northeast, are readily delineated. These subdivisions are significant because they are associated with differences in land use and farming practice.

### Ratio colour composite image (bands 4:5, 4:7, 5:7)

This image (Plate 7(b)) was found to contain more detail than the reference material—the NIRAD vegetation and land use map. Unlike the standard colour composite where active vegetation or crop growth appears in bright magenta hues, the false colour ratio composite shows active plant growth in blue. The density of the vegetation or crop cover is indicated by the colour saturation in the image. Although the ratio colour composite does not yield a classification of the land use types in the test area, it does provide a rapid method for locating and estimating the areas of active vegetation growth. Another test of this procedure in a different ecological zone would be particularly useful.

### Digital classification

The application of clustering techniques in an unsupervised classification resulted in a thematic map of land use and land cover in the test area. Comparison with the NIRAD vegetation and land use map of the subscene reveals many features of general agreement. The great advantage of digital classification is that the number of classes can be changed to take into account other sources of information. The procedure is interactive, and so the analyst can group classes to produce a smaller number of themes or subdivide to produce a larger number.

Initially, a total of 48 classes was identified. However, it was realized that many of these were artifacts in that variables, such as soil type or colour, moisture content, local relief, and differences in tillage, were 'creating' separate themes in an otherwise homogeneous vegetation or land use type. With relevant local field knowledge the analyst can eliminate these effects and recombine particular classes. In the digital classification shown in Plate 7(c), only ten themes are identified to permit easy comparison with the general land use and vegetation categories used in the NIRAD mapping project (Plate 7(d)).

Supervised digital classification would be even more effective. This requires the acquisition of field data at the time of the satellite overpass, in order to provide a training set for establishing the classifiers. Techniques of this type have great potential for land use and vegetation mapping in Nigeria and other parts of West Africa.

## CONCLUSION

In Nigeria, remotely sensed data have not been extensively used for resource inventory. However, reliable inventories, especially of agricultural land use, are urgently needed to provide a base for spatially balanced and ecologically controlled development. Given the size of Nigeria and its diverse ecological and cultural environment, it is clear that conventional methods of field survey cannot provide an adequate data base. Satellite remote sensing can meet this challenge.

The major contribution of this study has been the demonstration of the potential value of digital image analysis for mapping agricultural land use in Nigeria. It has been shown that thematic information about forest and agricultural lands derived from Landsat is compatible with 1:250 000 scale mapping in terms of geometric accuracy and information content. This is in agreement with the findings of Kalensky *et al.* (1980) for other parts of the developing world.

The study was experimental and no attempt was made to demonstrate cost effectiveness in comparison with other methods of land use mapping. However, it seems very likely that digital image analysis methods could result in significant savings. The current cost of a Landsat full scene computer compatible tape is $US730 and it is estimated that the costs of digital analysis are approximately $US3000 per scene. Thus, for the whole of Nigeria (50 Landsat scenes) the approximate costs of a full digital analysis would be $US186 500. Per unit area, this is substantially less expensive than existing, conventional methods of land use mapping.

It is hoped that further tests of ratio and digital clustering techniques can be carried out, using a subscene from the forest region of southern Nigeria. In addition, experimentation with more flexible digital processing subroutines is necessary in order to respond to the complexity of the national land use patterns.

One major problem facing the continent of Africa (with the exception of South Africa) is the lack of real time data. In order to derive optimum benefit from the application of satellite remote sensing, there is a need for the establishment of one or two regionally based, multifunctional ground receiving and data processing stations in tropical Africa. Although remote sensing is by no means a complete solution to the problems of land management in Nigeria, it seems to offer the best possibility for rapid inventory and mapping of land uses and land resources.

## ACKNOWLEDGEMENTS

This study was funded by the University of Lagos. The digital computer facilities were provided by the United States Geological Survey, Flagstaff, Arizona. The author is particularly grateful to L. Berlin, Pat Chavez, and Ellen Sanchez for their assistance in processing the Landsat data.

## REFERENCES

Adeniyi, P. O., 1979. Some fundamental issues in land and land use planning in Nigeria. *Niger. geogr. J.*, **22**, 59–79.

Agboola, S. A., 1979. *An Agricultural Atlas of Nigeria*. Oxford: Oxford University Press.

Anon., 1978. *NIRAD Project Interpretation Phase Summary: Report Prepared for the* Federal Department of Forestry, Federal Government of Nigeria by Hunting Technical Services Ltd. Boreham Wood, U.K.: Hunting Technical Services.

Anon., 1979. *Third Cocoa Project: Pilot Study to Evaluate the Role of Aerial Photography in Cocoa Resource Mapping*. Unpublished Rept. for the Federal Department of Agriculture, Nigeria, Hunting Technical Services Ltd., Boreham Wood, U.K.

Anon., 1981. *Feasibility Study for a Nigerian Natural Resources Information Centre*. Unpublished Rept. for the Federal Ministry of National Planning, Government of Nigeria, Hunting Technical Services, Boreham Wood, U.K.

Chavez Jr, P. S. and Bauer, B., 1982. An automatic optimum kernel-size selection technique for edge enhancement. *Remote Sensing Environ.* **12**, 23–38.

Chavez Jr, P. S., Berlin, G. L. and Sowers, L. B., 1982. Statistical method for selecting Landsat MSS ratios. *J. appl. photogr. Engng* **8**, 23–30.

Condit, C. D. and Chavez Jr, P. S., 1979. *Basic Concepts of Computerised Digital Image Processing for Geologists*, Bull. 1462, U.S. Geol. Surv., Washington.

Dey, B., 1978. Some applications of remote sensing in agricultural resources with reference to India. In Collins, W. G. and Van Genderen J. L. (Eds), *Remote Sensing Applications in Developing Countries*. Reading: Remote Sensing Society, pp. 35–42.

Gleave, M. B. and White, H. P., 1969. Population density and agricultural systems in West Africa. In Thomas, M. F. and Whittington, G. W., (Eds), *Environment and Land Use in Africa*. London: Methuen, pp. 273–300.

Hay, C. M., 1982. Remote sensing measurement techniques for use in crop inventories. In Johannsen, C. J. and Sanders, J. L., (Eds), *Remote Sensing for Resource Management*. Ankeny: Soil Conservation Society of America, pp. 420–433.

Kalensky, Z. D., Darmoyuwono, K. and Kaser, J., 1980. Resource mapping by Landsat in developing countries. *Proc. 14th Congr. int. Soc. Photogramm.*, Hamburg, 367–379.

Larin-Alabi, F. B., 1978. Problems of remote sensing in the tropics: an appraisal of the Nigerian situation with regard to forest resources. In Collins, W. G. and Van Genderen, J. L., (Eds), *Remote Sensing Applications in Developing Countries*. Reading: Remote Sensing Society, pp. 57–61.

Odenyo, C. P. and Negash, M., 1981. Use of Landsat imagery for mapping land use/land cover type of Ethiopia. Paper presented to *UN/ECA Regional Seminar on Remote Sensing Applications and Satellite Communications for Education and Development*, Addis Ababa, March 1981.

Oyelese, J. O., 1968. The mapping of land use patterns from air photographs in the forest zone of Ibadan Division. *Niger. geogr. J.*, **11**, 27–37.

Prothero, R. M., 1952. *Some Results from an Investigation of Land Use in the Cocoa Growing Region near Ibadan*. Research Notes, Geography Dept., University of Ibadan, 21–25.

UNDP/FAO, 1979. *Land Resources Survey, Sierra Leone: Review of Remote Sensing Techniques for Multidisciplinary Land Evaluation Purposes*. Tech. Rept. 4 (AG:DP/SIL/73/002), UNDP/FAO, Freetown.

Wikkamatileke, R., 1959. Problems of land-use mapping in the tropics. *Geography*, **44**, 79–95.

Remote Sensing and Tropical Land Management
Edited by M. J. Eden and J. T. Parry


# CHAPTER 10

# *Swamps and Rice Cultivation in Sierra Leone: Digital Landsat Techniques*

SUSAN M. RINGROSE
*Department of Environmental Science, University of Botswana*

Satellite remote sensing has been used to advantage for the management of natural resources in Third World countries where base maps are not available, precise topographic survey is lacking and aerial photographic coverage is incomplete. In this study the specific benefits of Landsat digital data for resource survey will be illustrated. A variety of restoration, enhancement, and information extraction techniques are now available and with these the Landsat analyst can obtain useful information for the management of tropical mangrove swamps and the associated rice-growing areas. Such areas occur along the coasts of West Africa, southeast Asia, Australia, and Central and South America (Schnell, 1971). In certain areas, optimal resource management requires resolving the conflict between the need for expansion of the area of productive rice land and the need to preserve the mangrove swamps as a unique habitat for marine life. In this study, which focuses on the mangrove swamplands of Sierra Leone in West Africa, consideration is also given to alternative rice production areas in the *bolilands* of the interior.

## SIERRA LEONE: THE REGIONAL SETTING

Sierra Leone lies between 7° 00′ N and 10° 00′ N, and 10° 00′ W and 14° 00′ W, and covers an area of 72 300 km$^2$. According to the Köppen system of climatic classification, the area lies in the Aw zone — tropical savanna with distinct wet and dry seasons. In common with West Africa as a whole, these seasons are caused by the migration of the Intertropical Convergence Zone (ITCZ), which separates warm, dry air originating over the Sahara from warm, moist air moving northwards from the Gulf of Guinea. Wet season flooding, which may last from

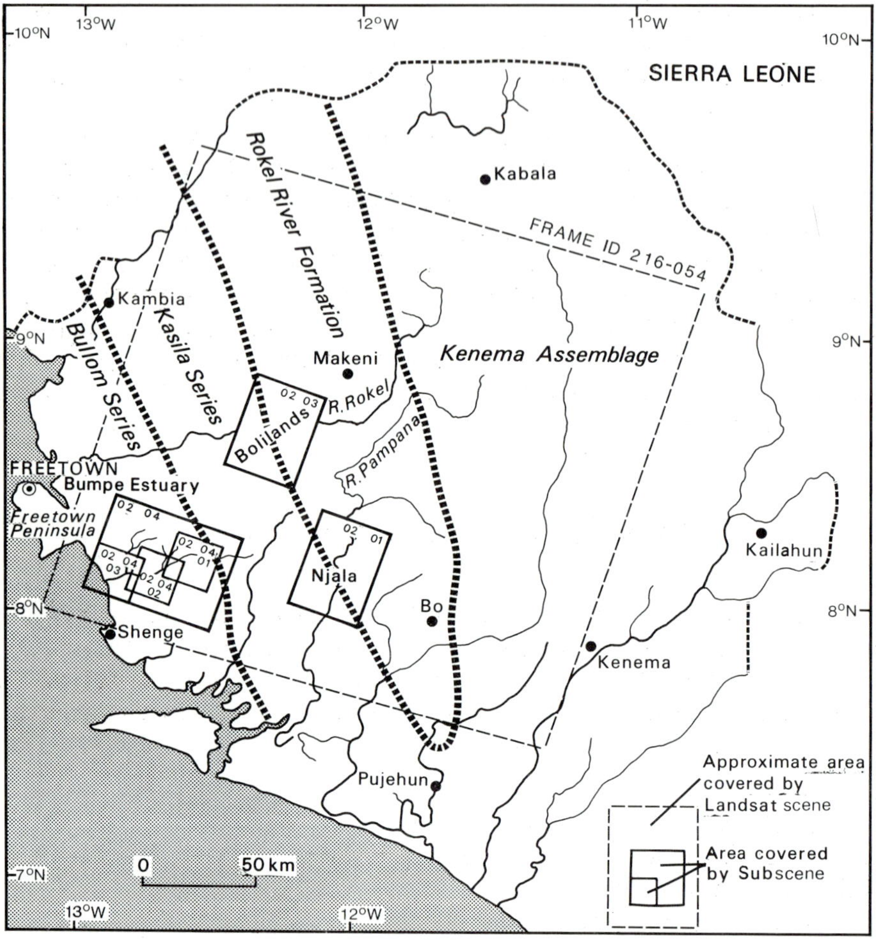


Fig. 1 Sierra Leone: location of Landsat scene and subscenes, and major geological units.

May till November, has considerable impact on the estuarine mangroves and inland depressions. The coast and the area of inland depressions, or *bolis*, experience an annual precipitation of about 3000 mm. These areas, which have the highest rainfall in Sierra Leone, are the country's most productive rice-growing lands.

Sierra Leone has a coastal plain which extends about 110 km inland and rises to 150 m above sea level. The plain is broadly coincident with the Bullom Series, which consists of Tertiary to Recent interbedded sands and clays over which the mangrove belt is established (Fig. 1). The zone of inland depressions, occurring between 150 and 300 m above sea level, is largely coincident with the outcrop area of the Rokel River Series which comprise both metasediments and volcanic rocks. Granites and granite gneisses of the Kasila Series and the Kenema Assemblage form undulating terrain adjacent to the main rice-growing areas (Morel, 1979; Macfarlane *et al.*, 1980; Ringrose, 1983).

Most of the drainage in Sierra Leone flows in a southwest direction from higher land in the north and east. River basins are relatively small, ranging from 7511 km$^2$ to 14 140 km$^2$, but discharge is high because of the heavy rainfall. Hydrographs on the major rivers show discharge peaks during the four-month period August to November. In the estuaries rice planting begins in October and the crop is harvested during January and February. Flooding in the *bolilands* area is less straightforward. The area is a complex of seasonal swamps that appear to have resulted from the silting up of a former, composite delta of the Mabole, Rokel, and Pampana rivers. The present drainage system is extremely variable, as is the extent and degree of flooding.

## SWAMPLAND MANAGEMENT FOR RICE CULTIVATION

In Sierra Leone rice is the staple crop. Twenty years ago, the country was a net exporter of rice. At the present time, rice is imported from the United States and southeast Asia. However, the Ministry of Agriculture and Forests is making determined efforts to re-establish self-sufficiency in rice production. The crop is grown in five major situations:

1. mangrove swamps in brackish water environments;
2. riverain *bolis* adjacent to major river valleys;
3. inland *bolis* or seasonally-flooded depressions;
4. inland valley swamps;
5. upland interfluve areas (upland rice).

The current policy of the Ministry of Agriculture and Forests in Sierra Leone is to extend the area of rice cultivation and to shift the emphasis from relatively low-yielding, upland rice to more productive swamp rice. There are numerous difficulties in achieving this transition, including the conservatism of the farmers and a stated preference for the taste of upland rice. However, through the Integrated Agricultural Development Projects (IADPs) and the Adaptive Crop Research and Extension (ACRE) project, farmers are being encouraged to move from the upland interfluve areas into the swamps. In certain areas, the process is being aided by extensive irrigation and water control projects. In addition, the Land and Water Development Division of the Ministry of Agriculture and Forests is engaged in appraising inland valley swamps for more intensive rice cultivation. Here, further water control measures, involving dams and sluice gates, will allow multiple cropping during the year.

## LANDSAT IMAGE ANALYSIS OF MANGROVE ZONES AND RICE PRODUCTION AREAS

Some digital analysis of Landsat imagery has been undertaken for the identification and assessment of mangrove environments and rice-growing lands in the humid tropics. In Taiwan Chen *et al.* (1982) attempted to use multispectral data for agricultural resource assessment. They achieved an 88% accuracy level in the estimation of paddy area. However, because of complex land use patterns in Taiwan, the spatial and spectral resolution of the multispectral scanner (MSS) data was found to be inadequate for detailed land use mapping.

Berg and Grégoire (1982) have used remote sensing techniques for rice production forecasting in Mali. Rice yield was estimated on the basis of digital image processing at three periods of growth, namely, the end of sowing, the end of heading and flowering, and the end of maturation. Four land cover types were defined using the maximum likelihood classifier: water bodies, rice, dry crop areas, and savanna. It was found that the visual interpretation of Landsat colour composite images yielded more accurate estimates of crop area than digital analysis.

Geiser *et al.* (1982) attempted rice yield forecasting in Sri Lanka using a model based on the area of paddy and the average yield figures. Using Landsat MSS data for two well-surveyed test sites, they concluded that the best time to detect paddy was six weeks before harvest, when a dense green carpet existed. No clear separation between paddy rice and other land cover types, such as sugar cane, brushland, grassland and homestead gardens, was established.

Blasco *et al.* (1982) mapped the Sundarbans, the mangrove zone at the margin of the Ganges delta. Landsat 2 digital data were used to differentiate five mangrove-related, land cover types, namely: areas cleared for agriculture and under cultivation, water areas, weakly saline wooded terrain, highly saline densely forested areas of low trees, and areas of sparse vegetation. The relative extent of these land cover types was reported to have changed as a result of subsidence of the delta.

In the present study, an assessment of several different types of Landsat imagery was made in order to determine the optimal type for mapping mangroves and rice-producing lands. This followed an earlier study of band 7 imagery, at 1:1 000 000 scale, which was interpreted with some success at the ecosystem level (Ringrose and Large, 1983).

Across West Africa, and Sierra Leone in particular, only limited dry season imagery is available from Landsat 1, 2 and 3. In the present study, cloud-free imagery was obtained for 29 February 1976. The imagery (frame 216-054, ID 2403-10204) is here referred to as frame 02 (Fig. 1). Black and white prints and colour composite images were examined prior to acquiring the computer compatible tape. Consideration has been given to both whole scene and subscene data. The whole scene was depicted on the Canada Centre for Remote Sensing

Image Analysis System (CIAS), and subscenes were chosen to show the mangrove zone with existing mangrove forest and land under rice cultivation, and other areas of potential rice production. Various subscenes, outlined in Fig. 1, were selected for detailed examination:

1. the Bumpe estuary (0204),
2. the *bolilands* (0203),
3. the Njala area (0201).

The subscenes were displayed at the $512 \times 512$ pixel level. More detailed work in the Bumpe estuary required additional subscenes at the $256 \times 256$ pixel level. These are referred to as 020401 for the upstream area of the Bumpe river, 020402 for the midstream portion, and 020403 for the downstream portion (Plate 9). The enhancement and classification techniques used, with the exception of the Martin Taylor transformation, are generally available in remote sensing centres throughout the world. In the study, each subscene was subjected to a series of enhancement and classification techniques aimed at differentiating rice-producing lands from mangrove forests in order to determine the area of each. An initial analysis of the separability of these land cover types was accomplished using discrimination matrices. The estimates obtained for the extent of land in each class were compared with previous Land Resource Survey mapping, based on aerial photographs (Gordon *et al.*, 1975–76), and with fieldwork undertaken as part of the present study.

## IMAGE ENHANCEMENT PROCEDURES

The data from Landsats 1, 2 and 3 consist of a two-dimensional array of measured incident radiation values in four spectral bands, namely, band 4 (0.5–0.6 $\mu$m—green), band 5 (0.6–0.7 $\mu$m—red), band 6 (0.7–0.8 $\mu$m—near infrared), and band 7 (0.8–1.1 $\mu$m—near infrared). The incident radiation is influenced by many factors other than surface reflectance. The most significant of these are sun angle, viewing angle, aspect angle of the terrain, and atmospheric absorption and scattering. The Landsat MSS data for Sierra Leone were not radiometrically corrected. However, the area is relatively flat and consistency is maintained because the three main subscenes are derived from the same satellite pass at a constant solar zenith angle. The histogram of radiance values from tropical scenes displays a high frequency of pixels within a limited intensity range, and so enhancement and classification techniques are essential in order to maximize feature separation. Each enhancement or classification display was photographed directly from the CIAS video screen using a tripod-mounted 35 mm camera with Ektachrome ASA 200 film. Black and white paper prints were made on a Versatec printer plotter and colour maps were made with an Applicon plotter.

The CIAS display combines in an optimal way as much data as possible from three of the four MSS bands. The image display can be transformed using a linear contrast stretch so as to provide greater feature separation. This transformation is based on the frequency distribution histogram of the pixel radiance intensities in the area. Unwanted data, such as the pixels representing water bodies, are excluded. The remaining data are then spread in a linear distribution over the whole range of radiance values present (Lillesand and Kiefer, 1979). A maximum of 64 radiance levels was used throughout this work, rather than the 258 possible values. This was advantageous in that the time required to process the transformations was reduced, while the data were adequately displayed. Linear contrast stretching was performed on all four channels, three of which (bands 4, 5, and 7) were combined to form a stretched composite image (Strome, 1976).

Image ratioing procedures, specifically the biomass ratio (bands 5:7), were applied to accentuate differences in vegetation cover (forest areas, *bolilands*, *Lophira* tree savanna, etc.) and minimize the shadow effects of low sun angle in more mountainous areas. Illumination differences on slopes of different aspect occur because of the relatively low solar illumination angle at the time of the satellite overpass (0930 hours). East-facing slopes are strongly reflecting whereas west-facing slopes are either in shadow or weakly reflecting. Because of the clear differentiation of vegetation types and their close association with the geology, geomorphology, and soils, more data were obtained in some cases by applying a biomass composite transformation in which the contrast-stretched data of bands 5 and 7 were ratioed. In other cases, the ratioed data were contrast-stretched in an attempt to provide maximum feature separation.

A form of edge enhancement using a high pass box car filter was applied to accentuate linear features. This technique is used to emphasize the boundaries between features. All radiance values that fall within the specified values for the box car are significantly increased or decreased so that the zone of threshold boundaries is sharpened. By varying the size of the box, features of different sizes can be enhanced. This technique was also applied to the bands 5:7 ratioed data in order to determine whether significant additional information could be obtained.

Principal component transformation was attempted in order to derive the maximum amount of data from all the spectral bands, and eliminate redundancy of data between the spectral bands. Principal component analysis computes the variance of the spectral reflectance data for each band in terms of covariance among the spectral bands. The resultant parameters—covariance matrix, eigenvalues, and eigenvectors—are used to produce four new sets of values for all pixels, such that the first principal component (PC1) includes 70% to 90% of the variance of the original values, and the second principal component (PC2) includes a large proportion of the remaining variance (Baltaxe, 1977). Normally PC1 and PC2 account for 95% to 99% of the variance in the original data. In general, the first and second principal components were used in the present analysis. When maximum information was obtained with PC1, only this was used. When information was available from both principal components, these data were stretched and then portrayed in black and white or with suitable colour added interactively.

The Martin Taylor enhancement (Taylor, 1974) is designed to maximize the variation in the data while absorbing any high degree of redundancy. The eigenvectors are determined by computing the variance matrix for the data in all channels. A coordinate rotation is then performed to produce the same number of channels with features represented by intensities along the eigenvectors. In selecting the three channels with maximum variation in rotated space, a new colour image is produced containing the maximum amount of information available. The reformed data on each channel can then be contrast-enhanced. Thus, the final image displays greater discrimination between terrain types with different spectral characteristics than is obtained through contrast stretching alone (Strome, 1976). An additional characteristic is that certain colours are added to improve visual feature separation. Emphasis is placed on the red-blueness and green-yellowness ascribed to different combined channel reflectance data.

The non-parametric unsupervised classification technique used at the Canada Centre for Remote Sensing is not interactive. It was applied to the West African data before field checking had been undertaken. Statistics for whole scenes and subscenes are examined and the data points are formed into distinguishable clusters or grouped together with gaps between the groups, each of which is arbitrarily defined as a class. The algorithm used for four-channel data is one that searches for isolated peaks in the four-dimensional histogram (Strome, 1976). Each of these peaks is identified as a centre for a class or cluster. Once pixel intensities have been ascribed to a class, new classes generated and new statistics devised, some exchange may take place until all classes are complete and statistically defined. Because of the low range of radiance values in the West African data, it was determined through trial and error that the optimum clustering achievable would yield six distinct land cover classes.

## LANDSAT SCENES AND SUBSCENES

Landsat whole scene data, at 1:1 000 000 scale, do not have sufficient resolution to provide information on all the swampland areas in Sierra Leone. The riverain *bolis* and inland valley swamps are frequently too small to be identified at that scale. Hence, there is a need for enlargement, which is met using digitized subscene data. With this approach, the location and mapping of swamp areas in the country appear to be feasible.

### Subscene: the Bumpe estuary (0204)

The Bumpe river flows through western Sierra Leone, south of the Freetown peninsula (Fig. 1). At the limit of navigation of the Bumpe estuary lies Rotifunk, a relatively large chiefdom centre. The main source of income in Rotifunk and its environs is the cultivation of rice, some of which is shipped to Freetown.

The Bumpe river system, which forms one of three minor catchments between the Freetown peninsula and Shenge, consists of minor waterways 40–60 km in length. Mangrove swamp development is believed to have begun with the rise in sea level during the late Pleistocene (Flandrian) transgression, which resulted in flooding of the lower valleys creating broad estuaries.

Four genera are common in the mangrove swamps of Sierra Leone. The most common colonizing genus is *Rhizophora*, a stiltroot mangrove. Three species of *Rhizophora* normally occupy that part of the intertidal zone which is flooded daily. *R. racemosa* prefers a silty substratum and colonizes newly emerged tidal flats. *R.mangle* and *R. harrisonii* are found on higher ground. Nowhere does *Rhizophora* reach its maximum height, but *R.racemosa* grows to 20 m in favourable locations in the Sherbro river. *Avicennia nitida* prefers a sandy substratum and generally occupies locations higher in the subtidal zone. It grows to 5 m in height. Two other species, *Conscarpus erectus* and *Laguncularia racemosa*, grow as scattered individuals around the high spring tide mark in semi-mangrove thickets (Fig. 2).

Mangrove zonation is controlled by the nearshore beach profile which determines the extent of tidal submergence, the tolerance of the species to daily and seasonal variations of salinity, and the character of the soil material. Mangroves rarely extend along an exposed coastline and mainly occupy sheltered areas such as estuaries. In addition to the lateral zonation, illustrated in Fig. 2, mangrove areas also display upstream variation in species, consequent upon the decrease in salinity. In upstream areas, in the presence of fresh water, *Raphia* swamps predominate, particularly along smaller tributaries. The swamps contain trees up to 20 m tall with a closed canopy, including both erect and climbing

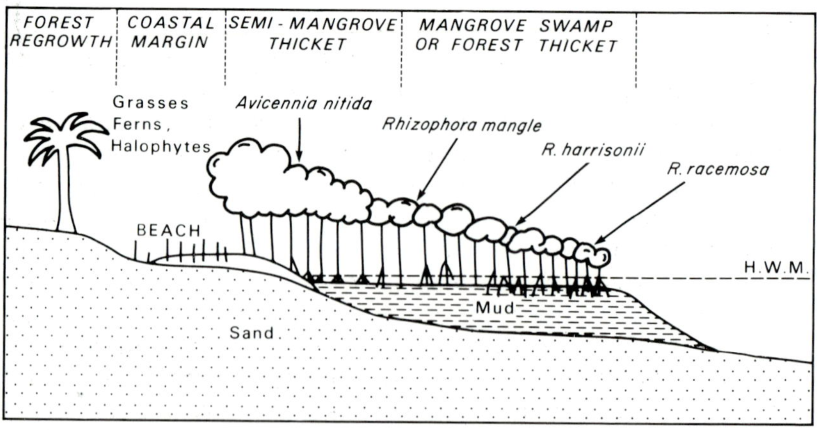


Fig. 2 Vegetation sequence in the mangrove zone of the Bumpe estuary, Sierra Leone.

rattan palms, which also occur in adjacent areas of forest regrowth. The undergrowth is mainly ferns.

Within the main mangrove swamp, Cole (1968) recognizes two types:

1. Estuarine mangroves, dominated by *Rhizophora racemosa*, which are found on silt- and clay-rich fibrous soil (30% silt and clay, 50% organic matter).
2. Saline mangroves, which occur in small estuaries and tidal fringes with little or no influx of fresh water. Here the soil tends to be sandy and more saline. The plant communities are dominated by open woodland with shrubby *R. mangle* and *R. harrisonii*.

In general, a rice crop will only grow in estuaries with an adequate fresh water influx. The dynamics of an estuary control the seasonal fluctuation in the position of the salt-water wedge, which appears to correspond to the transition zone between dominant *R. racemosa* and dominant *R. mangle* and *R. harrisonii*. Because the fresh-water influx during the rainy season maintains brackish conditions within the *R. racemosa* mangrove zone, it is in these areas that rice production has been established, whereas the *R. mangle* and *R. harrisonii* areas tend to be preserved. The fresh water swamps, characterized by *Raphia* forests, at the head of the estuaries and their tributaries, also tend to have been removed for rice production. In the more saline mangroves, dominated by *R. harrisonii* and *R. mangle*, experiments with poldering have been undertaken. In the Freetown peninsula, problems of sulphate toxicity have been encountered.

### *Spectral reflectance data of the Bumpe estuary subscene*

The reflectance data of the Bumpe river exhibit a limited spectral range, especially in the visible portion of the spectrum (Fig. 3). Very little information is obtained in band 4, probably because of interference from haze. Little additional information occurs on band 5. The higher reflectance values relate to areas of cultivated rice and the exposed coastal margin, with peak values representing forest regrowth and mangroves. The band 7 data have a wider range of spectral values. The high digital values represent the inland valley swamps, found in isolated channels beyond the the main mangrove belt. Most of the area under the curve represents forest regrowth within the coastal plain where the main agricultural activity is shifting cultivation. The lowest radiance values are obtained from the ocean, tidal and non-tidal mangrove thicket, and mangrove rice swamps.

The results of the analysis of the subscene are shown in Table 1 in the form of a discrimination matrix. Schematic sections illustrating the changing character of the mangrove zone are presented in Fig. 4. The rice-growing lands adjacent to the main channel can be discerned with relative ease on contrast stretched data, the principal component transformation, and the Martin Taylor transformation. In no instance are the rice-growing lands clearly separated from the main river channel, presumably because the radiance values of wet clay soils are similar to those of turbid river water (Plate 10).

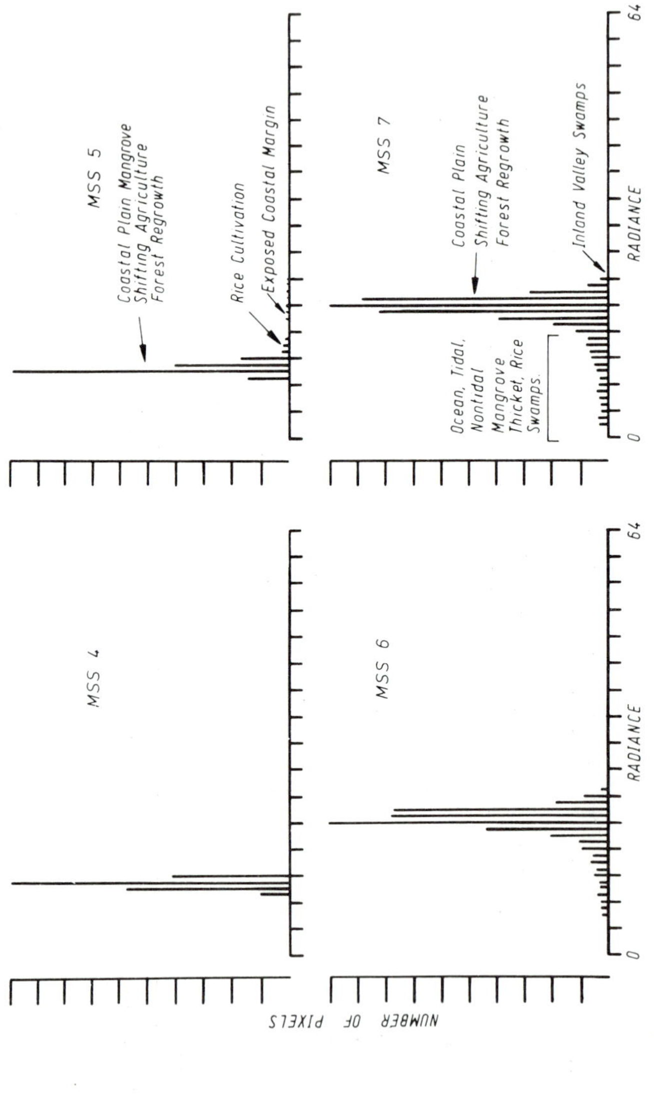


Fig. 3 Spectral radiance histograms for Landsat MSS bands 4, 5, 6, and 7 (image ID 2403-10204, 29 February 1976), showing feature segregation and response patterns in the Bumpe estuary, Sierra Leone (subscene 0204).

Table 1. Discrimination matrix for the Bumpe estuary subscene 0204.

| | Colour composite | Linear contrast stretch | Bands 5:7 ratio | Bands 5:7 ratio stretch | Principal component transformation | Martin Taylor transformation | Unsupervised classification |
|---|---|---|---|---|---|---|---|
| Rice growing lands: upstream section | P | F | P | N | F | F | N |
| Rice growing lands: midstream section | P | F | P | N | F | F | N |
| Remaining natural mangrove | G | G | G | G | G | G | G |
| Estuarine vs tidal mangroves | N | F | G | F | F | F | G |
| Inland valley swamps | G | G | P | P | F | N | F |
| Forest regrowth | F | F | G | G | F | N | F |

Discrimination levels achieved for the features under consideration: G good, F fair, P poor, N nil.

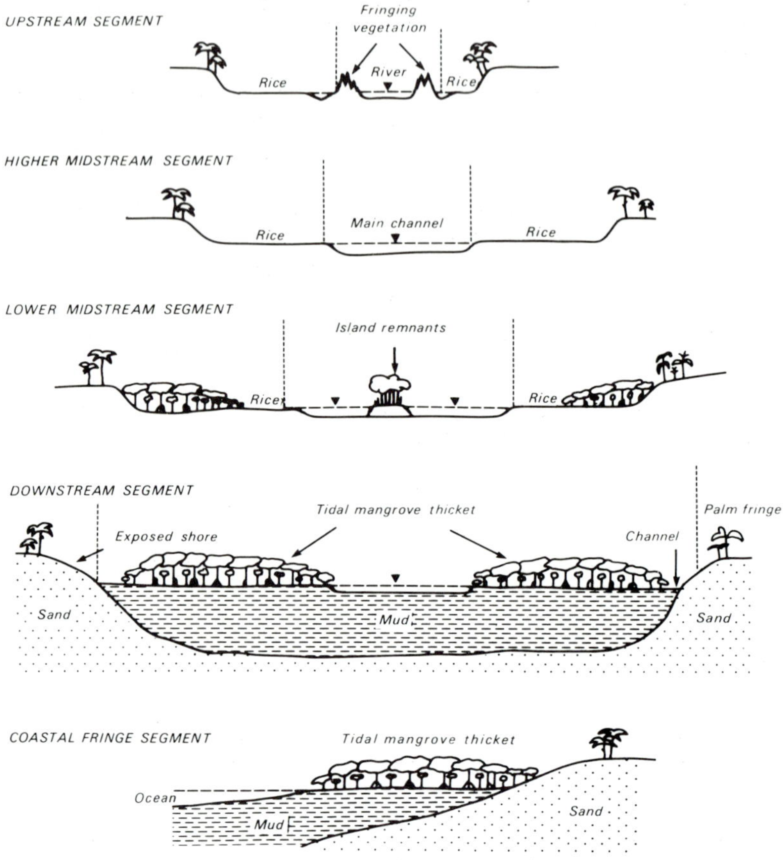


Fig. 4 Downstream changes in estuarine morphology and vegetation, identified from Landsat imagery in the main Bumpe estuary and adjacent fringing coast, Sierra Leone.

The enhancement and classification techniques are more successful in delimiting the extent of the remaining natural mangroves. In most of the enhancements, discrimination between estuarine and tidal mangroves is also reasonably satisfactory. The best separation of these features is achieved with the bands 5:7 ratio and unsupervised classification techniques.

Alternative rice-growing lands in the inland valley swamps stand out on colour composite and contrast stretch data because of their higher reflectance compared to adjacent areas of secondary forest. Reasonable recognition can also be obtained using the principal component transformation and unsupervised classification techniques.

Additional information on the interpretation of the unsupervised classification is given in Table 2. This technique allows good separation within the mangrove

Table 2. Unsupervised classification of the Bumpe estuary subscene 0204.

| Theme | Colour | Interpretation |
|---|---|---|
| 1<br>2 | Purple<br>Orange } | Riverain mangroves, *Raphia* fringe and tidal mangroves, coconut fringe, rice areas |
| 3 | Blue | Estuarine mangroves and cleared or burnt areas in forest regrowth |
| 4 | Green | Forest regrowth on interfluves, including areas of upland rice |
| 5 | Pink | Transitional, partially cleared areas |
| 6 | Light blue | Inland valley swamps |
| Unclassified | Black | Estuarine rice growing areas and main river channel |

Table 3. Areas of rice cultivation and natural mangrove in the Bumpe estuary, estimated from Landsat data and air photo interpretation.

| Land type | Airphoto analysis (Gordon *et al.* 1975–76) km$^2$ | Unsupervised classification km$^2$ | Unsupervised classification %error | Bands 5:7 ratio stretch km$^2$ | Bands 5:7 ratio stretch %error |
|---|---|---|---|---|---|
| Rice cultivation areas and natural mangrove | 130 | 105 | −19 | 148 | +14 |
| Swamp under rice cultivation | 25 | 20 | −20 | — | — |
| Tidal mangrove<br>Estuarine mangrove } | 105 | 26<br>59 | −19 | —<br>— | —<br>— |

zone. Most *Rhizophora racemosa* in the less saline, upstream and midstream areas has been cut down to allow rice production, leaving only small remnants of mangrove which appear as theme 1 (Plate 11). Estuarine mangroves, consisting of relatively dense stands of *R.mangle* and *R. harrisonii*, are identified mainly within theme 3, while the extremely saline, tidal areas along the coast, which support more open stands of these species, occur as theme 2.

The benefit of Landsat enhancement and classification techniques in relation to Land Resource Survey mapping was investigated as a major part of the study. As is indicated in Table 3, the unsupervised classification provides comparable results to those established by conventional land use mapping techniques (Gordon *et al.*, 1975–76) with a relatively low margin of error. Hence, it appears that rapid, reliable information can be obtained, using digital Landsat data, which would establish the areas under rice cultivation and those still under mangrove for any given estuary. For the Bumpe estuary, the areas are 20 km$^2$ and 85 km$^2$, respectively. The mangrove can be further subdivided into about 26 km$^2$ of tidal mangrove and 59 km$^2$ of estuarine mangrove using the themes from the unsupervised classification (Table 3). The boundary between the rice

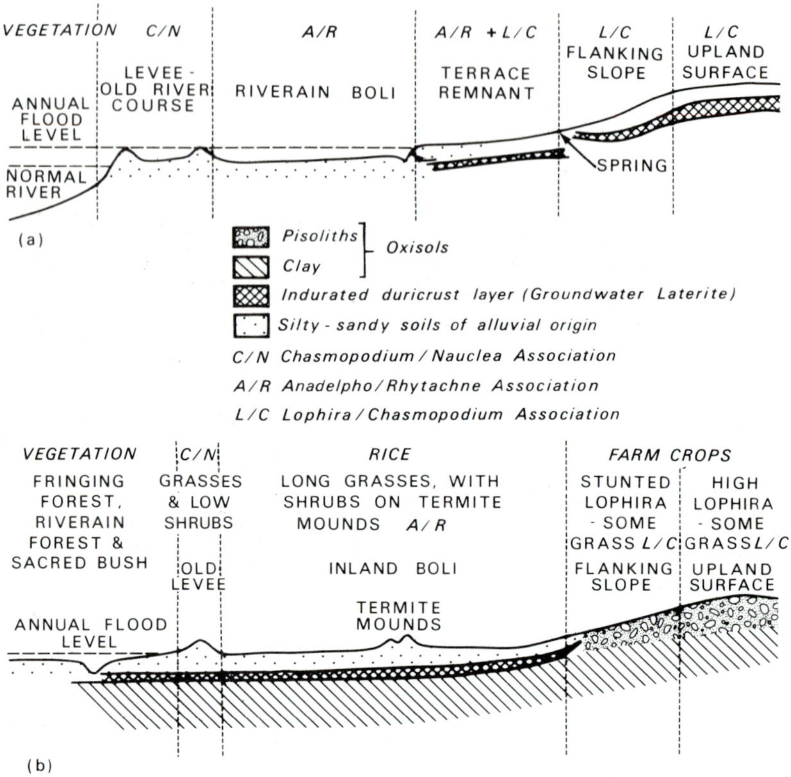


Fig. 5 Morphology and vegetation sequence in (a) a riverain *boli* and (b) an inland *boli* (after Stobbs, 1963).

cultivation zone and the mangrove is currently fixed by the position of the saline wedge that extends about one-third of the distance up the 24 km long estuary. The tidal and estuarine mangroves are not likely to be cleared for rice cultivation unless water control measures and poldering practices are introduced. Such schemes are expensive and would require a major investment of external financial aid.

### Subscene: the *bolilands* (0203)

This area lies in the central portion of Sierra Leone adjacent to the Rokel river. The Malal Hills (385 m) form a prominent feature in the subscene. The village of Roten Malal lies at the foot of the hills on the main road linking Freetown to Bo.

Stobbs (1963) defined a *boli* as an individual swamp suitable for rice cultivation. Differentiation is made between riverain *bolis* which are annually

Fig. 6 An inland *boli* during the dry season.

flooded areas on contemporary river floodplains and inland *bolis* which are areas flooded during exceptionally high floods by water draining over the flanking slopes. The *boli* types are illustrated in Fig. 5(a) and (b). According to Cole (1968), water begins to collect in the *bolis* in April. By mid-July, they are flooded and water cannot escape because of the levees with the result that flooding continues well into the dry season. This favours intensive leaching and the formation of groundwater laterites and acidic gleisols. These soils have low exchangeable base levels and little available phosphorus. The organic matter and nitrates normally increase during flooding, but they are lost through oxidation and leaching when the flood waters subside. The Malal district has been under mechanical rice production since 1951, and yields are moderately good if the crop is planted early in the wet season and superphosphate fertilizers are used. A typical inland *boli* in the dry season is illustrated in Fig. 6.

### *Spectral reflectance data of the* bolilands *subscene*

In general, the spectral reflectance data show high peaks in the visible portion of the spectrum with a low range of radiance values. In the near infrared portion of the spectrum, the converse is true. The pixel frequency peaks are noticeably lower, while the range of radiance values is double that of bands 4 and 5. However, there is a high degree of data redundancy. The histograms showing

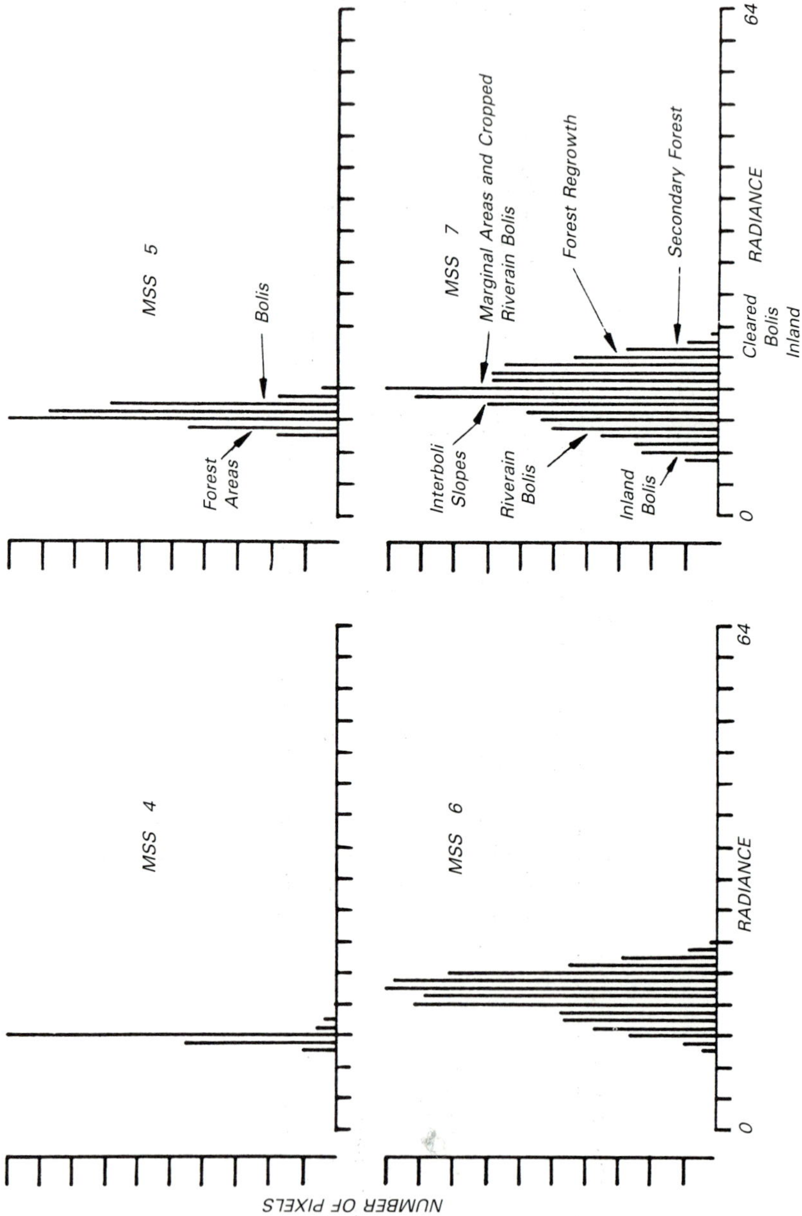


Fig. 7 Spectral radiance histograms for Landsat MSS bands 4, 5, 6, and 7 (image ID 2403-10204, 29 February 1976), showing feature segregation and response patterns in the *bolilands*, Sierra Leone (subscene 0203).

Table 4. Discrimination matrix for the *bolilands* subscene 0203.

| | Colour composite | Linear contrast stretch | Bands 5:7 ratio stretch | Principal component transformation | Martin Taylor transformation | Unsupervised classification |
|---|---|---|---|---|---|---|
| Inland *bolis*—seasonally flooded depressions | F | F | G | F | F | F |
| Riverain *bolis* | F | P | G | F | F | P |
| Secondary forest | F | G | F | F | F | N |
| Forest regrowth | G | G | F | G | F | F |

Discrimination levels achieved for the features under consideration: G good, F fair, P poor, N nil.

spectral reflectance characteristics are presented in Fig. 7. The histogram for band 7 is annotated to show inland and riverain *bolis*. The high frequency classes represent the moderate radiance values of inter-*boli* slopes, marginal areas, and cropped riverain *bolis*. The higher radiance values include areas of forest regrowth and secondary forest. Comparable subdivisions cannot be made on band 5 data, which only allow the separation of forested areas on the left and *bolis* and related features on the right.

Enhancement and classification techniques have been used to examine the swamp environments in the area. The different transformations were compared visually to provide an assessment of feature separation, as shown in Table 4. Most of the inland *bolis* stand out reasonably well as separate features on all the transformations, but they show up particularly well on the bands 5:7 ratio stretch. The low reflectance of the *boli* soils on the dry season imagery appears to be related to their high moisture content and the well-developed laterization. Riverain *bolis*, on the other hand, are indistinguishable from the forested and farmed inter-*boli* flanking slopes and upland surfaces because the *bolis* often support 'upland' crops during the dry season. Discrimination is only fair to poor, except for riverain *bolis* along the Rokel river, which are very evident after bands 5:7 ratio stretching (Plate 12). Areas of secondary forest and forest regrowth are also clearly differentiated on the enhanced data, particularly on contrast stretch and principal component transformations. It is in these forest areas that upland rice crops are grown. Clearings in the forest can be discerned on the enhanced data, but no discrimination was achieved between clearings under rice and those under other crops.

The results of the unsupervised classification of the *bolilands* are shown in Table 5, and depicted visually on Plate 13. The unsupervised classification, while partly successful, provides less clear results here than in the Bumpe estuary. Most of the inland *bolis* are discernible as themes 1 and 2, which represent pixels with the lowest radiance values. Themes 3 and 4 represent riverain *bolis* and adjacent slopes; in the dry season, it is very difficult to differentiate between the riverain *boli* itself, especially when under crops, and the farmed inter-*boli*

Table 5. Unsupervised classification of the *bolilands* subscene 0203.

| Theme | Colour | Interpretation |
|---|---|---|
| 1 and 2 | Black and Orange | Inland *bolis* |
| 3 and 4 | Dark blue and Yellow | Riverain *bolis* and farmed inter-*boli* flanking slopes |
| 5 | Pink | Mainly clearings under upland crops, thinly forested areas, and inland valley swamps |
| 6 | Light blue | Forest regrowth and secondary forest |

Table 6. *Boli* areas under cultivation in the *bolilands*, estimated from Landsat data and airphoto interpretation.

<table>
<tr><th>Land type</th><th>Airphoto analysis (Gordon et al., 1975–76) km²</th><th>Bands 5:7 ratio stretch km²</th><th>Unsupervised classification km²</th></tr>
<tr><td>Inland bolis</td><td rowspan="2">437</td><td>379</td><td rowspan="2">405</td></tr>
<tr><td>Riverain bolis and farmed inter-boli flanking slopes</td><td>189</td></tr>
</table>

slopes and upland surfaces. Clearings and areas under upland crops are represented in theme 5, while theme 6 depicts forest regrowth and secondary forest as one category.

An attempt was made to determine the extent of inland *bolis* and riverain *bolis* within the subscene. This was undertaken by comparing the results of the Vegetation and Land Use map (Gordon *et al.*, 1975–76), backed up by field checking, with the bands 5:7 ratio stretch and unsupervised classification data (Table 6). The results are unclear. Whereas a distinction can be made between definite (i.e. ground-truthed) inland and riverain *bolis* using bands 5:7 ratio stretch data, comparable results were not obtained from either the unsupervised classification or the air photo analysis undertaken by Gordon *et al.* (1975–76). In the latter cases, it appears that the inland *bolis* and riverain *bolis* have been combined as a single mapping unit. In general, there is also uncertainty over the extent to which the adjacent inter-*boli* flanking slopes have been differentiated. It is estimated that approximately 400 $km^2$ of potential rice-growing lands of the inland and riverain *boli* type exist within the area, with a higher proportion of land being ascribed to inland *bolis* and a lower proportion to riverain *bolis*.

**Subscene: the Njala area (0201)**

The area covered by the Njala subscene is indicated in Fig. 1. It lies in south central Sierra Leone, adjacent to the Taia river. Njala University College, in the southern portion of this area, is the centre of activities for the Adaptive Crop Research and Extension (ACRE) project. Farmers in the area are generally being encouraged to produce rice in both riverain and inland valley swamps.

Both types of *bolis*, in addition to significant inland valley swamps, are located in the area. Background data on the *bolis* have already been provided. The inland valley swamps are typically developed over granitic terrain associated with the Kasila Series (Thomas and Thorp, 1985), while larger linear swamps occur over the southern Rokel River Series. Only the larger valleys contain permanent

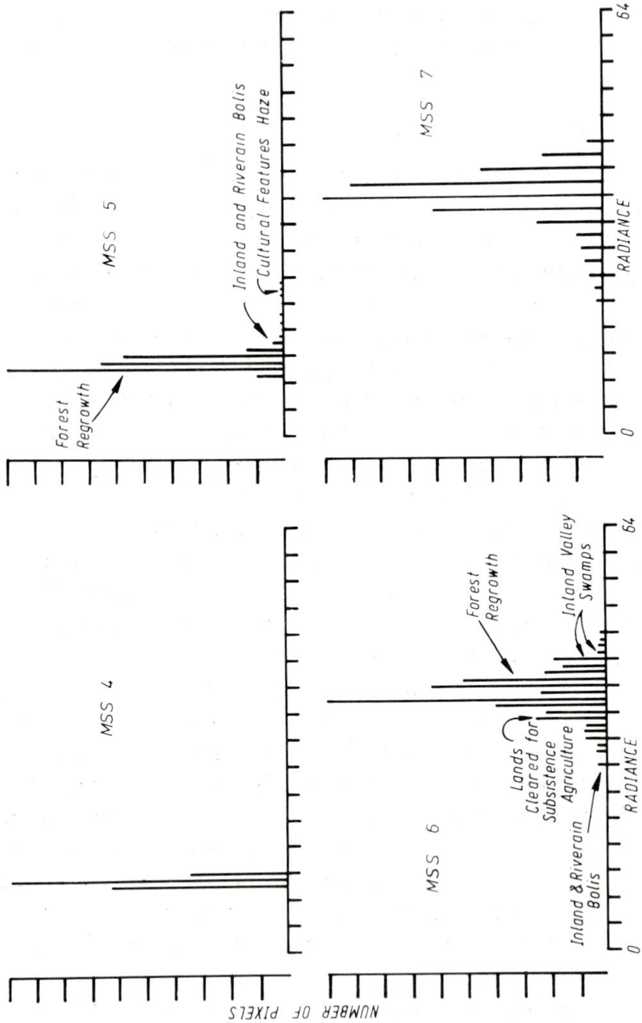


Fig. 8 Spectral radiance histograms for Landsat MSS bands 4, 5, 6 and 7 (image ID 2403-10204, 29 February 1976), showing feature segregation and response patterns in the Njala area, Sierra Leone (subscene 0201).

streams, but water tables are generally high, and flooding is likely to occur during the rainy season. A typical profile through a valley swamp reveals black clay loam, overlying mottled clay and grey clay, fine to medium sand and gravel, and finally bedrock. However, the valley floors are not always consistent, displaying considerable variation in width, sediment thickness, sediment profile, and bedrock topography. The problem with the valley swamps is that they dry up during the dry season when the land may support small crops, but not rice.

### *Spectral reflectance data of the Njala subscene*

The spectral reflectance data of the Njala area (Fig. 8) contrast with those of the Bumpe estuary and the *bolilands* in two respects. In the Njala data, there is complete separation in radiance values between the visible and infrared portions of the spectrum sensed. Also, there is a wider range in radiance values in the band 7 data. On band 5, inland *bolis* are distinguishable by their intermediate reflectance between the areas of settlement and the forest regrowth. The inland valley swamps cannot be distinguished. Greater feature separation is apparent on bands 6 and 7. For example, the band 6 histogram is polymodal, with peaks to the left representing such features as inland and riverain *bolis* and lands cleared for subsistence agriculture, and peaks to the right representing areas of forest regrowth and inland valley swamps.

The discrimination matrix (Table 7) presents the results of the different transformations. The intent here is to differentiate between inland valley swamps underlain by the Kasila Series and those underlain by the Rokel River Series. The swamps occupy numerous, small, interconnecting valleys in the Kasila Series and larger, discrete valleys in the Rokel River Series. Most of the transformations permit the identification of some inland valley swamps. The linear contrast stretch, bands 5:7 ratio stretch, and principal component transformations provide the clearest images. Examples of these features as they appear on the bands 5:7 ratio stretch are shown in Plate 14. The swamps in the Rokel River Series are generally less obvious than those in the Kasila Series, as they tend to be light-toned and indistinct with low contrast in relation to the surrounding areas of cleared farmland. Riverain and inland *bolis* are clearly depicted on all the transformations except the linear contrast stretch and the Martin Taylor enhancement, where they merge with the surrounding land uses. Linear contrast stretch is the only enhancement procedure that provides clear delimitation of the areas of forest regrowth and secondary forest. On other transformations, the distinction between areas of forest and cleared areas is considerably less marked.

Using bands 5:7 ratio and ratio stretch data, a comparison was made between the areas of riverain and inland *bolis* and inland valley swamps (Table 8). No direct comparison with the Vegetation and Land Use map of Gordon *et al.* (1975–76) was possible because different land use categories are employed.

Table 7. Discrimination matrix for the Njala area subscene 0201.

| | Colour composite | Linear contrast stretch | Bands 5:7 ratio | Highpass filter of bands 5:7 ratio data | Bands 5:7 ratio stretch | Principal component transformation | Martin Taylor transformation | Unsupervised classification |
|---|---|---|---|---|---|---|---|---|
| Inland valley swamps in Kasila Series | F | G | F | F | G | G | N | F |
| Inland valley swamps in Rokel River Series | F | F | F | F | F | F | N | P |
| Riverain and inland *bolis* | G | P | G | G | G | G | F | G |
| Forest regrowth and secondary forest | P | G | P | P | F | F | P | P |

Discrimination levels achieved for the features under consideration: G good, F fair, P poor, N nil.

Table 8. Area of potential lowland rice cultivation in the Njala subscene, estimated from Landsat data.

| Land type | Bands 5:7 ratio $km^2$ | Bands 5:7 ratio stretch $km^2$ | Mean value $km^2$ |
|---|---|---|---|
| Riverain and inland *bolis* | 132 | 170 | 151 |
| Inland valley swamps (Kasila Series) | 34 | 48 | 41 |
| Total potential lowland rice cultivation area | 166 | 218 | 192 |

Inland valley swamps in the Kasila Series can be readily mapped on both bands 5:7 ratio and ratio stretch enhancements: the total area of swamp was estimated at some 40 $km^2$. It was not possible to estimate the area of the Rokel River swamps, because of the poor discrimination levels achieved with the enhancements, but the extent is small. The area under riverain and inland *bolis* is relatively extensive in the Njala area. Field checking suggests that the *bolis* cover about 150 $km^2$ (Table 8).

## CONCLUSION

In this study, an attempt has been made to provide baseline data for vegetation and land use mapping in the humid tropics, using digital Landsat imagery and 'off the shelf' enhancement and classification techniques. Typical humid areas in Sierra Leone were chosen to exemplify the contrast between rice-growing lands and natural mangrove forest. In addition, techniques for estimating the extent of alternative rice production areas have been described.

A Landsat CCT, covering dry season conditions in February 1976, was used in the study. At this season the interior rice-growing lands (*bolis*) are not under cultivation. Their dark bare soils, with a relatively high moisture content, absorb infrared radiation in contrast to the surrounding areas of forest regrowth which are strongly reflective. A similar contrast exists between the swamp rice areas in the mangrove zone and the adjacent forest. However, because of the moist nature of the mangrove soils, it is difficult to differentiate between the highly absorptive, turbid river water and the equally absorptive rice swamps. In contrast, inland valley swamps are more reflective because they are continuously cropped and support growing vegetation, even in the dry season. Upland rice areas cannot be differentiated by digital enhancement techniques because their spectral radiance is similar to that of areas of subsistence agriculture.

The analysis showed that no single enhancement procedure was superior for all purposes. Discrimination matrices were used to assess the relative merits of each procedure with the following results:

1. Rice-growing swamps in a mangrove environment are best differentiated (from surrounding land uses) using linear contrast stretch and principal component transformations.
2. Estuarine and tidal mangroves are best differentiated using bands 5:7 ratio techniques or unsupervised classification.
3. The identification of natural mangrove areas can be achieved with all enhancement and classification techniques.
4. Inland *bolis* and riverain *bolis* are best differentiated using bands 5:7 ratio stretch enhancement.
5. Inland valley swamps can best be identified using linear contrast stretch, bands 5:7 ratio stretch, and principal component transformations.

It was found that there are about 20 $km^2$ of swamplands under rice cultivation in the Bumpe estuary and 85 $km^2$ of surviving mangrove. In parts of the *bolilands*, an estimated 400 $km^2$ of potential rice-growing lands of the inland and riverain *boli* type were identified, although the riverain type cannot be precisely determined because of its similar spectral reflectance to the surrounding terrain. In the Njala area, about 40 $km^2$ of inland valley swamps and 150 $km^2$ of riverain and inland *bolis* were identified.

An assessment of rice-growing lands is particularly critical in developing countries where rice is a staple food. Landsat imagery can provide a rapid and relatively inexpensive technique for obtaining an overview of the rice-growing lands, and digital analysis allows a degree of detailed assessment; the level achieved depends on the spectral characteristics of the crops and natural vegetation. The use of appropriate enhancement procedures considerably improves the effectiveness of Landsat data for the management of rice-growing swamplands. Areas under cultivation or potentially available for rice cultivation can be measured, and it is feasible to make an estimate of rice production, both actual and potential, using yield data for the different swamp types.

Digital Landsat imagery has also proved to be useful in determining the area of mangrove swamp. Using enhancement and classification techniques, the various types of mangrove can be differentiated. These techniques can be used as a management tool to provide baseline data and to achieve longer-term monitoring of the mangrove cover.

## ACKNOWLEDGEMENTS

The author gratefully acknowledges funding and support from the following sources: the International Development Research Centre and the Canada Centre for Remote Sensing, Ottawa, Canada; the Land and Water Development

Division, Ministry of Agriculture and Forests and the Njala University College, Sierra Leone. The University of Botswana facilitated the compilation of the final manuscript. Acknowledgement is made to the National Oceanic and Atmospheric Administration (NOAA), USA for the use of original Landsat imagery.

## REFERENCES

Baltaxe, R., 1977. *Preliminary Investigation of Mapping Land Use and Vegetation by Computer Processing of Landsat Data.* Technical Note 14, UNDP/FAO, Rome.

Berg, A. and Grégoire, J. M., 1982. Use of remote sensing techniques for rice production forecasting in West Africa (Mali and Guinea: Niger Bani Project). In *Satellite Remote Sensing for Developing Countries.* European Space Agency, SP-175, pp. 161–168.

Blasco, F., Deshayes, M., Le Toan, T., Chaudhury, M. U. and Favard, J. C., 1982. Cartographie des mangroves du Gange à partir des données Landsat. *Act.Symp.int. de la Commission VII de la Société Internationale de Photogrammetrie et Télédétection,* Toulouse, vol. 24-VII/1, 945–946.

Chen, A. J., Wang, H. T. and Lin, C. C., 1982. Temporal spectral measurements of rice crop with the GSFC Mark-II three band hand-held radiometer. *Act.Symp. int. de la Commission VII de la Société Internationale de Photogrammetrie et Télédétection,* Toulouse, vol. 24-VII/2, 81–93.

Cole, N. H. A., 1968. *The Vegetation of Sierra Leone.* Njala: Njala University College.

Geiser, U., Sommer, M., Haefner, H. and Itten, K. I., 1982. Multistage landuse mapping and change monitoring in Sri Lanka. In *Satellite Remote Sensing for Developing Countries.* European Space Agency, SP-175, pp. 153–160.

Gordon, O. L. A., Kater, G. and Schwaar, D. C., 1975–76. *Vegetation and Land Use of Sierra Leone. Land Resources Survey Map 1:500 000.* Freetown: FAO/UNDP—MANR.

Lillesand, T. M. and Kiefer, R. W., 1979. *Remote Sensing and Image Interpretation.* New York: John Wiley.

Macfarlane, A., Crow, M. J., Arthurs, J. W., Wilkinson, A. F. and Avcott, J. W., 1980. *The Geology and Minerals of Northern Sierra Leone.* London: Institute of Geological Sciences.

Morel, S. W., 1979. The geology and mineral resources of Sierra Leone. *Econ. Geol.,* **74**, 1563–1576.

Ringrose, S. M., 1983. Geological exploration and land use using Landsat data, Sierra Leone, West Africa. *Advd Space Res.,* **3**(2), 81–89.

Ringrose, S. M. and Large, P., 1983. The comparative value of Landsat print and digitized data and radar imagery for ecological land classification in the humid tropics. *Can.J.Remote Sensing,* **9**, 45–60.

Schnell, R., 1971. *Introduction à la Phytogéographie des Pays Tropicaux. Troisième partie: Les Plantes et les Milieux Intertropicaux,* vol. 2. Paris: Gauthier-Villars.

Stobbs, A. R., 1963. *The Soils and Geography of the Boliland Region of Sierra Leone.* Freetown: Government Printer.

Strome, M., 1976. *Progress towards Operational Use of Remote Sensing in Canada.* Unpublished Rept., Canada Centre for Remote Sensing, Ottawa.

Taylor, M. M., 1974. Principal component colour display of ERTS imagery. *Proc. 2nd Can. Symp.Remote Sensing,* 295–314.

Thomas, M. F. and Thorp, M. B., 1985. Environmental change and episodic etchplanation in the humid tropics of Sierra Leone: the Koidu etchplain. In Douglas, I. and Spencer, T., (Eds), *Environmental Change and Tropical Geomorphology.* London: Allen and Unwin, pp. 239–267.

Remote Sensing and Tropical Land Management
Edited by M. J. Eden and J. T. Parry


# CHAPTER 11

# *Agricultural Land Use Mapping in Lesotho: Problems and Limitations of Landsat*

EDWARD M. MAKHANYA
*Department of Geography, National University of Lesotho*

About 85% of the population of Lesotho depends on agriculture, yet to date there has been no attempt to carry out a comprehensive programme of mapping and monitoring agricultural land use. The first aerial photography of the territory was not obtained until 1950. This covered only the western region and the rest of the territory was not covered until two years later (March 1952). The Directorate of Overseas Surveys in the United Kingdom used this photography (at a nominal scale of 1:30 000) to produce the first topographic maps of Lesotho at a scale of 1:50 000. The main purpose of these maps was to provide topographic information and so the amount of socio-economic detail including land use is very limited.

A second aerial photographic coverage of the territory, again at a scale of 1:30 000, was obtained in May 1961. Using this photography Bawden and Carroll (1968) produced a fairly comprehensive review of the land resources of the country including a soil map, a land systems map and an agricultural potential map (all at the 1:250 000 scale). This is one of the few attempts to make use of aerial photography in resources surveys in Lesotho. However, it should be noted that the actual aerial photographic interpretation and map preparation were carried out at the DOS offices in the United Kingdom and so there was little or no opportunity for local government staff to learn the appropriate techniques.

The existing aerial photographs have been used in several urban planning projects. Other projects involving the use of aerial photography have been few. As far as agricultural land use is concerned, a pilot project was undertaken by the Department of Agriculture for the Tebe-tebeng Valley Agricultural Development Scheme in northwest Lesotho using the 1950 photography. Another study was undertaken by the author for the Thaba-Bosiu Rural

Development Project in west central Lesotho using the early aerial photographs, 1950 and 1961, and two later coverages, 1971 and 1976 (Makhanya, 1979).

From this brief review of attempts at mapping land resources in general and agricultural land resources in particular, it is clear that there are several constraints which limit the use of aerial photographs for land resources surveys in Lesotho. These same constraints apply with even greater effect to other types of remote sensing:

1. The high cost of obtaining aerial photography or other types of imagery which limits the frequency and scale of the coverage.
2. The lack of photogrammetric and interpretive equipment in Lesotho.
3. The lack of trained interpreters.
4. The lack of funds for this type of work.

When Landsat went into orbit in 1972 the United States government agreed to supply, free of charge, photographic prints of scenes in which Lesotho was imaged. The whole territory of Lesotho, some 30 000 $km^2$, is covered in four Landsat scenes. These materials were at first kept by the Lesotho Department of Geology, but since no one there could find use for them they were transferred to the National University. At that time only one or two people in the government of Lesotho had any idea of how to use Landsat data, and since no meaningful reports on Landsat applications were forthcoming the United States government discontinued the policy of free supply of Landsat data products in 1976.

Since that time there has been a growing interest in the use of Landsat for gathering agricultural land resources data. The Food and Agriculture Organization (FAO) of the United Nations, as well as United States agencies, have played an important role in changing government attitudes towards remote sensing by providing training programmes and technical aid.

The study described in the next few pages is an example of the type of analysis that can be carried out relatively cheaply in spite of the constraints mentioned above. The purpose of the study is to examine the degree to which rural land use detail can be extracted from digitally enhanced Landsat images. The area of interest is lowland Lesotho where crop raising is dominant. A comparable study, also based on Landsat imagery, is being undertaken for highland areas in Lesotho where pastoralism is dominant. The latter study is being carried out by the Lesotho Department of Agriculture with the help of American specialists in pasture management.

## LAND TENURE AND LAND USE IN THE STUDY AREA

The study area lies in northwestern Lesotho (Fig. 1). It is rural, with an undulating to rugged topography punctuated by scattered sandstone buttes and mesas. The natural vegetation is predominantly grassland and there is an almost complete absence of natural tree growth.

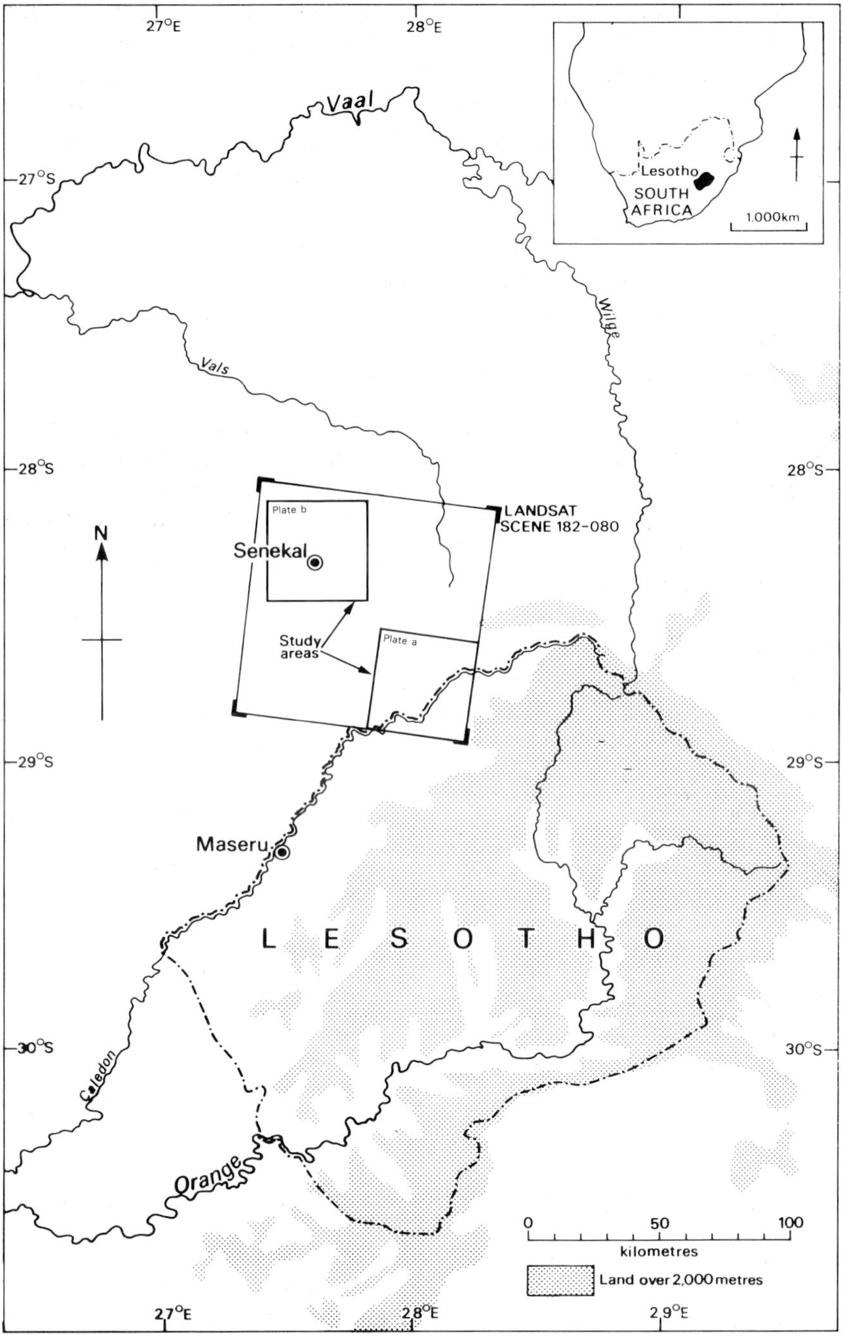


Fig. 1 Lesotho, showing location of Landsat subscene study areas.

The traditional system of land tenure in Lesotho is communal in the sense that all land belongs to the nation with the king as guardian of the lands of his people. The power to allocate land is vested in the monarch who discharges this responsibility through his local chieftains (Anon., 1973). Traditionally land was granted to individuals solely for the subsistence of the family group. It included a residential site with or without a garden depending on local conditions, and three parcels of land for cultivation. There was no standard area laid down with regard to the size of these parcels and they differed according to topographic conditions.

Grazing land and lands containing commodities such as building material, fuel, wild plants, etc. are used communally. Such lands are not allocated to individuals, but individuals can obtain the right of access to them. After the harvest, the cultivated lands are also opened to communal stubble grazing. During this period the individual's rights over his arable land are temporarily suspended.

As already pointed out, the traditional object of peasant agriculture in Lesotho is the production of food for direct family consumption. The manner in which the land can be used is greatly influenced by the communal nature of the land tenure system as well as by the chieftainship system. Three broad types of land use can be identified: areas of settlement, cultivated land, and permanent rough grazing land.

Most of the settlements are of the traditional type, sparsely distributed, and situated on the steeper and more rugged slopes, allowing the more level areas to be used for cultivation. On average, the villages in the study area contain about 45 homesteads, most of which are thatch-roofed.

Each household was by tradition allocated three parcels of land. In 1936, when a survey was carried out by the Colonial government (Anon., 1937), each land parcel was about 1 ha in size. Owing to the increase in the number of landholders who are sharing such an inelastic resource, the average size of land parcels has been reduced to about 0.8 ha at the present day, and it is now impossible to allocate the traditional three land parcels to each householder. As a result, there is an increasing number of landless, rural households. The introduction of legislation governing the prevention of soil erosion has lead to further subdivision of land parcels into strips of about $12 \times 100$ m (Anon., 1959).

The landholdings of many farmers have been further fragmented in response to two additional factors: firstly, the fact that traditionally a farmer in Lesotho was allocated land parcels at different sites to ensure equity in the distribution of good, mediocre, and poor land; and secondly, the fact that land shortage at the present day is chronic resulting in further subdivision with each successive generation.

During the period of crop growth, grazing is restricted to areas that are not used for cultivation, including the mesas, plateaux, and the *maboella* (special areas of protected grazing land). After the harvest the arable land is also opened to communal grazing.

## LAND USE CLASSIFICATION

Approximately 90% of the cultivated land in Lesotho is allocated to grains, mainly maize, sorghum, and wheat. As a result of the type of land use and management practices discussed above, there has been very little diversification of crops. The main crops grown in the summer (rainy) season are maize, sorghum, and, to a small extent, beans. In winter, wheat and peas are grown. A basic crop calendar for northwestern Lesotho is presented in Fig. 2. There is hardly any irrigation, and very few farmers raise crops in winter. This is especially true in areas where the chiefs are strict in the application of the practice of communal stubble grazing.

The general practice is to classify land use simply into three broad categories: areas of settlement, cultivated land, and permanent rough grazing land. Mapping of the land use on the basis of these broad categories can be carried out with consistent accuracy using medium scale air photographs (1:40 000 and larger).

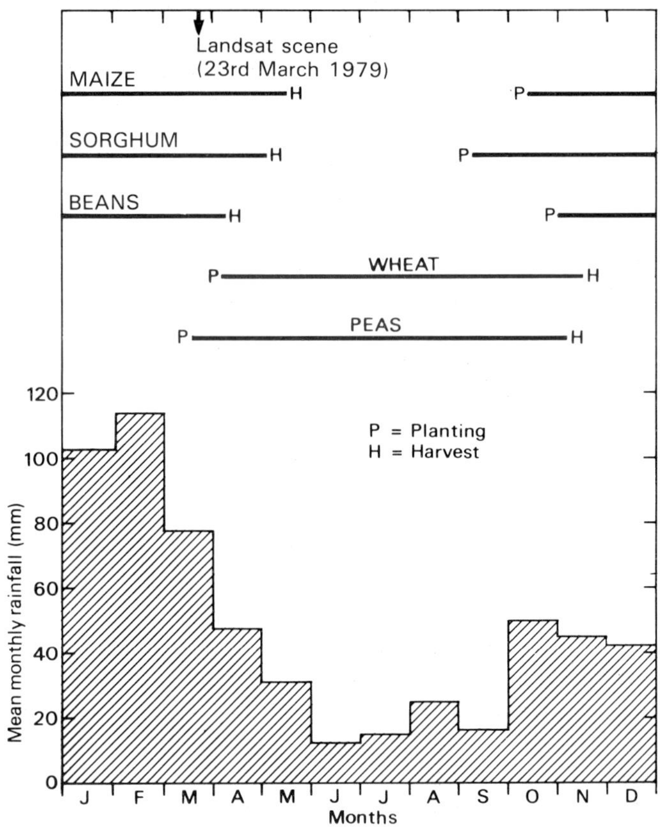


Fig. 2 Crop calendar and mean monthly rainfall for northwestern Lesotho.

Even at a scale of 1:60 000 the boundary line between cultivated and non-cultivated land can still be drawn with ease, but the villages are less easy to identify (Makhanya, 1979).

In view of the simplicity of the land use pattern and the relative ease with which land use mapping can be carried out with medium and small scale photography it was decided to investigate the potential of Landsat imagery, digitally enhanced for maximum contrast, to determine if this could provide additional detail, such as crop type and area of fallow.

## LANDSAT DATA AND INTERPRETATION

The Landsat product used in this study was an MSS computer compatible tape (CCT) of scene 182-080 acquired on 23 March 1979 and supplied by the Eros Data Center. Unfortunately, the scene contained about 10% cloud cover, most of which was situated over the Lesotho study area thus greatly reducing its size. Nevertheless, there was still a considerable area suitable for the purpose of this study.

The original CCT was digitally processed by the United States Geological Survey Image Processing Facility at Flagstaff, Arizona. In the first phase of processing, image restoration techniques were carried out, then the image was enhanced using contrast stretching, band ratioing, and filtering. Paper prints at a scale of 1:500 000 were generated at each stage of the processing.

The contrast stretched colour composite image (bands 4, 5, 7), obtained from the United States Geological Survey at Flagstaff, covered mainly the Republic of South Africa with part of Lesotho to the southeast. The image has optimum contrast and fairly good resolution. It was subjected to a detailed visual examination. Very little sign of land parcelization on the Lesotho side of the frontier could be discerned. In marked contrast, the land parcels in the neighbouring region of the Republic of South Africa were clearly imaged (Plate 26(a)).

An examination of the individual bands that had undergone contrast stretching was equally disappointing; land parcels in Lesotho were not readily identifiable. Band ratioing, which was useful in revealing some land use contrasts on the South African side of the border was less effective on the Lesotho side. Even where land use differences could be identified the land parcels were too small to be effectively mapped.

The marked difference in the detectability of land parcels on the two sides of the border is a direct function of the spatial resolution of the Landsat system. It has already been noted that an average land parcel in Lesotho is only 0.8 ha, and this may be further subdivided into strips of land about $12 \times 100$ m in size. These land parcels are too small to be identified individually by the Landsat MSS which has a spatial resolution of $79\,m \times 79\,m$. In the Republic of South Africa, where farms are producing cash crops rather than subsistence crops, farm sizes range from 100 ha to 5000 ha and individual fields are more readily recognized. In this situation it was found that the Landsat imagery (digitally enhanced colour composites of bands 4, 5, 7) did provide sufficient information to produce land use maps at the scale of 1:500 000 (Plate 26(b)). A sample land use map of the Senekal area of the Republic of South Africa is shown in Fig. 3. The land use types were identified on the Landsat image and transferred onto a 1:250 000 topographic map containing cadastral detail. This allows direct comparison of land parcels and actual land use. The mapping was carried out with a simple detail-transfer instrument, the sketchmaster, which proved very effective for this task.

It is possible that Landsat images acquired during the middle of the summer growing season (December to February) rather than the end (March) would be more useful for rural land use mapping. The problem with images obtained

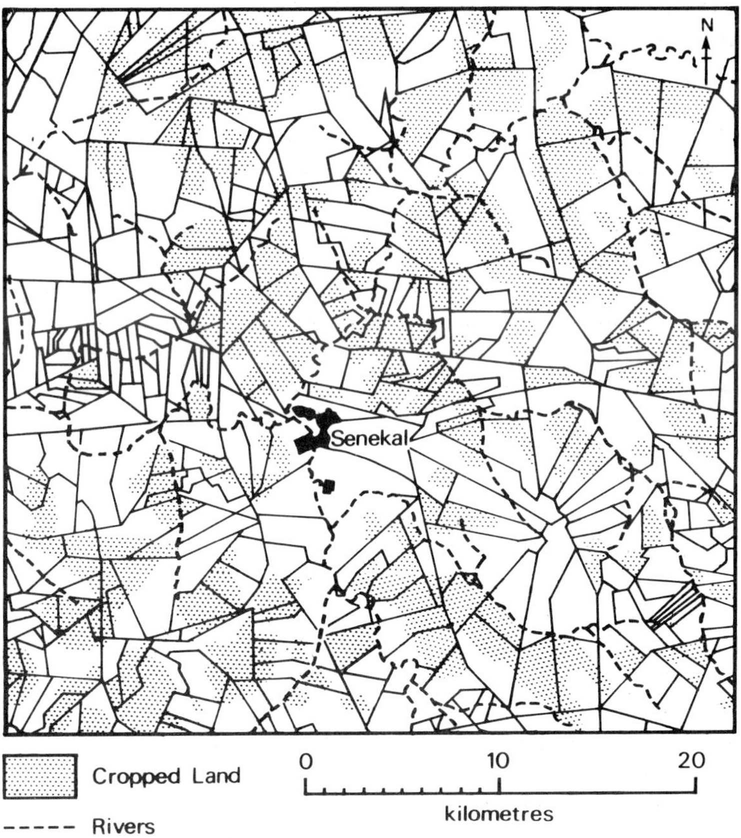


Fig. 3 Land use of the Senekal region, Republic of South Africa mapped from Landsat colour composite.

at the end of the growing season is that harvesting is in progress resulting in numerous mixed pixels which makes discrimination very difficult.

## CONCLUSION

In conclusion it can be stated that although digitally enhanced Landsat images could be used effectively in rural land use mapping in the Republic of South Africa where farm sizes are large and farming is commercialized, and therefore also specialized, they had limited applications in Lesotho. The limiting factor is the traditional subsistence system of rural land use and land management that results in the allocation of very small land parcels to individual homesteads. This problem is not unique to Lesotho. It is a limiting factor in the application of Landsat to land use studies over much of Africa.

Although it was not possible to prepare detailed land use maps of Lesotho from the available Landsat imagery, images acquired during the summer growing season should be more effective for distinguishing cultivated land from non-cultivated land. Such information would be useful in making general crop estimates and identifying those areas that have experienced crop failure as a result of drought, insect damage or blight.

## REFERENCES

Anon., 1937. *Annual Report on Basutoland 1936*. London: HMSO.

Anon., 1959. *The Laws of Lerotholi* (revised). Maseru: Morija.

Anon., 1973. *Government Gazette*. Maseru: Government Printer.

Bawden, M. G. and Carroll, D. M., 1968. *The Land Resources of Lesotho. Land Resources Study No. 3*. Tolworth: Directorate of Overseas Surveys.

Makhanya, E. M., 1979. *The Use of Land Resources for Agriculture in Lesotho*. Maseru: Mazenod.

Remote Sensing and Tropical Land Management
Edited by M. J. Eden and J. T. Parry


# CHAPTER 12

# *Multistage Remote Sensing, Land Systems, and Soil Erosion in Central Tanzania*

HEZEKIEL M. MUSHALA
*Department of Geography, University of Dar es Salaam, Tanzania*

Land is a critical resource base in developing countries and if mismanaged it creates an array of self-perpetuating problems, among which decline in agricultural production is often the most severe. Land mismanagement manifests itself most dramatically in the form of badlands, areas of severe, accelerated soil erosion. Badlands are essentially unproductive, most of the soil nutrients having been lost, and they are difficult to rehabilitate because of their physical configuration. Many countries in the tropics are severely affected, and one of the major problems is how to monitor the rates at which badlands and other less severe forms of soil erosion are spreading, and at the same time devise effective means to halt or retard the process.

In this study, the procedure adopted in using remote sensing techniques to identify and map areas of accelerated soil erosion in central Tanzania are described. The mapping exercise is part of a general attempt to generate a data base for use in the control of accelerated soil erosion (Mushala, 1986). An essential part of the project is the combination of Landsat imagery and aerial photography in a multistage mode to assess the development and spatial extent of accelerated soil erosion relative to land use practices in the Kondoa district of central Tanzania.

## THE KONDOA STUDY AREA

As indicated in Fig. 1, Kondoa is located in central Tanzania. The district is characterized by rolling topography with rocky hills interspaced by broad river valleys. The area is underlain by Pre-Cambrian crystalline rocks with a predominance of felspathic gneisses. Fault lines of the East African central rift zone are marked by two parallel escarpments running in a south-southwest to

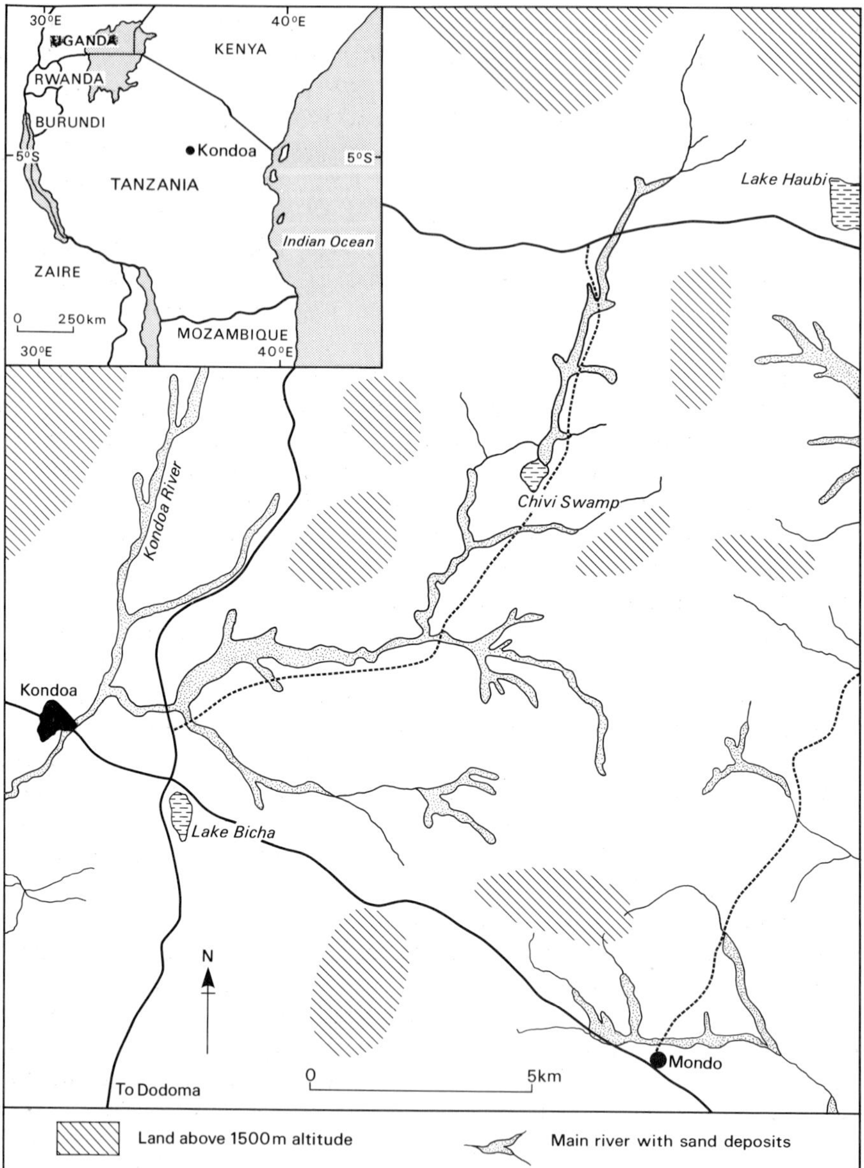


Fig. 1 The Kondoa study area, Tanzania.

north-northeast direction. The general relief of the escarpments is about 300 m. A few inselbergs with elevation in excess of 1500 m are characteristic features. Their height above the local terrain is about 300 m. While contemporary soil erosion is most obvious along footslopes, earth pillars and buried horizons of laterite and quartz pebbles are evidence of earlier erosion surfaces (Christiansson, 1972).

The landscape is dissected by numerous, intermittent streams which form part of the larger Mkuku-Kondoa drainage network. Kondoa district lies within an area of internal drainage with Lake Haubi in the northeast forming the largest basin. Smaller basins include Lake Bicha and the Chivi swamp.

The district is semi-arid, receiving rainfall for only a short period of the year. The average annual rainfall is below 600 mm. Most of the rainfall occurs between December and April, but is concentrated in late December and January. Total precipitation varies from year to year and its reliability is very low. Rainfall commonly occurs as scattered local showers or cloudbursts of short duration. Due to the intensity of falls, the erosion risk is extremely high.

The vegetation consists of dry montane forest, woodland, bush, and wooded grassland. Its distribution reflects the influence of local relief and previous trends in vegetation clearance. During the 1930s and 1940s, many wooded areas were cleared as a result of tsetse control, agricultural expansion, new settlement, and possibly also fuelwood collection. Sites of ritual significance were generally protected, and these still retain the characteristics of the original vegetation cover. Under present conditions, scattered trees and grassland generally occupy the flat terrain, woodland and bush are found on moderate slopes, while the dry montane forest is on more elevated slopes.

The soils of the Kondoa district are coarse-textured and friable, and are vulnerable to erosion by water. Reddish-grey sandy soils are common along pediment slopes and on alluvial fans, but most of the soils are reddish-brown coarse sandy loams and dark yellowish-brown loamy sands. The regional climate is such that relatively slow rates of soil development occur, and the soils themselves are commonly shallow and lacking in organic matter. Also, many of the soils have been truncated as a result of accelerated erosion. As well as sheetwash processes, rill and gully erosion and stream channel erosion are commonly encountered in the study area.

The Irangi people occupy most of the Kondoa district. Their major economic activities are mixed farming and animal grazing. The staple food crops are maize, millet, and sorghum. Oilseeds and vines are the main cash crops.

Land preparation and tillage are achieved by family labour with simple tools, usually a machete and a hand hoe. Only a few ox-drawn ploughs are employed. With the growth of population, the traditional system of shifting cultivation has ceased to be sustainable, and fallow periods in the region are gradually being reduced. A zonation of land use is apparent around individual villages. Small farm plots are located around the central homesteads, and livestock enclosures are in close proximity. The larger *ujamaa* fields, where labour input and production are communal, are located farther away from the settlements. The agricultural cycle in the area is adjusted to climatic conditions, with the growing season extending from the beginning of the rains, in December, into the early part of the dry season. At the start of the rainy period, the agricultural plots are bare, and intense rainstorms tend to erode the soil before a protective crop cover is established. Much of the land is today marginal in character, and cultivation occurs across less than 25% of the area.

## SOIL CHARACTERISTICS AND REMOTE SENSING

It is generally accepted that soil characteristics cannot be interpreted directly from either Landsat imagery or aerial photography, but have to be inferred from other directly observable criteria. The use of satellite imagery in soil survey is limited by the relatively poor resolution and the weak stereoscopic capability. Even obvious soil degradational features, such as gullies, cannot be detected on unenhanced Landsat imagery (Riquier, 1978). For this reason, satellite imagery must be used with aerial photographs, in a complementary manner, in order to obtain useful results.

In spite of its limitations, satellite imagery has certain unique features which are particularly useful for developing countries. A major advantage is the large-area coverage of the Landsat system, which provides a good synoptic view for general assessment of the physical environment, especially geological structures, drainage patterns and vegetation characteristics. Another feature is the repetitive coverage which is useful in monitoring rates of change of particular features. In addition, satellite imagery can be an effective tool in the planning of the fieldwork programme.

The surrogate features used by the remote sensing analyst in the assessment of soil conditions are land use and land cover. The land use of an area refers to the type of human activity that is dominant—agriculture, forestry, etc.—whereas land cover refers to the biomass veneer that obscures the land surface from the direct view of the aircraft or spacecraft. The tonal and textural variations in the land use and land cover provide the analyst with the essential clues for the interpretation of soil conditions. Armed with these, the skilled interpreter can generally delimit soil associations and soil series with a reasonable degree of accuracy (Evans *et al.*, 1976).

The acquisition of basic data on land and soil conditions using medium-scale to very small-scale imagery is generally undertaken using a land systems approach as the basic framework. The land systems are identified and defined by a combination of imagery interpretation and field survey. The main advantages of this approach are that the constituent land units are easily recognized and mapped, they provide a convenient storage system for a range of terrain information, and it is possible to extend the limit of the survey by identifying analagous conditions in adjacent or more distant areas (Parry and Beswick, 1973; Webster and Beckett, 1970).

Land classification requires familiarity with several of the physical and biological sciences, because land systems represent an integrated approach to the environment involving the assessment of individual factors and also their interrelationships (Fezer, 1971).

The land systems approach introduces a useful hierarchical structure to land classification which permits the application of the system at a range of different levels of generalization. A land system is defined as a recurrent pattern of genetically linked land units. For practical purposes a land system is the highest level of classification used in resource surveys. A land unit is the next lower level in the classification hierarchy, and comprises an area of terrain that is essentially homogeneous in terms of either morphology, lithology, or amplitude of relief. At the next level are land components, which are areas exhibiting a single, dominant land cover type. At the lowest level in the hierarchy are land elements. These are the simplest constituents of the landscape with uniform lithology, morphology, soil, and vegetation (Cook, 1981).

The land systems methodology is widely used in reconnaissance surveys of all types, and it provides a useful framework for the assessment of soil erosion—either the actual incidence and severity of soil loss or the potential for soil degradation. The land systems approach has been successfully applied in many developmental projects in the tropics where the assessment of soil conditions was one of the major requirements (Cook, 1981; Hill, 1978; King, 1978; Young and Goldsmith, 1977). It should be noted that in each of these studies the analysis went beyond site description. Imagery and aerial photographic interpretation was supplemented by field checks for soil–vegetation–land use linkages and by laboratory analyses of the physical and chemical properties of the soils.

## AERIAL PHOTOGRAPHY AND LANDSAT IMAGERY

The aerial photographic coverage of the Kondoa study area was obtained in September 1977 at the end of the dry season. The nominal scale of the photography was 1:15 000 and the normal 60% overlap provided stereoscopic coverage of the whole area.

The selection of suitable Landsat images was based on the quality of the image, the amount of cloud cover, and the date. Details of the Landsat images selected are given in Table 1. It was essential to have multi-date imagery because both the land use and the land cover detail exhibit major differences between wet and dry seasons.

Table 1. Landsat imagery used in the study.

| Scene ID | Date | Season | Cloud cover % | Image quality (scale 1→5) | Selected |
|---|---|---|---|---|---|
| E-1192-07185 | 31 Jan 1973 | Wet | 30 | 2 | Yes |
| E-1426-07171 | 22 Sep 1973 | Dry | 5 | 4 | No |
| E-30229-07081 | 20 Oct 1978 | Dry | 30 | 2 | Yes |
| E-30301-07075 | 31 Dec 1978 | Wet | 15 | 2 | Yes |
| E-30481-07064 | 29 Jun 1979 | Dry | <10 | 1 | Yes |
| E-30840-06532 | 22 Jun 1980 | Dry | <10 | 5 | No |

The multispectral scanner in Landsats 1 and 2 operated in four spectral bands, namely: band 4—0.5–0.6 $\mu$m, band 5—0.6–0.7 $\mu$m, band 6—0.7–0.8 $\mu$m, band 7—0.8–1.1 $\mu$m. Each band contributes specific details in an interpretation. For example, drainage lines and water bodies are best identified on band 7 because at these wavelengths water absorbs nearly all the incident radiation, giving a very dark signature. Vegetation is most readily identified on band 5, where it is recorded in various dark tones. The contrast between vegetated and unvegetated areas is particularly dramatic in this band, which is thus useful for mapping agricultural land use and distinguishing between crops at different stages of growth. For the examination of lithology and structure, band 5 is found to be the most useful (Kondratyev *et al.*, 1973; Colvocoresses, 1977).

The visual interpretation of Landsat imagery was undertaken using single band transparencies at 235 × 235 mm and 70 × 70 mm. The latter were examined with an additive colour viewer, which allows the generation of false colour composites. The technique can sharpen the spatial resolution of some image characteristics by varying the intensity of illumination of particular bands in the composite; it also allows the overlaying of images obtained at different dates in order to provide a multi-temporal comparison for change detection.

Stereoscopic analysis of Landsat imagery was only possible with two scenes for which there was sidelap. Vertical exaggeration is optimal in tropical latitudes, but the area of overlap is very restricted and so little information was derived from this procedure.

The interpretation of the land system boundaries and the characterization of each land system were guided by the *a priori* selection of certain criteria, including the vegetation and land use patterns, the drainage characteristics and the topography. The vegetation and land use details were found to be strongly influenced by the season, as also were the drainage patterns. Drainage details provided useful indications of soil conditions and susceptibility to erosion. Topography and slope were interpreted mainly from the aerial photographs, and these two elements were found to exert the dominant control on the extent and severity of soil erosion.

## SECONDARY SOURCES OF INFORMATION AND FIELD SURVEY

All the available published materials on the Kondoa Highlands were consulted in the course of the study. These included the meteorological data, census data, government documentation on soil and water conservation, and the soil and vegetation classification schemes elaborated for the area in previous studies (Banyikwa *et al.*, 1978).

Field surveys were carried out with several objectives in mind; firstly, to verify land system boundaries at various levels in the hierarchy and to obtain on-site

Table 2. The Kondoa land system

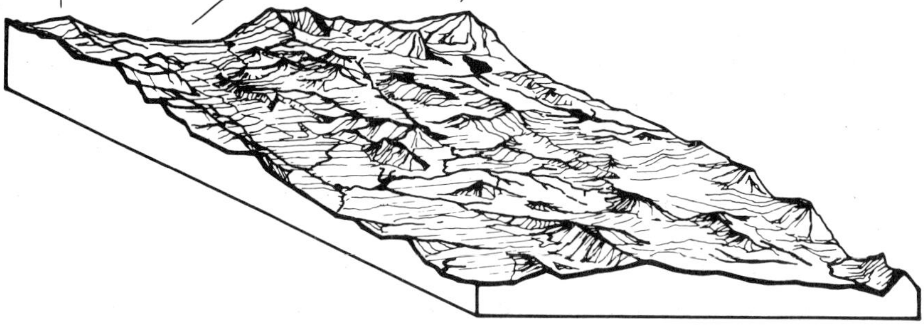


*Kondoa escarpment*

| | |
|---|---|
| Land unit: | Kondoa escarpment |
| Parent material: | Usagaran metasedimentary series (Pre-Cambrian) |
| Average elevation: | 1500–1700 m |
| Morphology: | East-facing concave slopes of 20–25° |
| Vegetation: | Grassland and shrub cover (< 40% canopy closure) |
| Land use: | Grazing |
| Land components: | Fault line scarp, footslopes, alluvial fans (termitaria form conspicuous land elements) |
| Soil erosion: | Severe to moderate sheetwash, local gully erosion |

*Kondoa river floodplain*

| | |
|---|---|
| Land unit: | Kondoa river floodplain |
| Parent material: | Neogene deposits |
| Average elevation: | 1350–1400 m |
| Morphology: | Level |
| Vegetation: | Bush and shrub cover (50% canopy closure) |
| Land Use: | Cereal crops and horticulture |
| Land components: | Alluvial fans, dry washes, stream terraces |
| Soil erosion: | Stream channel erosion |

*Inselberg*

| | |
|---|---|
| Land unit: | Inselberg |
| Parent material: | Usagaran gneiss (Pre-Cambrian) |
| Average elevation: | 1600-1650 m |
| Morphology: | Concave slopes of 30–40° |
| Vegetation: | Open woodland (20% canopy closure) |
| Land use: | Mixed horticulture, cereal crops |
| Land components: | Pediments, footslopes, alluvial fans |
| Soil erosion: | Severe sheetwash and gully erosion |

details of terrain conditions; secondly, to obtain information on agricultural practices and crop types; thirdly, to obtain soil samples and profile descriptions from the various land units; and fourthly, to identify suitable sites for intensive study at later stages in the investigation.

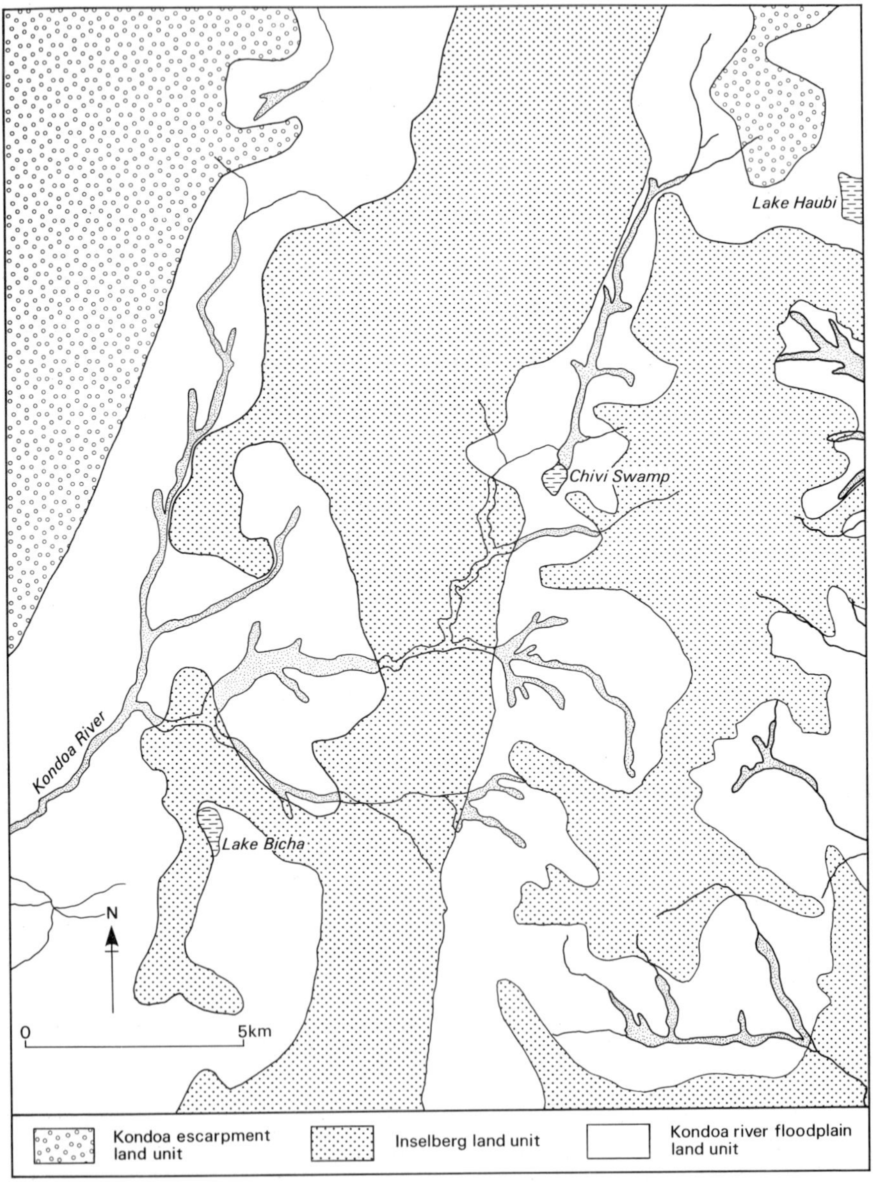


Fig. 2 Land units of the Kondoa land system, Tanzania.

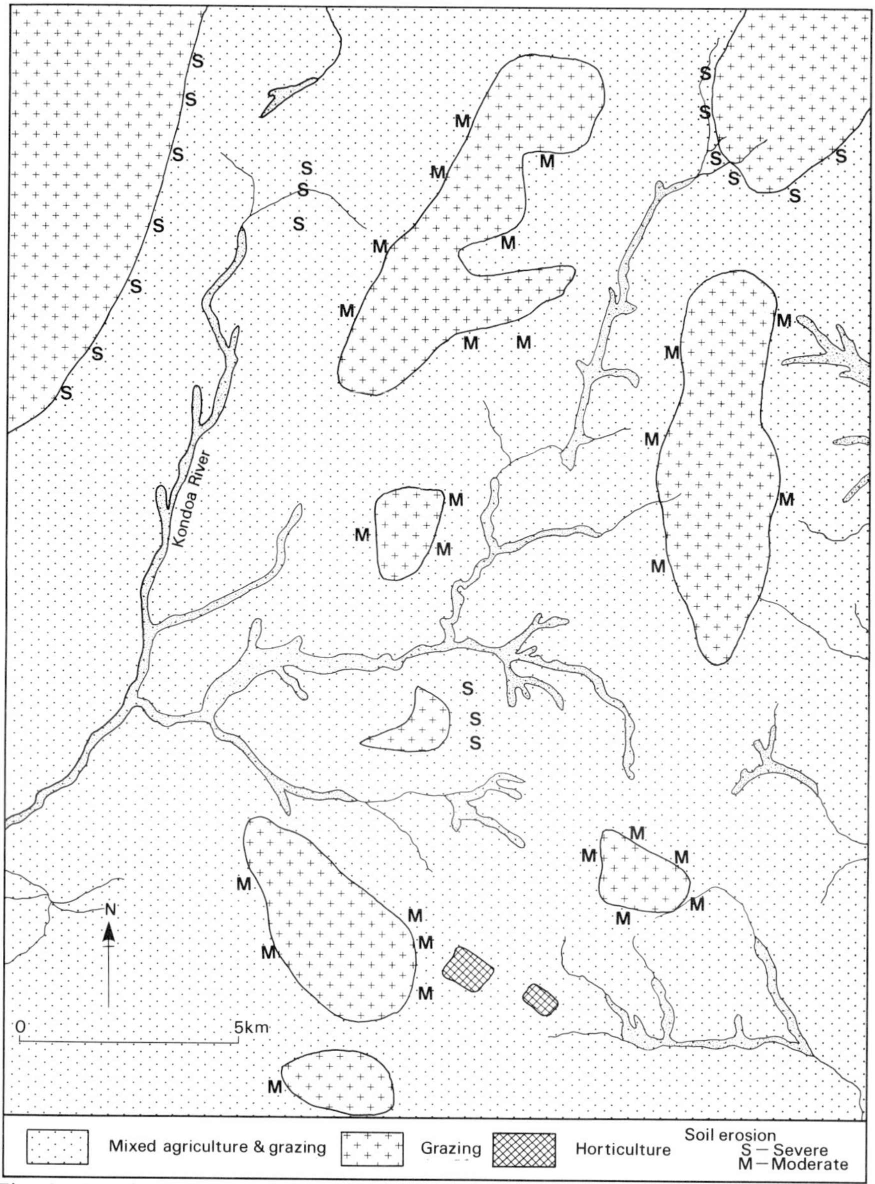


Fig. 3 Land use/land cover map of the Kondoa study area, Tanzania.

The field programme was limited by a variety of problems, many of which are familiar to researchers in the tropics: the inaccessibility of areas of interest because of their distance from the road network, the limited field-time available because of the high cost of rental vehicles, and the physical characteristics of the soils in the dry season, which made augering and pit digging both exhausting and time-consuming.

## SUMMARY REVIEW OF THE RESULTS

The major product of the aerial photograph and imagery interpretation was a land systems map of the Kondoa study area (Fig. 2). The details of the Kondoa land system are listed in Table 2, and the characteristic interrelationship of the constituent land units is illustrated in the block diagram. In addition, a land use/land cover map was prepared (Fig. 3), according to the classification scheme outlined by Anderson (1971) for use with aerial imagery.

Land systems were easily identified and delineated on both the Landsat images and the aerial photographs. Land units could be recognized on the Landsat imagery, but the small scale prevented their delineation. On the aerial photographs, at the larger scale, the units could be mapped without difficulty.

Land cover and land use were identified on the Landsat colour composites by careful assessment of the different shades of magenta, and on the aerial photographs by examination of the details of tone and texture: the darker the tone and the coarser the texture, the greater the vegetation or crop density.

The most important outcome of the study was the identification of the areas directly affected by soil erosion and the categorization of the effects—severe, moderate or weak. An indication of the pattern of erosion impact is given in Fig. 3. As shown, severe erosion is commonly located along footslopes and is associated with wooded grassland and scrub where cultivation has ceased and the land is used for grazing. The commonest form of erosion is gullying. The gullies range from small features, less than 0.5 m in depth, to large incisions that reach to 6 m or more. Moderate erosion is common along lower pediments, alluvial fans, and floodplains. Such land supports annual crops, with fallowing for a few months each year. Sheetwash erosion is dominant, with slight rill development occurring during the periods of cultivation. Stream channel erosion is usually moderate, except during the heavy rains when stream banks may collapse.

The remainder of the area is locally affected by weak erosion, although an increasing erosion severity may develop in the future, if current levels of land pressure persist. By extrapolation, the erosion hazard can be determined for areas as yet unafflicted, depending upon the land unit and the land cover and land use. At this stage it is not possible to identify the specific factors that control the erosion process. It is assumed that soil structure has a direct effect on the erodibility of soils, but this will require verification and field testing.

In conclusion, the following points should be emphasized. Firstly, although remote sensing is not the optimum source of data on soil conditions, it can assist in generating a useful base for mapping and for reconnaissance classification. Secondly, multi-scale imagery provides considerably more detail than a single scale coverage and is better adapted to a hierarchical classification scheme such as the land systems approach. Thirdly, the usefulness of Landsat imagery and aerial photography is controlled not only by the scale and resolution, but also by the season and the conditions under which the image is obtained. Fourthly, remotely sensed data are generally inadequate for making recommendations on alternative land uses and land management practices. Management decisions require considerably more detailed information than is available through image interpretation. Fifthly, the land systems approach has only partial value for soil erosion assessment. It provides a framework for the delimitation of affected areas and a ranking of the erosion severity. It also has some predictive value, but it does not allow the identification of the specific factors responsible for the erosion itself, nor does it contribute to the selection of appropriate soil conservation practices or land use alternatives.

## ACKNOWLEDGEMENT

Fieldwork for the study was supported by a grant from the Research and Publications Committee of the University of Dar es Salaam, for which the author is grateful.

## REFERENCES

Anderson, J. R., 1971. Land use classification schemes used in selected recent geographic applications. *Photogramm. Engng*, **37**, 379–387.

Banyikwa, F. F., Kikula, I. S., Mohamed, S. and Rugumamu, S., 1978. *Studies on Soil Erosion Control and Land Reclamation in Kondoa District.* Unpublished Paper, University of Dar es Salaam.

Christiansson, C., 1972. Notes on morphology and soil erosion in Kondoa and Singida Districts, Central Tanzania. *Geogr. Annlr* **54**, 9A, 319–324.

Colvocoresses, A. P., 1977. Proposed parameters for an operational Landsat. *Photogramm. Engng Remote Sensing*, **43**, 1139–1145.

Cook, A., 1981. Integrated resource survey as an aid to soil survey in a tropical flood plain environment. In Townshend, J. R. G. (Ed.), *Terrain Analysis and Remote Sensing*. London: Allen and Unwin, pp. 184–203.

Evans, R., Head, J. and Didkzwager, M., 1976. Air photo-tones and soil properties: implications for interpreting satellite imagery. *Remote Sensing Environ.*, **4**, 265–280.

Fezer, F., 1971. Photo interpretation applied to geomorphology—a review. *Photogrammetria*, **27**, 7–50.

Hill, I. D. (ed.), 1978. *Land Resources of Central Nigeria. Agricultural Development Possibilities*, vols 1–7. Surbiton: Land Resources Development Centre, Ministry of Overseas Development.

King, R. B., 1978. *Land Resources of the Rukwa Region: a Reconnaissance Assessment*, vol. 2, *Land Systems*. Dar es Salaam: BRALUP, University of Dar es Salaam.

Kondratyev, K. ya, Vassilyev, O. B., Grigoryev, A. A. and Ivanian, G. A., 1973. An analysis of the ERTS-1. *Remote Sensing Environ.*, **2**, 273–283.

Mushala, H. M., 1986. The influence of agricultural practices on soil erosion processes in Central Tanzania. *Taamuli*, Dar es Salaam (in press).

Parry, J. T. and Beswick, J. A., 1973. The application of two morphometric terrain classification systems using air-photo interpretation systems. *Photogrammetria*, **29**, 153–186.

Riquier, J., 1978. *A Methodology for Assessing Soil Degradation in FAO/UNEP*. Rept. on FAO/UNEP Expert Consultation on Methodology for Assessing Soil Degradation, World Assessment of Soil Degradation Project, No. 1106-75-5, 25–59, Rome.

Webster, R. and Beckett, P. T., 1970. Terrain classification and evaluation using air photography: a review of recent work at Oxford. *Photogrammetria*, **26**, 51–75.

Young, A. and Goldsmith, P. F., 1977. Soil survey and land evaluation in developing countries: a case study of Malawi. *Geogrl J.*, **153**, 407–438.

# PART IV

# Monitoring Agricultural Resources and Land Cover

Remote Sensing and Tropical Land Management
Edited by M. J. Eden and J. T. Parry


# CHAPTER 13

# *Change Detection in the Tropical Forest Environment of Northeastern India using Landsat*

ASHBINDU SINGH
*Indian Forest Service, Government of Manipur, Imphal, India*

India is the seventh largest country in the world, covering an area of 3 288 000 km$^2$. The population of India constitutes nearly 15% of the world's population, while its territory accounts for only 2% of the world's land area. Forestry is a major land use in India, being second in importance only to agriculture. The designated forest area in the country amounts to 750 000 km$^2$, which is 22.8% of the land area (Anon., 1976). Sixteen major forest types have been identified within the country (Champion and Seth, 1968), and the forest flora and fauna are rich and varied in character.

One of the most serious environmental problems in India is the accelerating degradation and depletion of its forest cover. Undisturbed forests are virtually non-existent. Changes have occurred as a result of various activities including shifting cultivation, permanent cultivation, logging, and grazing. In many areas, clearance of the forest is creating serious socio-economic problems, particularly involving shortages of timber, fuelwood, and fodder. The impact of clearance is also evident in the more frequent occurrence of landslides in the Himalayas, of flooding in the Gangetic plain, and in reservoir siltation (Anon., 1976).

It should be emphasized that the official designation of 750 000 km$^2$ of land as forest indicates no more than that this area is under the control of government forest departments. There is no certainty that the land itself is actually forest. Indeed, it has been estimated that during the period 1947–76 some 41 500 km$^2$ of forest have been cleared for agriculture and other purposes (Anon., 1978). These figures, which relate purely to planned and authorized encroachment on forest land, represent an average annual loss of some 1500 km$^2$ of forest.

Meanwhile, no systematic investigation has been undertaken to determine the real status of the forest cover of India.

In the present study, the potential of satellite imagery for investigating vegetation cover is examined. In particular, an attempt is made to assess the value of Landsat data for determining forest cover changes in northeastern India. The synoptic view and temporal coverage of the Landsat system make it invaluable for monitoring deforestation over large areas.

## NORTHEASTERN INDIA

The northeastern region of India covers 6.7% of the national territory, and contains 3.9% of the population. The region currently contains the most valuable forest reserves in the country. The forest displays luxuriant growth, and contains major species like *Shorea*, *Dipterocarpus*, *Bauhinia*, *Terminalia*, and *Bombax*, together with tropical pines, oaks, and various types of bamboo. The region appears to have the richest reserve of orchids and ferns in India. Among the regional fauna are tigers, panthers, and other smaller carnivores. The region also contains elephants, deer, antelopes, and langurs, and has the world's major reserve of the one-horned rhinoceros.

One of the primary causes of deforestation in the tropical world is shifting cultivation (Hadley and Lanly, 1983), and northeastern India is no exception in this respect. The precise area under shifting cultivation is not known, because of the lack of resource surveys, an absence of official land records, and the fact that clearance for cultivation is continually occurring. An estimate from FAO (1981) suggests that an area of approximately 27 000 km$^2$ of northeastern India is currently affected by shifting cultivation, of which about one-sixth may be cultivated at any one time.

## SHIFTING CULTIVATION IN NORTHEASTERN INDIA

Shifting cultivation is a system of cyclic cultivation practised in tropical and subtropical countries. A patch of land is first cleared of trees and other plants, the debris is burnt, and crops are then established. Shifting cultivation, known as *jhuming*, is the primary occupation of almost all the tribes in northeastern India. Although the system is relatively uniform across the area, local variations occur in respect of crop composition, the period of cultivation, and the length of fallowing. Of late, some general changes have occurred within the region. The cultivation interval in individual plots has traditionally been 15–20 years or more, but, as a result of increasing population, this interval has now shortened to 5–10 years in many areas. A vicious circle of declining fertility and crop yield has developed as a result of the abbreviation of the fallow period, necessitated by the growing food requirements of the increasing population.

Fig. 1 Forest cover in Manipur State, northeastern India.

Fig. 2 Degraded forest cover in Manipur State, northeastern India.

Fig. 3 Cleared forest land in Manipur State, northeastern India.

It is widely accepted that shifting cultivation can have a modifying or destructive impact on the tropical forest (Denevan, 1975). As cultivation continues, closed forests are often replaced by more open, degraded forests or by bamboos or grassland (Figs. 1–3). Destruction of the forest by shifting cultivators, especially in high rainfall areas, frequently results in heavy soil erosion.

## DETECTION OF LAND COVER CHANGES

In order to develop suitable land management strategies for northeastern India, it is essential to inventory land cover types and monitor the changes caused by shifting cultivation. Existing methods of estimating the area affected by shifting cultivation are based on physical surveys and socio-economic approximations, but serious discrepancies arise (Sharmah, 1979). According to the Forest Survey of India, in 1975, Manipur State with a total area of 22 300 $km^2$ contained 1800 $km^2$ under shifting cultivation. Individual fields are normally cropped for a period of approximately two years, and so the extent of annual clearance was assumed to be 900 $km^2$. According to census figures, there were about 300 000 tribal people in the State who were dependent on shifting cultivation. Demographic data indicate that a tribal family averages five members and that each family needs to clear one hectare annually. Thus, the total annual clearance was estimated at only 600 $km^2$, some 300 $km^2$ less than the reported clearance.

More precise information on the extent of forest clearance can be obtained by the use of aerial photographs but the cost of acquiring the necessary repetitive cover is high and much time is spent in the process of interpretation. In any case, aerial photographs in India generally have a 'restricted' status, and access to them is difficult. Against this background, the potential of other forms of remote sensing needs to be explored. In particular, data from the Landsat multispectral scanner (MSS), with its favourable synoptic view and repeat coverage, appear to offer possibilities for the assessment and mapping of forest degradation (Aldrich, 1975). There are three aspects to the problem. The first is that of detecting changes in land cover; the second lies in the identification of the nature of the changes; and the third involves estimating the areal extent of the changes. Landsat has the potential to contribute to all three, but at different levels of accuracy and completeness (Thomson *et al.*, 1979). The present study is primarily addressed to the first aspect, namely, the detection of land cover changes. Before discussing the methodology of change detection, the nature of Landsat data and the associated processing will be reviewed.

## LANDSAT IMAGERY AND PROCESSING

Landsat data are available in photographic form, either as black and white images or as false colour composites at scales of 1:1 000 000, 1:500 000 and 1:250 000. Landsat data in digital form are available as computer compatible tapes (CCTs). The data, whether in visual or digital form, provide a synoptic view of very large areas under uniform illumination conditions. The repetitive coverage provided by the system offers the possibility of obtaining data from large areas at different dates or seasons. Subject to cloud cover limitations, the selection of imagery from optimal periods for land cover analysis can be achieved.

Landsat data can be interpreted from colour composites (usually MSS bands 4, 5, and 7), or from single-band black and white images. Landsat imagery is essentially distortion-free, so that the process of transferring the boundaries of thematic classes from an image to a base map is relatively straightforward. Considerable information can be derived from Landsat data in photographic format. However, the major advantage of the Landsat multispectral scanner is that it generates digital numbers on magnetic tape which are amenable to computer-assisted analysis (Lillesand and Kiefer, 1979). Consequently, since the launch of the Landsat series of satellites in 1972, there has been a progressive shift from conventional photo-interpretative procedures to computer-aided analysis. This has been stimulated by parallel developments in digital image processing systems, which once appeared to be the rich man's pastime (Townshend and Justice, 1981). In general, digital image processing systems require various input/output devices together with system consoles for man-machine interaction. Each image processing task usually involves relatively simple, but highly repetitive operations on vast amounts of image data (Ince, 1983).

Initially, digital analysis was undertaken using large, mainframe computers and operated solely in batch mode. The second generation of data processing systems for remote sensing was also designed for large, mainframe computers, but took advantage of interactive demand mode processing. The third generation of systems was developed for minicomputers and specifically designed to speed up certain algorithms, with a consequent improvement in the interactive capabilities of the systems. In general, mainframe systems and dedicated image processing systems are prohibitively expensive for the majority of remote sensing centres in developing countries.

In recent years, as a result of the revolution in microchip technology, computers and their peripherals have become much cheaper, and microprocessor-based systems are being developed that allow interactive analysis of digital data at much lower cost. With interest increasing in the use of microcomputers for image processing, software systems are now becoming commercially available, e.g. APPLEPIPS Apple II Personal Image Processing System, developed by the Telesys Group, and RIPS Remote Image Processing System, designed by the Spectral Data Corporation of New York. Microcomputer-based systems of this kind are most suitable for applications in developing countries (Harrison and Singh, undated), and such a system has been employed in the present investigation.

## CHANGE DETECTION TECHNIQUES

Various procedures have been proposed and used to detect changes in land use or land cover from Landsat multispectral data. The fundamental assumption is that any changes in land cover will result in changes in reflectance values which are sufficiently large to register despite other variables. Both photographic and digital methods can be employed for change detection. Photographic methods are identified as follows (Crapper and Hynson, 1983):

1. A side-by-side comparison of two photographic images can be carried out with annotation of the changes that are detected.
2. Transparencies of two images can be overlaid, registered, and viewed over a light table, with similar marking of areas where changes have occurred.
3. One negative and one positive transparency can be overlaid. Areas in which no change has occurred will appear as medium density tones and areas in which change has occurred will appear either darker or lighter toned.

### Digital change detection techniques

Digital change detection was employed in the present study, and the principal techniques are outlined below. All these procedures require the precise spatial registration of Landsat images from different dates. A residual image is produced by various procedures, and then a threshold value, determined empirically or statistically, is applied in order to define areas of change. Other available techniques involve the independent classification and subsequent comparison of images from different dates.

In practice, these digital techniques have rarely been applied in areas of tropical forest, while few comparative evaluations of their effectiveness have been made for any environment (Weismiller *et al.*, 1977; Howarth and Wickware, 1981; Nelson, 1983). However, it is apparent that the method of change detection employed can have a significant effect on estimates of the amount of change (Colwell and Weber, 1981). In the present study, the potential of these techniques for the analysis of digital Landsat MSS data is examined for forest areas of northeastern India. The specific aim is to compare the relative effectiveness of the techniques in detecting land cover changes induced by shifting cultivation.

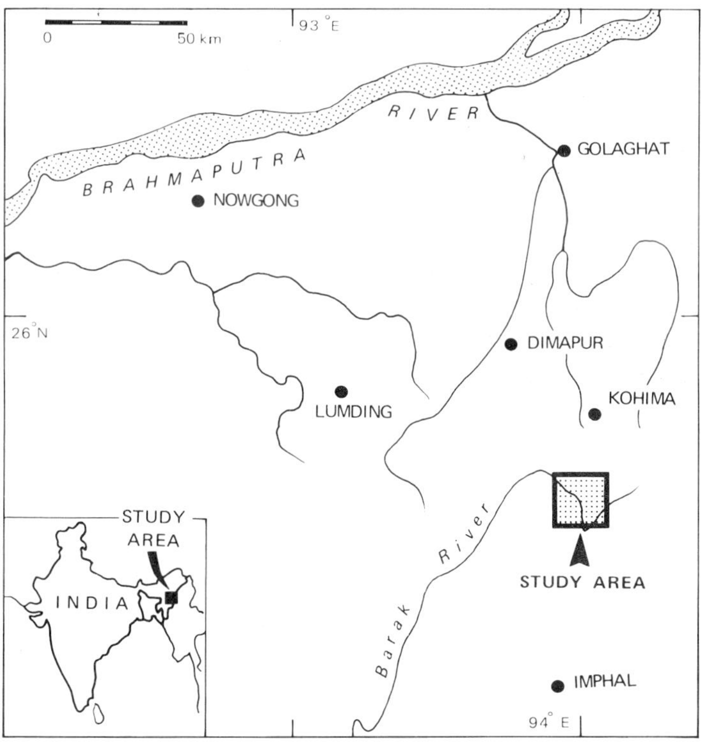


Fig. 4 Study area, Manipur State, northeastern India.

## THE STUDY AREA

The study area is located in the northern district of Manipur State in northeastern India (Fig. 4). The major considerations in selecting the area were as follows:

1. it was undergoing rapid change as a result of shifting cultivation;
2. forest maps from earlier dates were available;
3. the author was already familiar with the area;
4. it was relatively accessible for fieldwork.

The area selected for investigation represents 256 × 256 pixels on the Landsat imagery, and covers approximately 280 km$^2$ on the ground. It is hilly, with moderate to steep slopes, and ranges in altitude from 800 to 1700 m above sea level. Rainfall is within the range 1200–2000 mm per annum. The forest cover along valleys and on moderate slopes is currently being cleared and cultivated by the tribal people. The area supports mainly tropical, moist deciduous forest, although at higher altitudes the latter is replaced by tropical to subtropical evergreen forest. Among the dominant forest trees are *Quercus* spp., *Castanopsis* spp., *Schima wallichi*, *Terminalia myriocarpa*, *Michelia champaca*, and *Bombax ceiba*.

## DATA SOURCES AND METHODOLOGY

Two cloud-free Landsat images of the study area were selected for digital analysis:

| Landsat | 1 | 2 |
|---|---|---|
| Date | 26 Jan 1974 | 6 Dec 1981 |
| Path/row no. | 145/042 | 145/042 |
| Sun elevation | 34° | 32° |
| Sun azimuth | 140° | 145° |

In addition, Landsat 3 imagery of the study area for 31 January 1980 was available for visual examination (Fig. 5).

Image processing and analysis were undertaken with a microcomputer-based, interactive image processing system in the Department of Geography, University of Reading, England, using locally developed software. A description of the hardware configuration and software system is available in Harrison and Singh (in press). Initial digital processing of the 1974 and 1981 images was undertaken to remove or reduce radiometric and geometric distortions. In order to make the radiometric correction, the data were rescaled to produce a consistent 8-bit scale for all four bands. Destriping, correction

Fig. 5 Landsat MSS band 5 subscene (ID no. 83069-03280, 31 January 1980) of northern Manipur State, northeastern India. The light tone patches in the dark tone forest cover are areas of recent clearance by shifting cultivators (*Reproduced by permission of NOAA.*)

for 'bit slips', and haze correction were carried out, and digital values were divided by the cosine of the sun elevation angle in order to normalize the data for an incidence angle of zero degrees. Geometric registration was undertaken so that it was possible to match the two images taken at different times. This may be performed by image-to-image or image-to-map registration (Hardy, 1978). In the present study, quadratic trend surface fitting and nearest neighbour resampling algorithms were used for image-to-image registration. The mean square errors in approximation for pixels and lines were 0.09 and 0.08, respectively.

In order to test the accuracy of change detection techniques on the selected Landsat subscene, reference ground data were required. General information on the area was available from forest-type maps (1:50 000 scale) prepared by the Forest Survey of India, and these were of assistance during fieldwork undertaken by the author in 1983. Land cover data were collected at sample points located on the forest-type maps by means of random spatial coordinates within the following land cover types: (1) forest, (2) cultivated forest land, and (3) regenerating forest. A five-hectare plot, representing a $3 \times 3$ pixel unit on the imagery, was selected as the nominal sampling point. At each point, land cover data for a representative one-hectare plot were then collected in the field, together with information on slope, aspect, and the density of forest cover. A total of 150 point samples was obtained, sufficient to ensure that cultivated areas were adequately represented and that inaccessible sites could be replaced by accessible ones. The point locations were transferred from the maps to photographic enlargements of the Landsat imagery, and the reference data collected at these sites were used to assess the relative accuracy of the digital change detection techniques.

As indicated above, three land cover classes were identified in the field, namely, forest, cultivated forest land, and regenerating forest. In the context of change detection over the time interval 1974 to 1981, only changes of the following kind were assumed to have occurred:

$$\text{Forest} \rightarrow \text{Shifting cultivation} \longleftrightarrow \text{Regeneration}$$

On this basis, forest cover at the time of field survey (1983) was designated for analytical purposes as a ‘no change’ category, whereas areas of clearance and regenerating forest were combined to form a ‘change’ category. Of the 150 sample points used to provide reference data for the digital analysis, 115 were classified in the ‘no change’ category and 35 in the ‘change’ category.

## Change detection analysis

The digital change detection techniques described below were employed to analyse the Landsat data for the 150 sample sites in the study area. The procedures and the results obtained are outlined below. Additional details are available in Singh (1984).

### *Image differencing*

In this method, geometrically registered images acquired at different times are subtracted to produce a residual or *difference image* which represents the change between the two dates. The resulting values for each pixel show positive or negative shifts or no change. Pixels of no radiance change are distributed around the mean, while pixels of radiance change are distributed in the tails of the distribution (Stauffer and McKinney, 1979). Single-band images or combinations

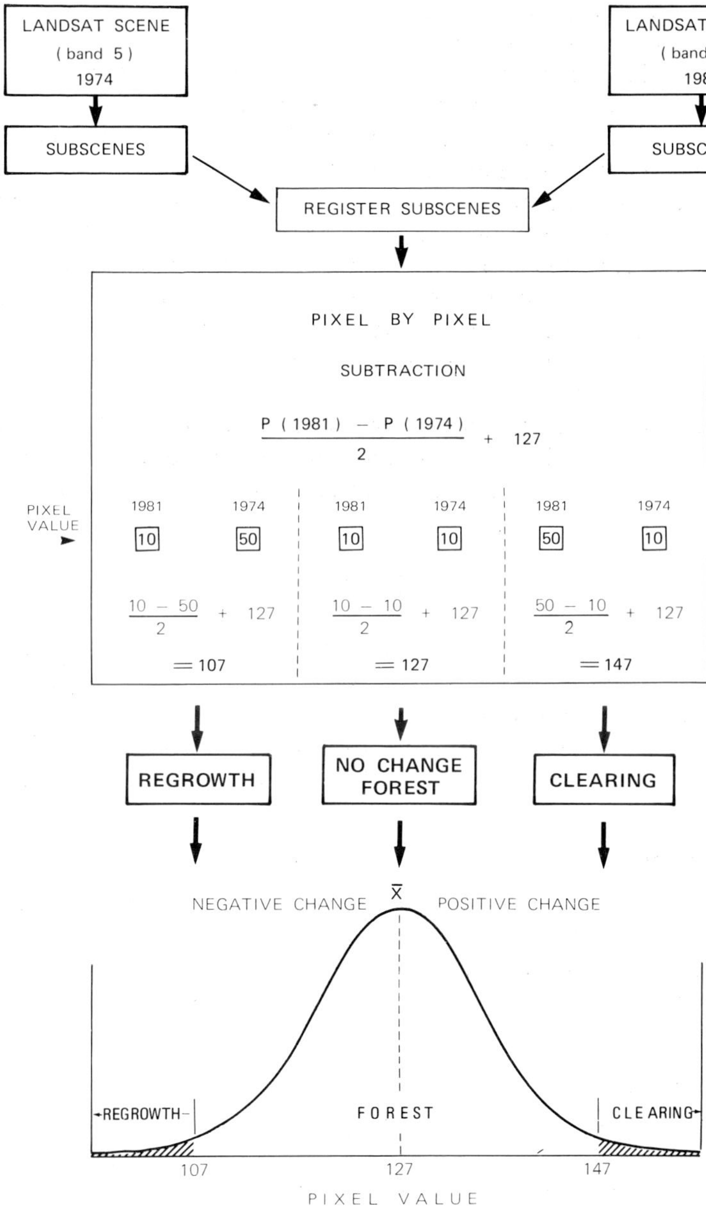


Fig. 6 Forest cover change detection using image differencing techniques.

of different bands can be employed in this technique. The procedure has been used for various topics with differing degrees of success (Angelici *et al.*, 1977; Weismiller *et al.*, 1977; Miller *et al.*, 1978; Ingram *et al.*, 1981).

The method employed in carrying out the differencing is given by the following equation and presented schematically in Fig. 6:

$$DX_{ij}^{k} = \frac{X_{ij}^{k}(t_2) - X_{ij}^{k}(t_1)}{2} + 127$$

where:

$X_{ij}^{k}$ = pixel value for band $k$
$i$ = line
$j$ = column
$t_1$ = first date
$t_2$ = second date
127 = a constant used in order to generate output values between 0 and 254.

Two *difference images* were created for bands 5 and 7, respectively. Data transformations were confined to these bands because they are considered to be the most useful for discriminating forest canopy alterations (Nelson, 1983).

*Ratio image differencing*

A difference/sum ratio image was calculated for each date, as follows:

$$\left(\frac{\text{MSS 7} - \text{MSS 5}}{\text{MSS 7} + \text{MSS 5}} + 1\right) \times 127$$

This index has been found to be strongly correlated with the amount of green biomass (Tucker, 1979), and is used here as a basis for comparing green biomass at different times. Image differencing procedures were again used to detect changes between the two ratioed images.

*Image regression*

In this method it is assumed that the pixel values at time $t_1$, $X_{ij}^{k}(t_1)$, are related to those at time $t_2$, $X_{ij}^{k}(t_2)$, by a linear function so that one image can be regressed against the other using a least squares method. An image differencing method (Ingram *et al.*, 1981) is then applied that subtracts the predicted values $\hat{X}_{ij}^{k}(t_2)$, based on the regression line, from the actual values in the second image, pixel by pixel, as follows:

$$RDX_{ij}^{k} = \frac{\hat{X}_{ij}^{k}(t_2) - X_{ij}^{k}(t_1)}{2} + 127$$

A regressed image of band 5 was calculated using this procedure and image differencing was performed. The method has the advantage of reducing atmospheric haze and sun angle effects.

### *Image ratioing*

*Ratio images* are prepared by dividing the pixel values in a particular band at time $t_1$ by the corresponding pixel values at time $t_2$. Pixel values greater or less than one are taken to indicate areas of change (Wilson *et al.*, 1976; Todd, 1977; Howarth and Wickware, 1981; Howarth and Boasson, 1983).

The following ratioed images were calculated:

(a) $$\frac{\text{Band 5 (1981)}}{\text{Band 5 (1974)}}$$

(b) $$\frac{\text{Band 7 (1981)}}{\text{Band 7 (1974)}}$$

### *Principal components analysis*

This multivariate technique is used to reduce the number of spectral components to fewer principal components, which account for most of the variance in the original multispectral images. Data from two or more dates are then compared, as in the image differencing or image regression techniques (Lodwick, 1979; Byrne *et al.*, 1980). Principal components analysis was performed separately on the 4-band Landsat data from each date. The transformation was investigated because it involves linear combination of data from all four bands. The analysis was carried out using unstandardized variables. Principal component two (PC2) from each date was used in the image differencing procedure. PC2 was chosen because it emphasizes the difference between the infrared and the visible bands, and is likely to be most useful for the enhancement of vegetation features. PC1 contains variability due to overall scene brightness, while PC3 and PC4 account for very little variance in the original data set.

### *Post-classification comparison*

The most obvious method of change detection is a comparative analysis of spectral classifications for times $t_1$ and $t_2$ produced independently. In this context it should be noted that the change map of two Landsat images will only be as accurate as the product of multiplying the accuracies of each individual classification, i.e. 80% × 80% = 64% accuracy (Stow *et al.*, 1980). The method is more demanding in terms of computer time than other methods because it requires complete, separate classifications for each image (Robinson, 1979).

In the present study, Landsat MSS data for dates $t_1$ and $t_2$ were independently classified using a minimum distance to means classifier. Sample sites representing 'change' and 'no change' cover classes were used in assessing the classification accuracy.

### *Direct multidate classification*

This approach is based on the single analysis of a combined data set of times $t_1$ and $t_2$, in order to identify areas of change (Swain, 1976). In a two-date Landsat sequence, eight bands are analysed at one time in either a supervised or unsupervised mode. Instead of using Landsat MSS bands from two dates, the first two principal components (PC1 and PC2) were used in a supervised classification. The use of the first two PCs in the analysis was justified on the grounds that they accounted for approximately 98% of the variance in the original data set.

Table 1. Standard deviation thresholds and classification accuracy achieved using the image regression method for band 5 data from Manipur State, northeastern India.

| Standard deviation threshold | Commission error* | Omission error* | % Correct classification accuracy† | | |
|---|---|---|---|---|---|
| | | | Average | Overall | Combined |
| 0.25 | 0.47 | 0.05 | 73.22 | 62.00 | 67.61 |
| 0.50 | 0.30 | 0.17 | 76.20 | 72.66 | 74.43 |
| 0.75 | 0.19 | 0.40 | 70.43 | 76.00 | 73.21 |
| 1.00 | 0.13 | 0.57 | 64.46 | 76.00 | 70.23 |
| 1.25 | 0.10 | 0.68 | 60.49 | 76.00 | 68.24 |
| 1.50 | 0.06 | 0.77 | 57.94 | 76.66 | 67.30 |
| 1.75 | 0.06 | 0.88 | 52.66 | 74.67 | 63.66 |
| 2.00 | 0.04 | 0.97 | 49.25 | 74.00 | 61.62 |

*Commission error: Landsat data incorrectly indicate a change in land cover; omission error: Landsat data fail to indicate a change in land cover.
†Classification accuracy:

$$\text{Average} = \frac{(\%\ \text{Correct change} + \%\ \text{Correct no change})}{2}$$

$$\text{Overall} = \frac{\text{Total no. of correctly classified pixels}}{\text{Total no. of pixels}}$$

$$\text{Combined} = \frac{(\text{Average} + \text{Overall})}{2}$$

In the application of these digital change detection techniques, with the exception of the post-classification comparison and the direct multidate comparison, it was necessary to establish a thresholding level in order to define land cover change. All histograms were examined and the mean and standard deviation values for each data set were calculated. Various standard deviation threshold levels were applied to the lower and higher tail of each distribution in order to find the threshold value that produced the highest change classification accuracy (Table 1). It has been pointed out by Nelson (1983) that two commonly employed measures for accuracy assessment, namely, overall accuracy (which weights each class according to the number of pixels in that class) and average accuracy (which weights each class equally), have inherent bias depending on the number of pixels in each class of interest. For this reason he suggested the use of a combined accuracy index (the average of the overall accuracy and the average accuracy), which effectively damps the biases of the individual indices. A summary of the results obtained by using these techniques is given in Table 2.

## SUMMARY REVIEW OF THE TEST RESULTS

The results of the study can be summarized as follows:

1. The highest percent correct classification in relation to the available reference

Table 2. Comparison of the performances of the differential digital change detection techniques applied in Manipur State, northeastern India.

| | | % Correct classification | | |
|---|---|---|---|---|
| Method | Standard deviation level | Average classification accuracy | Overall classification accuracy | Combined classification accuracy |
| Image differencing (band 5) | 0.50 | 70.99 | 75.33 | 73.16 |
| Image differencing (band 7) | 0.50 | 55.58 | 70.00 | 62.79 |
| Ratio image differencing | 1.00 | 62.11 | 80.00 | 71.05 |
| Image regression (band 5) | 0.50 | 76.20 | 72.66 | 74.43 |
| Image ratioing (band 5) | 0.75 | 71.42 | 76.00 | 73.71 |
| Image ratioing (band 7) | 0.50 | 57.32 | 72.66 | 64.99 |
| Principal components analysis (PC2) | 1.00 | 63.66 | 79.33 | 71.49 |
| Post-classification comparison | — | 51.38 | 51.33 | 51.35 |
| Direct multidate classification | — | 57.26 | 57.33 | 57.29 |

data was achieved using the image regression method with data from band 5. The image ratioing and image differencing techniques using band 5 data yielded comparable, albeit marginally poorer, results.
2. The post-classification comparison and direct multidate classification methods produced the lowest classification accuracy. The poor performance of these approaches can be attributed to the relatively weak classification levels achieved for each data set at the different dates.
3. In single band analyses, band 5 data proved to be superior to band 7 data for detecting changes in forest cover. The reason for the relatively poor performance using the band 7 data is to be found in the high infrared return from the herbaceous understorey in cleared areas which produces classification errors.
4. In order to achieve maximum combined classification accuracy, it was found that standard deviation thresholds must be set at a low level, i.e. in the range 0.5 to 1.0.
5. Commission error (the situation where Landsat data incorrectly indicate cover changes) was the major type of misclassification at lower threshold values. Conversely, omission errors (the situation where Landsat data fail to indicate a cover change) were greater at higher threshold levels.

It is clear that land cover changes caused by shifting cultivation can be detected in a tropical forest environment, using Landsat MSS data. The maximum change detection accuracy achieved in the Manipur study area was 74%. The shortfall can be explained by such factors as differences in illumination, differences in atmospheric conditions, variations in surface moisture, and the possible misregistration of images. It is unlikely that noise factors of this kind can ever be entirely eliminated. Also, it should be noted that land cover changes that do not significantly alter the spectral response will go undetected and these will contribute to the inaccuracy of change detection. This problem is particularly evident in areas of shifting cultivation, because of the spatial complexity of the system. In forests with shifting cultivation, cleared areas frequently tend to multiply rather than expand individually. Many cleared areas of irregular shape and limited size, representing different stages in the cultivation-regrowth cycle, become intermixed with mature forest, creating a complex land cover mosaic. In this respect, the cultivation system itself makes objective change detection from Landsat MSS data a somewhat imprecise task. It is expected that the improved resolution of the Thematic Mapper and the SPOT series will allow an increased precision of change detection in areas of shifting cultivation.

In spite of the current imperfection of digital change detection techniques in areas of tropical forest, they are useful in drawing attention to the magnitude of change over specific periods of time. This type of information may be sufficient to initiate appropriate, remedial measures (UNESCO, 1978). This consideration, rather than the specific attainment of accuracy, currently justifies the use of Landsat MSS data for forest survey purposes (Baltaxe, 1980).

## ACKNOWLEDGEMENTS

The present work was carried out during postgraduate studies in the Department of Geography, University of Reading, England; the author is grateful to Dr J. R. G. Townshend of that department for his valuable suggestions. The author also thanks the Commonwealth Scholarship Commission in the United Kingdom for their financial support, and the Government of Manipur, India, for their sponsorship.


## REFERENCES


Aldrich, R. C., 1975. Detecting disturbances in a forest environment. *Photogramm. Engng Remote Sensing*, **41**, 39–48.

Angelici, G. L., Bryant, N. A. and Friedman, S. Z., 1977. Techniques for land-use change detection using Landsat imagery. In *Proc. Fall tech. Meet.*, American Society of Photogrammetry, Falls Church, Virginia, pp. 217–228.

Anon., 1976. *Report of the National Commission on Agriculture 'Forestry'*, part 9. New Delhi: Ministry of Agriculture and Irrigation, Government of India.

Anon., 1978. *Indian Forestry 1978 Pamphlet*. New Delhi: Forestry Division, Ministry of Agriculture and Irrigation, Government of India.

Baltaxe, T., 1980. *The Application of Landsat Data to Tropical Forest Surveys*. Rome: Food and Agriculture Organization.

Byrne, G. F., Crapper, P. F. and Mayo, K. K., 1980. Monitoring land-cover change by principal component analysis of multitemporal Landsat data. *Remote Sensing Environ.*, **10**, 175–185.

Champion, H. G. and Seth, S. K., 1968. *Revised Survey of Forest-types of India*. New Delhi: Government of India Press.

Colwell, J. E. and Weber, F. P., 1981. Forest change detection. *Proc. 15th int. Symp. Remote Sensing Environ.*, **2**, 839–852.

Crapper, P. F. and Hynson, K. C., 1983. Change detection using Landsat photographic imagery. *Remote Sensing Environ.*, **13**, 291–300.

Denevan, W. M., 1975. *The Cause and Consequences of Tropical Shifting Cultivation*. Rome: Food and Agriculture Organization.

FAO, 1981. *India and Sri-Lanka: Agroforestry*. Rome: Food and Agriculture Organization.

Hadley, M. and Lanly, J. P., 1983. Tropical forest ecosystems, identifying differences and seeking similarities. *Nature and Resources*, **19**, 2–19.

Hardy, J. R., 1978. Methods and accuracy of location of Landsat MSS points on maps. *J.Br. interplanet. Soc.*, **31**, 305–311.

Harrison, A. R. and Singh, A., undated. Microcomputer based image processing systems in remote sensing. *Photogramm. Engng Remote Sensing* (in press).

Howarth, P. J. and Boasson, E., 1983. Landsat digital enhancements for change-detection in urban environment. *Remote Sensing Environ.*, **13**, 149–160.

Howarth, P. J. and Wickware, G. M., 1981. Procedure for change-detection using Landsat digital data. *Int. J. Remote Sensing*, **2**, 277–291.

Ince, F., 1983. Digital image processing systems and remote sensing. *Int. J. Remote Sensing*, **4**, 129–148.

Ingram, K., Knapp, E. and Robinson, J. W., 1981. *Change-detection Technique Development for Improved Urbanised Area Delineation*. CSC/TM-81/6087 Report prepared for NASA, Computer Sciences Corporation, Springfield, Maryland.

Lillesand, T. H. and Kiefer, R. W., 1979. *Remote Sensing and Image Interpretation*. New York: Wiley.

Lodwick, G. D., 1979. Measuring ecological changes in multitemporal Landsat data using principal components. *Proc. 13th int. Symp. Remote Sensing Environ.*, **1**, 1–11.

Miller, L. D., Nualchwee, K. and Tom, C., 1978. Analysis of the dynamics of shifting cultivation in the tropical forests of northern Thailand using landscape modelling and classification of Landsat imagery. *NASA/GSFC Tech. Mem.*, 79545, 1–19.

Nelson, R., 1983. Detecting forest canopy change due to insect activity using Landsat MSS. *Photogramm. Engng Remote Sensing*, **49**, 1303–1314.

Robinson, J. W., 1979. *A Critical Review of Change-detection and Urban Classification Literature*. CSC/TM-79/6235, Computer Sciences Corporation, Springfield, Maryland.

Sharmah, B. S. K., 1979. Shifting cultivation in Manipur and Nagaland. In *Proceedings of the Agro-forestry Seminar*, Imphal. New Delhi: Indian Council of Agricultural Research, pp. 132–137.

Singh, A., 1984. *Tropical Forest Monitoring Using Digital Landsat in Northeastern India*. Unpublished Ph.D. Thesis, Department of Geography, University of Reading, England.

Stauffer, M. L. and McKinney, R. L., 1979. *Landsat Image-differencing as an Automated Land Cover Change Detection Technique*. Rept. Contract NAS5-24350, General Electric Company, Beltsville, Maryland.

Stow, D. A., Tinney, L. R. and Estes, J. E., 1980. Deriving land use/land cover change statistics from Landsat: a study of prime agriculture land. *Proc. 14th int. Symp. Remote Sensing Environ.*, **2**, 1227–1237.

Swain, P. H., 1976. *Land Use Classification and Mapping by Machine-assisted Analysis of Landsat Multispectral Scanner Data*. LARS Information Note 112776, Laboratory for Application of Remote Sensing, Purdue University, Indiana.

Thomson, F., Wilson, C., Sadowski, F., Malila, W. and Dye, R., 1979. *Application and Further Development of Remote Sensing Technique for Forest Management*. Final Report Contract No. 53-3187-8-19, Forest Service USDA, ERIM Application Division, Ann Arbor.

Todd, W. J., 1977. Urban and regional land use change detected by using Landsat. *J.Res., U.S.G.S.*, **5**, 529–534.

Townshend, J. R. G. and Justice, C. O., 1981. Information extraction from remotely sensed data. *Int. J. Remote Sensing*, **2**, 213–229.

Tucker, C. J., 1979. Red and photographic infra-red linear combination for monitoring vegetation. *Remote Sensing Environ.*, **8**, 127–150.

UNESCO, 1978. *Tropical Forest Ecosystems*. Paris: UNESCO.

Weismiller, R. A., Kristof, S. J., Scholz, D. K., Anuta, P. E. and Momin, S. A., 1977. Change detection in coastal zone environment. *Photogramm. Engng Remote Sensing*, **43**, 1533–1539.

Wilson Jr, I. R., Blackman, C. and Spann, G. W., 1976. Land use change detection using Landsat data. *Proc. 5th Annual Remote Sensing Earth Resour. Conf.*, Tulhahama, 79–91.

Remote Sensing and Tropical Land Management
Edited by M. J. Eden and J. T. Parry


# CHAPTER 14

# *Monitoring Indigenous Shifting Cultivation in Forest Areas of Southwest Guyana using Aerial Photography and Landsat*

MICHAEL J. EDEN
*Department of Geography, Royal Holloway and Bedford New College, London University, U.K.*

Indigenous shifting cultivation has long been practised in the Amazon lowlands of South America. Some authors have suggested that shifting cultivation and associated burning have previously resulted in the conversion of forest to savanna (Sombroek, 1966; Prance and Schubart, 1978), but the cultivation has traditionally been based on localized exploitation and extended fallowing, and does not generally appear to be destructive of the forest (Denevan, 1978). In contrast, contemporary land colonization in Amazonia commonly results in widespread destruction of forest. Some land is cleared directly for permanent cultivation, but much is settled by migrant colonists, who crop the land for a few years and then move on to fresh, primary forest. The abandoned land is converted to low-grade pasture or degraded secondary growth. Pioneer shifting cultivation of this kind is inherently unstable in character, and contrasts with the more adaptive and sustainable indigenous cultivation system (Meggers, 1971; Eden, 1978).

The examination of indigenous shifting cultivation is of interest from an historico-cultural point of view, but also has relevance for contemporary land exploitation and management. In particular, such cultivation can provide an important model of appropriate technology for the rainforest environment. Its polycultural cropping characteristics, for example, provide specific advantages for the cultivator. These include protecting the soil, maintaining the nutrient cycle without chemical fertilizers, and also maintaining plant diversity which minimizes diseases and pests (Denevan, 1978). Agroforestry and similar

Fig. 1 Location of the southern Rupununi, Guyana (after Eden, 1964).

integrated systems involving livestock are currently being developed from this tradition and may provide a foundation for the more sustainable exploitation of rainforest land (Bishop, 1978; Eden, 1980, 1982; Denevan *et al.*, 1982; Hecht, 1982). Some scope may also exist for the introduction of intensive farming systems to Amazonia, but these have hitherto had such a poor record in the region (Eden, 1978) that alternative attention to the development of adaptive, traditional systems would appear to be desirable.

One source of information on indigenous cultivation in Amazonia is remote sensing. Aerial photography and Landsat imagery can both provide useful, sequential data on vegetation and land use in the area. In the present study, remote sensing has been used to acquire information on the cultivation activities of one indigenous group, namely, the Wapisiana Indians of the southern Rupununi region of southwest Guyana (Fig. 1). The impact of their cultivation is examined and the value of remote sensing, as an aid to land management, is assessed.

The study area, located 2°–3° N of the equator, is a savanna-forest boundary zone. The Wapisiana Indians, whose numbers have substantially increased in recent decades, total some 5000 or more people. This population is normally resident in the savanna, but its agricultural activities are located in the adjacent forest. Aerial photographs of the southern Rupununi are available from the period 1952–1970, and later information is obtainable from Landsat images acquired in 1972 and 1983. Visual examination of these data sources indicates that, in spite of an increasing intensity of Wapisiana shifting cultivation over recent decades, the cultivation system itself has been sustained and the forest cover generally preserved. This conclusion is at variance with earlier studies from the area, which have tended to assume that deforestation was occurring as a result of indigenous cultivation and burning (Waddell, 1963; Hills, 1964; Talbot and Kesel, 1975; Potter, 1978).

## REMOTE SENSING AND ENVIRONMENTAL MONITORING

Sequential aerial photography exists for many tropical areas, but its use for environmental monitoring has been relatively limited. Since the mid-1970s, interest in environmental monitoring has increased, but this principally reflects the growing availability of Landsat multispectral imagery. The latter has been adopted, for example, as a basis for the *Programa de Monitoramento da Cobertura Florestal do Brasil*, established by the Brazilian Institute of Forest Development (IBDF). In spite of cloud problems in Amazonia, repetitive Landsat cover is generally available for the region and can readily be used to monitor large-scale forest clearance, associated with the establishment of cattle ranches or with peasant colonization along new highways. In the case of Brazil, such monitoring is facilitated by the existence of a national Landsat receiving station.

In regions where aerial photographs are available, more detailed information on land exploitation can be obtained, especially in respect of traditional shifting cultivation. According to Myers (1980), aerial photographs can be used to differentiate between cultivated fields, relict grasslands including fallow patches, and secondary forest of various ages. If sequential photography is available, temporal changes in the pattern and status of shifting cultivation and of the forest matrix can be monitored with considerable precision. Existing Landsat imagery is less valuable in this respect on account of its relatively low resolution, but it can still provide useful information.

Some earlier studies of shifting cultivation undertaken with aerial photography illustrate the monitoring capability of the imagery. In Papua New Guinea, Taylor (1968) used aerial photographs dating from 1942–1953 to examine native cultivation in forested areas of the Popondetta plains. He showed that, although forest regrowth normally occurred in old fields, savanna was occasionally able to establish itself. On the ground such areas were found to be colonized by nearly pure stands of *Imperata cylindrica*. It was assumed by Taylor that these herbaceous communities would persist as a result of annual burning, although in time other grasses might spread at the expense of *I.cylindrica*.

Comparable conditions have been reported by Eden (1974) near Port Moresby in Papua New Guinea. In this area, the examination of aerial photographs from the period 1942–1968 showed that approximately one-fifth of native fields, originally established in seasonal forest, was converted to grassland after 3–4 years of cultivation. The aggressive pioneer species, *Imperata cylindrica*, was again dominant in the early stages of regrowth, being encouraged by the incidence of burning in newly abandoned fields around village settlements (Eden, 1974).

In the Gran Pajonal of eastern Peru, aerial photographs from the 1950s have been used, in conjunction with later fieldwork, to examine the impact of Amerindian shifting cultivation on the forest cover (Scott, 1978). In this area, abandoned fields commonly revert to forest but, if fire intervenes, a community dominated by bracken fern (*Pteridium aquilinum*) can become established. With continued annual burning, this gives way to communities dominated by *Imperata brasiliensis* and then *Andropogon* and *Cyperus*. In the Gran Pajonal, some forest invasion of established savanna has also been detected since the 1950s. This is encountered both in areas of increased herbivory by domestic cattle and in areas that are no longer used for indigenous settlement (Scott, 1978).

Previous use has been made of aerial photographs in southwest Guyana. Baseline investigations in the Rupununi region were undertaken by Waddell (1963) and Hills (1964), using the original 1952/1953 photographic coverage. These authors concluded that forest retreat was taking place in response to indigenous cultivation and burning, although in one unpopulated area the forest appeared locally to be extending into the savanna. The assumption of general forest retreat was confirmed by Potter (1978) on the basis of an examination of aerial photographs from 1952 and 1970.

## THE STUDY AREA

The southern Rupununi comprises an undulating, planar landscape, developed over weathered igneous rocks of the Guiana Shield. The eastern savanna-forest boundary, extending from Achiwib to Shea, roughly parallels the transition from a younger, lower land surface (140–200 m) in the west to an older, elevated surface (210 m) in the east. However, there is no precise coincidence of the relief and vegetation boundaries, and both surfaces have acid and infertile soils, with lateritic characteristics (FAO, 1966; Eden, 1971).

Annual rainfall in the region is 1500–1800 mm, with an extended dry season of 6–7 months. In the savanna, *Trachypogon plumosus* is the dominant grass and is associated with an open tree layer containing *Curatella americana* and *Byrsonima* spp. The savanna extends into the adjacent northern Rupununi and is continuous with the Rio Branco savanna of Brazil (Eden, 1964). The flanking

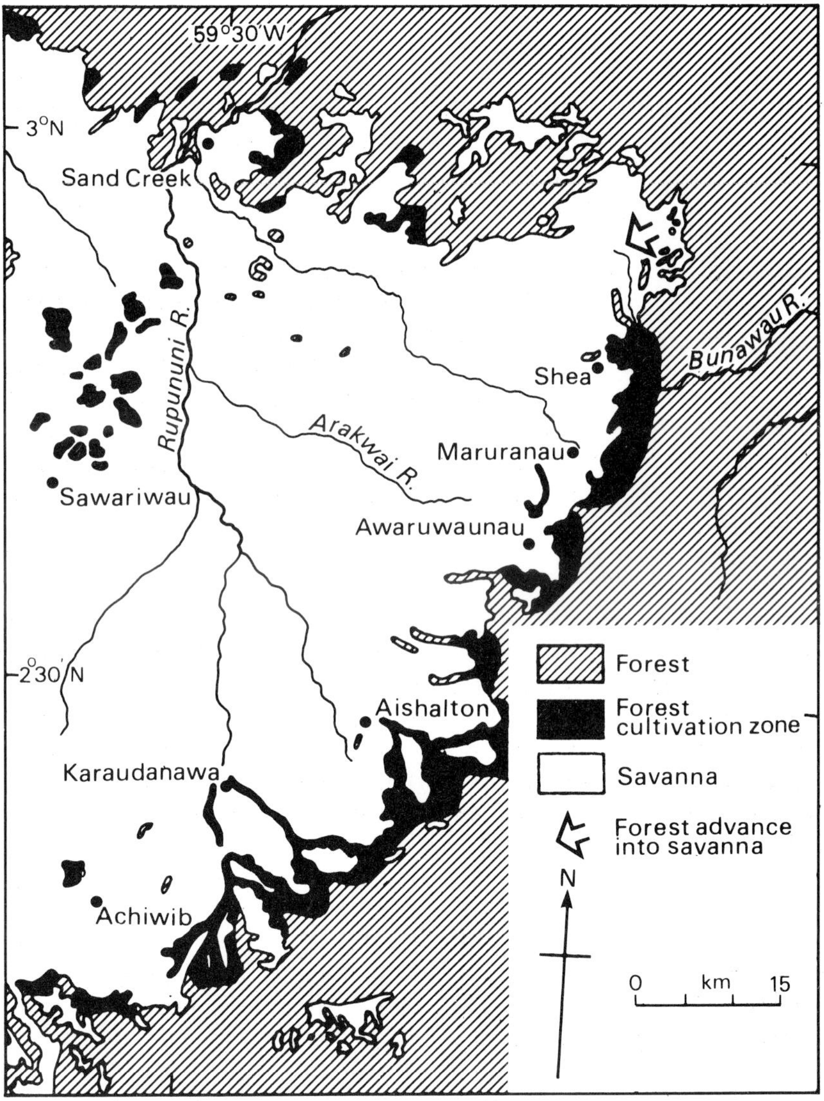


Fig. 2 The southern Rupununi showing the main Wapisiana villages. Cultivation data are derived from 1962 aerial photographs.

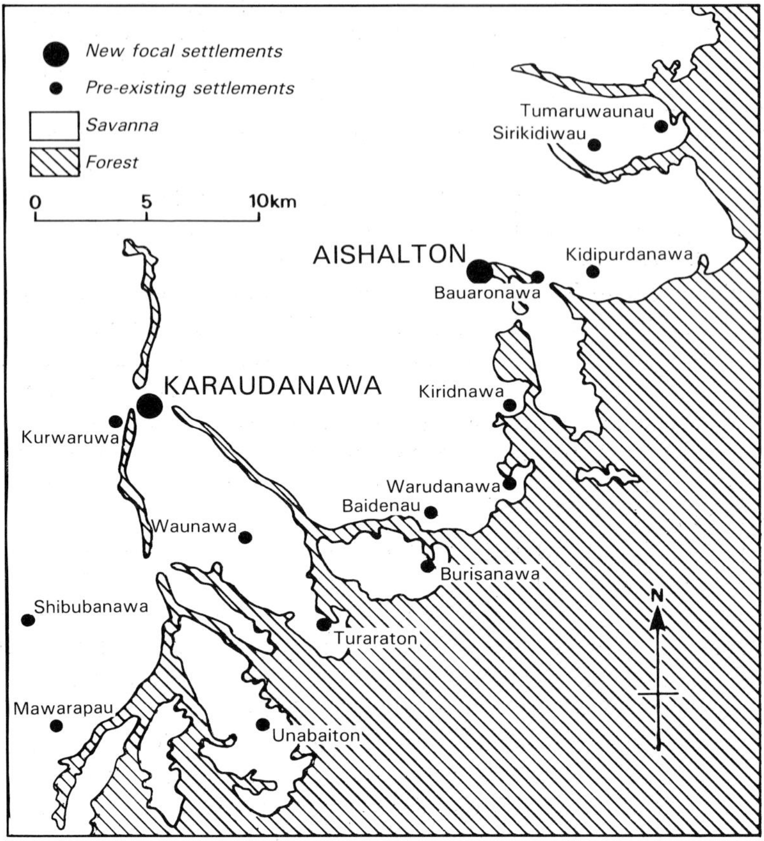


Fig. 3 Redistribution of settlement in the Karaudanawa-Aishalton area of the southern Rupununi, 1923–1960 (after Hills, 1964).

forest in the southern Rupununi is deciduous to semi-deciduous in character, forming part of the main Amazonian *hylaea* (Myers, 1936; Ducke and Black, 1953). Riparian forest is locally present within the savanna, which also contains scattered 'bush islands' of the kind occurring between Sawariwau and Sand Creek (Fig. 2). As in other areas, the savanna is burnt during dry seasons. The fires commonly extend to the margin of the forest, but rarely penetrate it.

Most of the population of the southern Rupununi are Wapisiana Indians. Some Macushi Indians are present west of Sand Creek, and there are a few ranchers and other settlers in the area. In earlier times, Indian settlements were mostly small and impermanent and, according to Evans and Meggers (1960), were to be found within both the savanna and the forest. Under European influence in the present century, larger and more permanent villages have developed in the savanna (Fig. 3). There are now eight major Wapisiana settlements of this kind which have grown up around introduced service facilities like water pumps, medical centres, schools, and churches. Some dispersion of the population persists, but Hills (1964) indicated that almost two-thirds of the Wapisiana were concentrated within the main villages.

In parallel with the development of village centres, there has been an increase in the population. In 1945, the Wapisiana totalled some 2000 people, but the number increased to 4300 by 1960 and approximately 5000 by the early 1970s. About 1000 people reside in the largest village, Aishalton, with the other main centres each containing 400–800 people. The population increase reflects improved medical services and in particular the control that has been achieved over malaria, tuberculosis and other diseases (Eden, 1966). The mean population density across the savanna is currently $1-2\,km^{-2}$, but a more realistic calculation for the settled marginal zone would be $15-20\,km^{-2}$ (Hills, 1964; Potter, 1978). In addition, several thousand Wapisiana reside in the adjacent Rio Branco region of Brazil, where many of the group have become 'integrated' (Migliazza, 1978). Over the years, periodic migration appears to have occurred between the two areas.

Occupance of the savanna-forest boundary zone in the southern Rupununi has advantages for the indigenous population, who today prefer the more open and accessible conditions of the savanna for their residence, but remain dependent on forest land for their shifting cultivation. This is mainly orientated towards manioc production, although intercropping of maize, sweet potatoes, yams, bananas and other crops takes place. Individual fields are commonly 1–2 ha in extent, and are mostly cut in secondary forest. After drying out, the tree debris is burnt and fields are cultivated for 2–4 years. In the past most fields were dispersed in the forest and, if adjacent plots were cleared, a band of tree growth was left separating the fields (Salisbury, 1968). This encouraged the regrowth of woody plants when field areas were abandoned and provided a fire barrier between newly cleared fields, requiring to be burnt, and older weedy plots that might be fire-deflected to grassland. As forest regeneration took place, nutrient capital was restored and weedy growth suppressed. After fallowing for a period of 10–20 years, the land was considered suitable for recultivation (Dummett, 1968; Hills, 1968). In many respects the traditional pattern persists, although in recent years more concentrated cultivation and abbreviated fallowing have developed as a response to population growth and the nucleation of of settlement.

Some fishing and hunting activities are undertaken by the Wapisiana, while balata (*Manilkara bidentata*) collecting from the forest and some cattle raising in the savanna have been undertaken in recent decades (McCann, 1972). Small yard gardens are present within savanna villages, and local 'corral cultivation' has developed (Dummett, 1968; Salisbury, 1968), but general cultivation in the savanna is impractical because of the low soil fertility.

## METHODS OF STUDY

The first systematic aerial survey of the southern Rupununi was undertaken by Hunting Surveys Limited of England in 1952/1953, at a scale of 1:30 000. A second survey was made in 1962, when the Royal Air Force obtained photographs at 1:60 000, and a third survey was made by Terra Surveys Limited of Canada in 1969/1970, at the scale of 1:40 000. Black and white panchromatic film was used in all cases. Usable Landsat imagery of the southern Rupununi is available from July 1972 and July 1983. The 1972 image (ID no. 1007-13402) is from the Landsat 1 return beam vidicon (RBV) camera. The image has 50% cloud cover, but provides good visibility over parts of the southern Rupununi. The 1983 image (ID no. 483190–134110) is from the Landsat 4 multi-

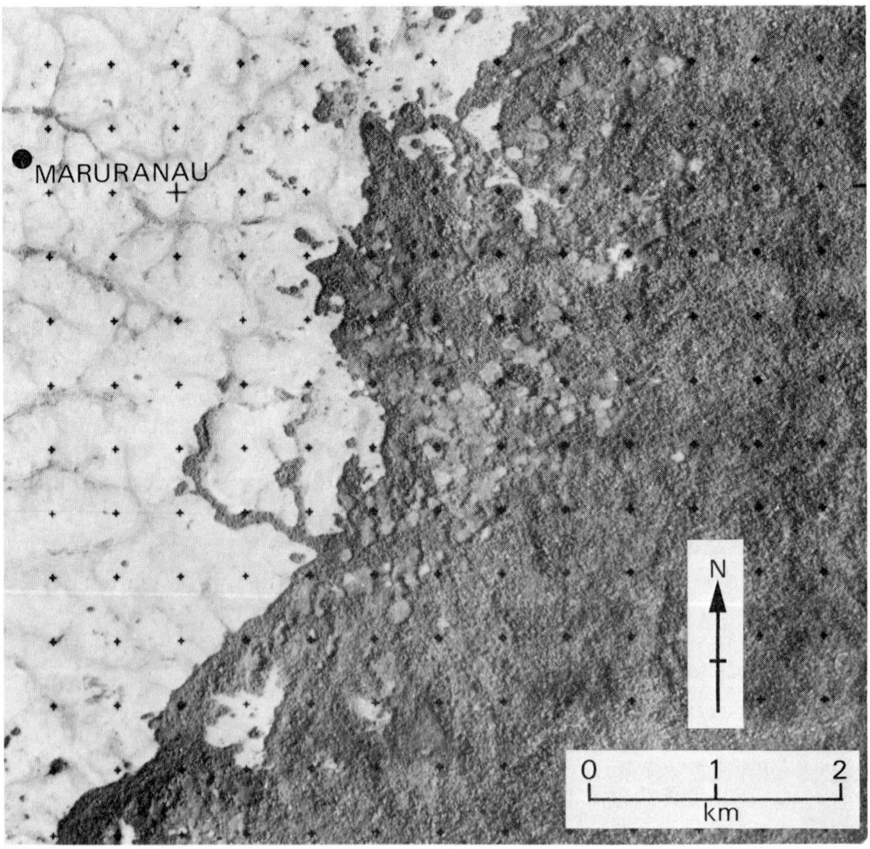


Fig. 4 Savanna-forest boundary near Maruranau, southern Rupununi. Panchromatic aerial photograph, 29 September 1962 (*British crown copyright reserved/RAF photograph*). + + camera reseau marks

spectral scanner (MSS). Cloud cover is minimal and haze effects appear negligible.

Two studies were carried out using the above imagery. Firstly, visual monitoring of the savanna-forest boundary was undertaken. An initial examination of the boundary was made on selected aerial photographs, dating from 1952, and comparisons were made with later aerial photographs and Landsat images. Photographic enlargements of the Landsat RBV band 2 (0.58–0.68 μm) from 1972 and MSS band 5 (0.6–0.7 μm) from 1983 were used in the study. Altogether some 50 km of the main, eastern savanna-forest boundary were examined, together with the complex of bush islands in the vicinity of Sawariwau. In general, the savanna-forest boundary in the area is abrupt, and is clearly identifiable on aerial photographs by virtue of contrast in colour and texture (Fig. 4). On Landsat imagery, the boundary is also clearly visible, although resolution is inferior to that of the aerial photographs.

The second study, based solely on the examination of aerial photographs, focused on individual cultivated fields. Wapisiana fields are located within the forest and, particularly at an early stage of cultivation, are clearly identifiable on the photographs by virtue of a pale colour, fine texture, and distinctive shape. As the field crop develops, its colour darkens and its texture coarsens. In general, in the southern Rupununi, it is not possible to separate mature crop growth, especially of manioc, from the early stages of forest regeneration. However, as forest regrowth develops, its texture continues to coarsen. On this basis, newly planted fields were identified on selected photographs, dating from 1952/1953 and from 1962. On photographs of the same areas taken 7–10 years later, i.e. in 1962 and 1969, respectively, the original field sites were located. These were re-examined and an attempt made to determine the cover type, as follows:

1. secondary forest,
2. derived grassland,
3. recultivated land.

The above classes were adopted on the assumption, based on fieldwork in the southern Rupununi, that fields are normally cut from forest, cultivated, and then abandoned within a period of 2–4 years. Altogether 128 fields from four areas were examined on the aerial photographs (Table 1). Although newly planted fields can be identified on the photographs, no estimates were possible of the total area under cultivation at any particular time, because of the difficulty in differentiating between mature crop growth and early regenerating forest.

## RESULTS

The examination of the savanna-forest boundary on sequential imagery focused on sections of the main eastern boundary from Karaudanawa to north of Shea, and on the complex of bush islands to the north of Sawariwau (Fig. 2).

Table 1. Sample fields in the southern Rupununi, Guyana.

| Location | No. of field sites examined | Year of photo coverage (a) | (b) |
|---|---|---|---|
| Karaudanawa | 17 | 1953* | 1962† |
| Karaudanawa | 13 | 1962† | 1969‡ |
| Aishalton | 11 | 1953* | 1962† |
| Aishalton | 12 | 1962† | 1969‡ |
| Maruranau | 34 | 1952* | 1962† |
| Maruranau | 11 | 1962† | 1969‡ |
| Sawariwau | 30 | 1952* | 1962† |

* 1:30 000 scale.
† 1:60 000 scale.
‡ 1:40 000 scale.

The Sawariwau area contains about a dozen large bush islands and numerous small ones, together with stretches of gallery forest. A comparison was made of aerial photographs, dating from 1952 and 1962, over an area of approximately 130 km$^2$. The bush islands are of irregular form and frequently give the impression of having been shaped by human activity (Fig. 5). However, over the period in question, no clear changes in the location of the savanna-forest boundary were detected. The most recent aerial photographs of the area (1970) were not available for detailed inspection, but the Landsat images for 1972 and 1983 were examined. While the resolution of the latter precluded precise monitoring of the boundary, there was no evidence of major changes in the extent of the bush islands over the period 1952–1983 (Fig. 6(a)–(c)).

Similarly, in the vicinity of Karaudanawa, Aishalton and Maruranau where approximately 50 km of the main savanna-forest boundary were examined, no major changes were apparent on the aerial photographs over the period 1952–1969. A few minor changes, involving extensions of both savanna and forest, were evident but no 'field size' changes were detected in spite of the often irregular shape of the boundary itself. Examination of the Landsat imagery confirmed the general stability of the savanna-forest boundary over the period 1952–1983 (Fig. 7(a)–(c)).

Finally, the Bunawau basin to the northeast of Shea was examined. This is an area at present lacking settlement, where Hills (1964) suggested that local advance of the forest was taking place. On the earlier aerial photographs the savanna-forest boundary in the area is diffuse and irregular and the regenerating forest is thin and locally overtopped by mature riparian forest (Fig. 8). On the 1983 Landsat image, the forest appears to have consolidated its earlier advance, but there is no sign of any further extension into the savanna since 1962.

The second study was of individual fields within the forest zone. In the vicinity of Sawariwau, fields are scattered throughout the bush islands and adjacent

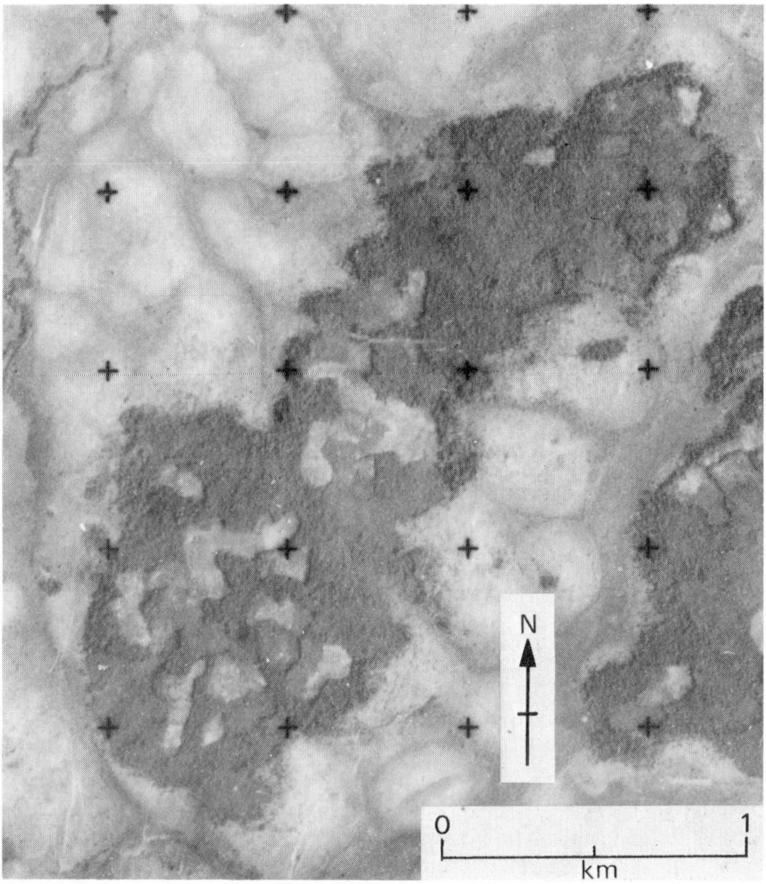


Fig. 5 Cultivated 'bush island' near Sawariwau, southern Rupununi. Panchromatic aerial photograph, 29 September 1962 (*British crown copyright reserved/RAF photograph*). + + camera reseau marks

gallery forest. Since even the larger bush islands are rarely more than 1–2 km across, their entire area is easily accessible from the savanna and usable for cultivation. Over the period 1952–1962, there was an increase in the area of cultivation and of immature forest regrowth within the bush islands and gallery forest. However, individual fields that were under cultivation in 1952 were found to be regenerating to forest on the 1962 aerial photographs. No evidence of deflection to grassland was apparent (Table 2).

In the main forest area to the east, fields are generally confined to a marginal zone some 2–3 km in breadth. A few fields are located 4 or 5 km into the forest, but this is rare. The marginal location of the fields reflects the fact that the

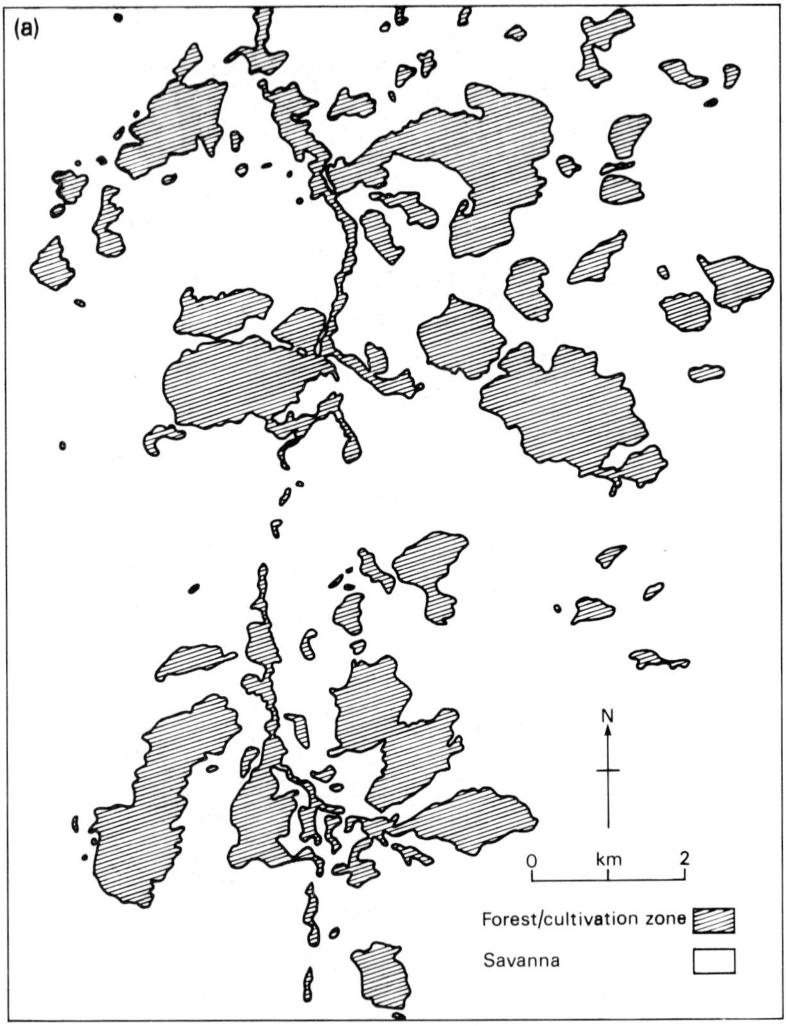


Fig. 6 'Bush islands' in the savanna north of Sawariwau, southern Rupununi; (a) distribution of 'bush islands' from 1953 aerial photograph, (b) Landsat RBV band 2 image subscene, July 1972, ID no. 1007-13402 (local cloud arrowed) (*Reproduced by permission of NOAA*), and (c) Landsat MSS band 5 image subscene, July 1983, ID no. 483190-134110. (*Reproduced by permission of CNPq/INPE, Brazil.*)

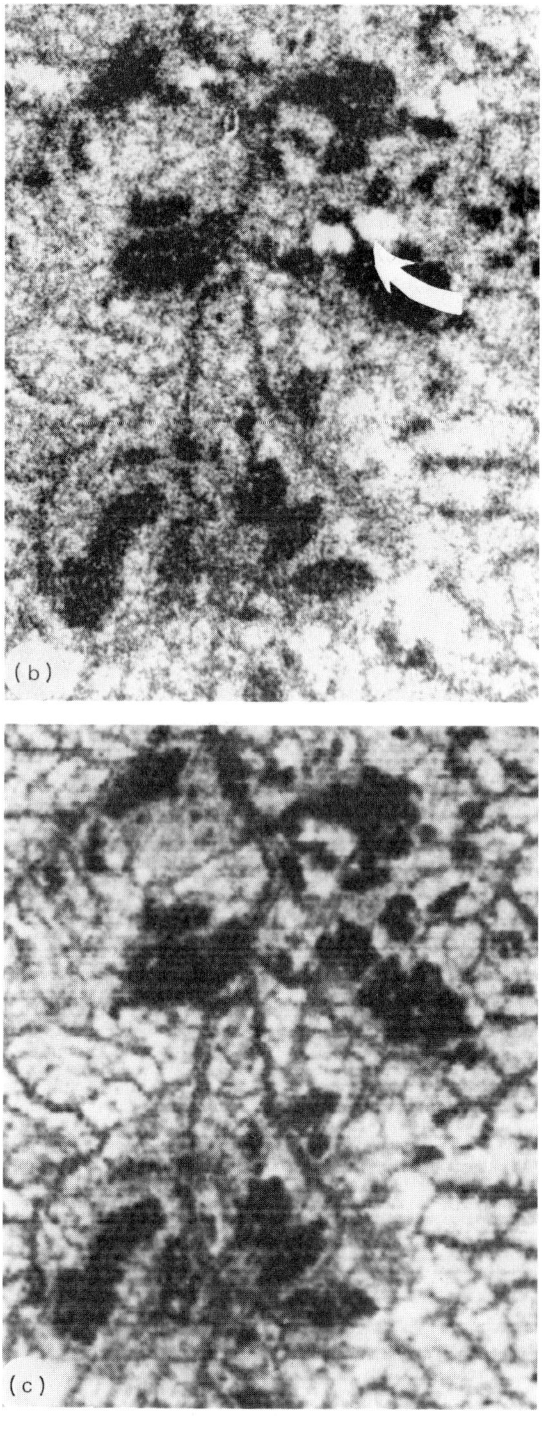
(b)
(c)

main savanna villages are already several kilometres from the forest edge, so that any great penetration into the forest makes visits to fields and the transport of produce therefrom increasingly arduous and time-consuming. Since the savanna villages are spaced out along the savanna-forest boundary, usually 10–20 km or more apart, the lateral concentration of fields along the forest margin is somewhat uneven. Areas of denser forest cultivation closer to the main villages tend to alternate with less densely cultivated zones between. As is suggested by the persistence of the broad cultivation pattern over recent decades (Figs 9 and 10), the effect of population growth and increasing land pressure is primarily expressed in the intensification, rather than extension, of forest land use. Even so, individual fields under cultivation in 1952/1953 and in 1962 were again found to be regenerating to forest on later aerial photographs (Table 2). The study revealed a particularly high level of field recultivation in the vicinity of Maruranau. This accords with an increasing intensity of land use, although some contraction of the cultivated zone has also occurred (Fig. 9). In the case of Aishalton, some extension of the cultivated area is evident (Fig. 10), but this is largely within the marginal forest zone and more concentrated exploitation within the original cultivation area is also evident. In none of the sample areas was there any indication of the deflection of old fields to grassland or savanna.

On the ground, it is evident from both the Sawariwau bush islands and the main forest zone that the intensification of land use is being achieved by the abbreviation of the fallow period. Individual fields are still being used for periods

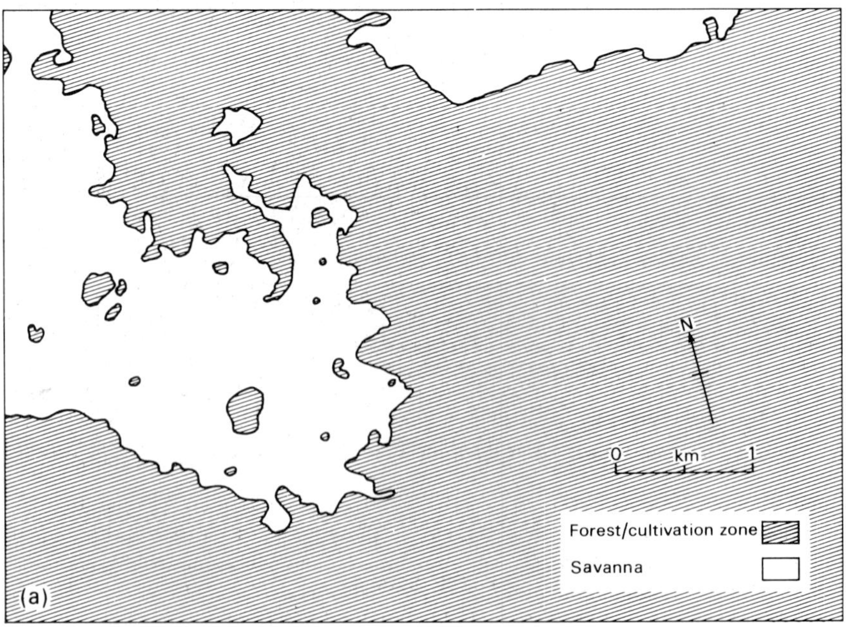

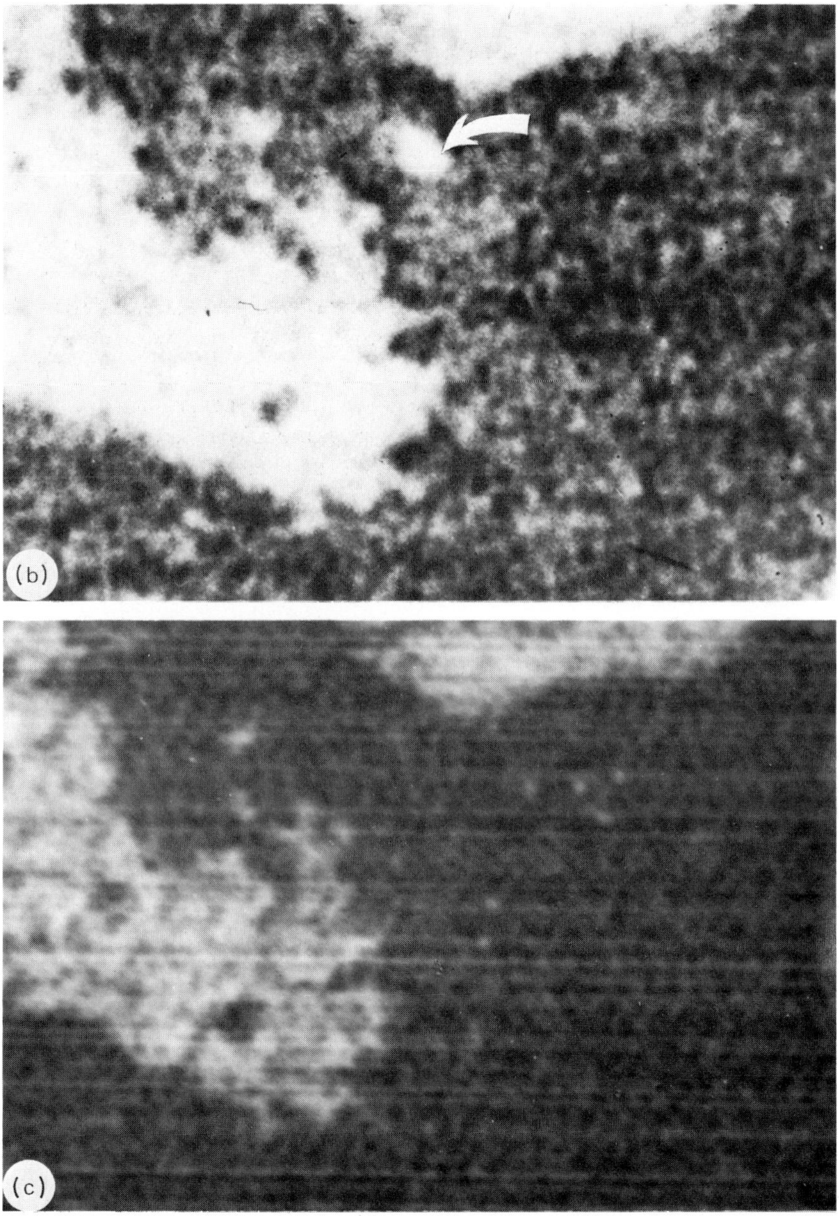

Fig. 7 Savanna-forest boundary zone near Karaudanawa, southern Rupununi; (a) forest (including cultivation zone) and savanna from 1953 aerial photograph, (b) Landsat RBV band 2 image subscene, July 1972, ID no. 1007-13402 (local cloud arrowed) (*Reproduced by permission of NOAA*), and (c) Landsat MSS band 5 image subscene, July 1983, ID no. 483190-134110. (*Reproduced by permission of CNPq/INPE, Brazil.*)

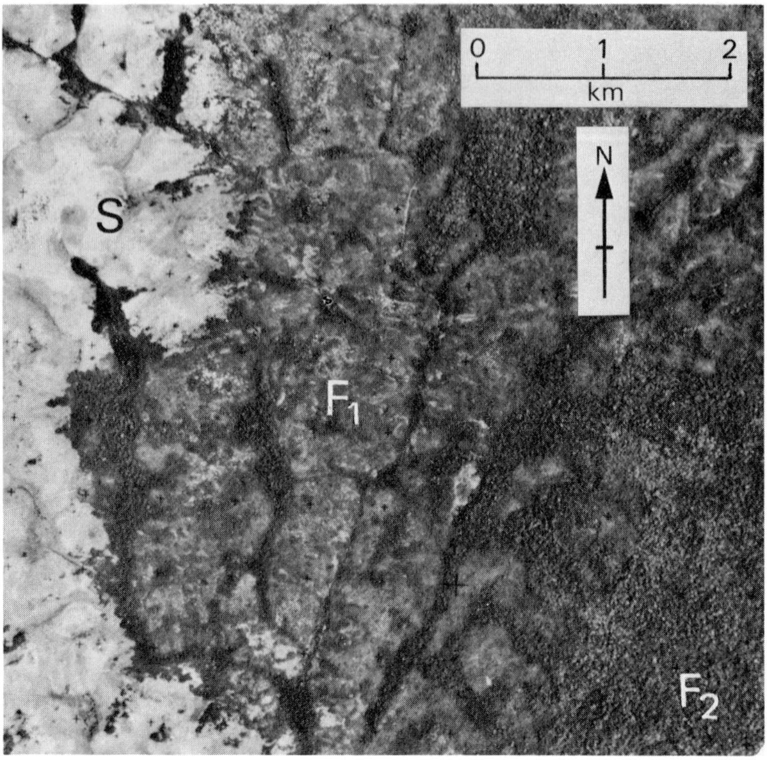


Fig. 8 Savanna-forest boundary in the upper Bunawau basin, southern Rupununi. S: savanna, $F_1$: regenerating forest advancing into savanna, $F_2$: mature forest. + + camera reseau marks. Panchromatic aerial photograph (1:60 000), 11 October 1962 (*British crown copyright reserved/RAF photograph*).

Table 2. Field regeneration in the southern Rupununi, Guyana.

| Location | No. of field sites | Sites regenerating to forest | Sites recultivated *in toto* |
|---|---|---|---|
| 1952/1953–1962 | | | |
| Karaudanawa | 17 | 14 (82%) | 3 (18%) |
| Aishalton | 11 | 11 (100%) | — |
| Maruranau | 34 | 25 (74%) | 9 (26%) |
| Sawariwau | 30 | 29 (97%) | 1 (3%) |
| 1962–1969 | | | |
| Karaudanawa | 13 | 9 (69%) | 4 (31%) |
| Aishalton | 12 | 12 (100%) | — |
| Maruranau | 11 | 4 (36%) | 7 (64%) |
| | 128 | 104 (81%) | 24 (19%) |

of only 2–4 years, and then abandoned. In the past, forest fallowing normally lasted for a period of at least 10–20 years, allowing the land to recover and the secondary forest to become sufficiently open at ground level for easy reclearance (Dummett, 1968; Hills, 1968). Latterly, abbreviated fallowing has become widespread. According to McCann (1972), recultivation of fallow land commonly takes place after 7–9 years, while Potter (1978) mentions 'drastic reductions' in the fallow period, sometimes to as little as one or two years. At Sawariwau, in 1979, cultivators informed the present author that old field areas were being recultivated after fallowing for 2–5 years.

## SUMMARY AND CONCLUSIONS

Field investigators in the early 1960s, with access to a single photographic coverage from 1952/1953, were aware of the scale of forest cultivation in the southern Rupununi, and were of the opinion that the savanna was generally extending into the forest (Waddell, 1963; Hills, 1964). The irregular and angular nature of the savanna-forest boundary and its lack of conformity with any clear soil or relief variables were considered indicative of this process (Eden, 1964). A similar view was held by Talbot and Kesel (1975), who argued that the effect of grass fires on forest margins, already 'weakened' by shifting cultivation, was to cause a gradual retreat of the forest. Potter (1978) also considered that in areas of shifting cultivation the savanna was invading land that was formerly under forest. According to Hills (1964) an exception to this pattern prevailed in the Bunawau basin where forest was thought to be locally advancing into the savanna. Hills (1973) concluded that during the present century the Rupununi had been a region 'of some ecological instability'.

In fact the examination of sequential imagery from selected areas of the southern Rupununi provides negligible direct evidence of changes in the savanna-forest boundary. This is perhaps surprising in view of the socio-demographic changes that have occurred in the area and the more concentrated cultivation that has latterly developed in parts of the forest. Admittedly, the increasing anthropic impact was only partly achieved at the time of the inital photographic cover, but, if forest retreat had previously been occurring, as earlier writers assumed, one would have expected some evidence of it to have been visible on the recent imagery. Even the Landsat images tend to confirm, within the limits of their resolution, the general persistence of the forest cover. At the very least, one must conclude that any changes in the savanna-forest boundary, if they have occurred, have been more gradual or spasmodic than was hitherto thought to be the case.

The apparent stability of the savanna-forest boundary presumably reflects the fact that Wapisiana shifting cultivation is sufficiently adaptive to sustain itself and the forest matrix in the face of increasing land pressure. Admittedly, the traditional advantage of field dispersal in the forest, aiding woody regrowth

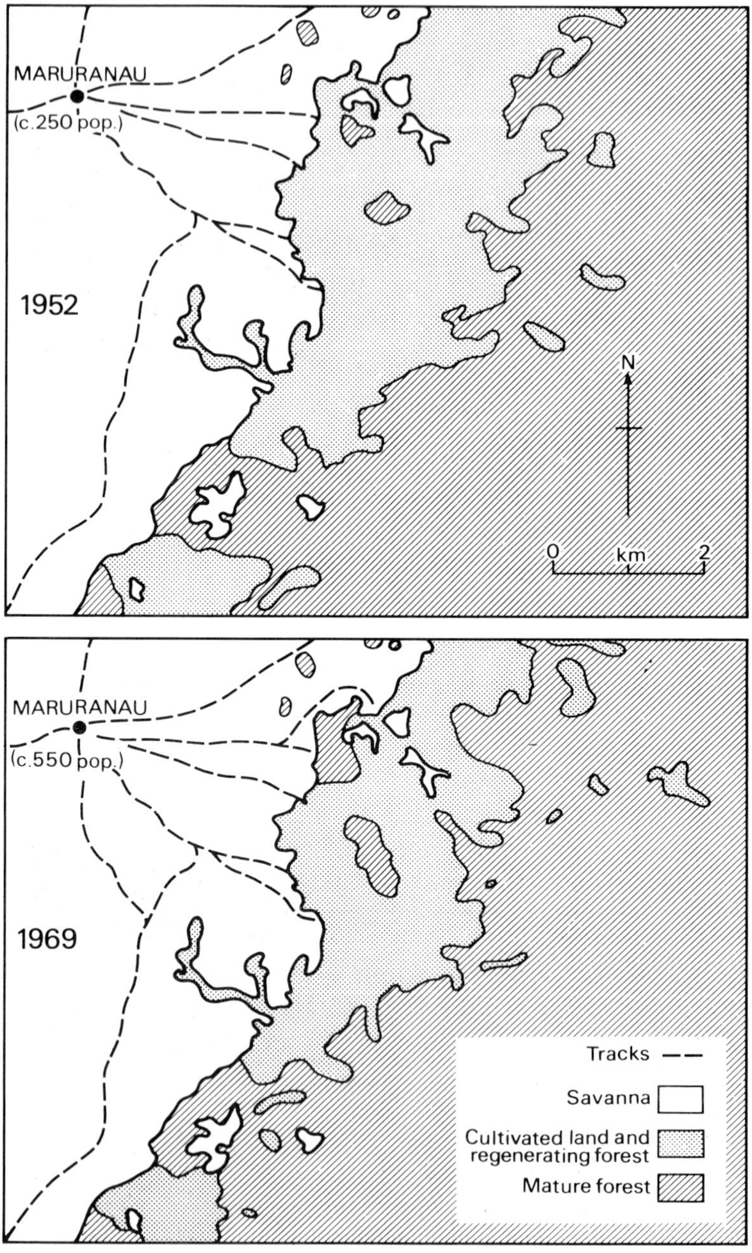


Fig. 9 Land cover near Maruranau, southern Rupununi, based on aerial photographs from 1952 and 1969.

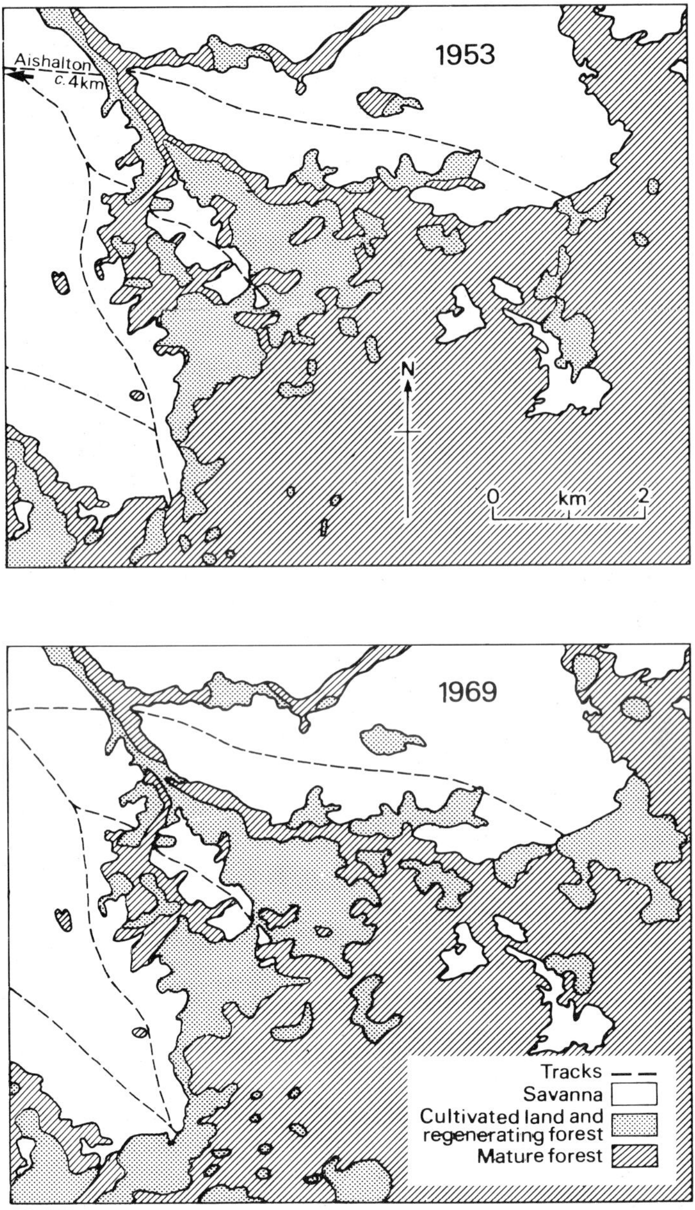


Fig. 10 Land cover near Aishalton, southern Rupununi, based on aerial photographs from 1953 and 1969.

in abandoned fields and minimizing the spread of fires from newly burnt fields to older weedy plots, has been diminished as the concentration of cultivation has increased. However, this has perhaps been compensated by some reduction in fire risk, as a result of the increasing physical separation of fields and villages over recent decades (Fig. 3). Herbaceous fires commonly occur around villages in the savanna, but, in the isolated field areas, burning is more easily controlled and generally restricted to the essential clearance of newly cut fields. This lowers the risk of fire-deflection of older fields to savanna.

An additional factor affecting the fire risk in the southern Rupununi is weed incidence. Although herbaceous weeds occur in cultivation areas, providing fuel for burning, their incidence is relatively low. This may be partly related to soil conditions, but also seems to reflect the dominance of manioc cultivation. As Dummett (1968) indicates, early weeding is required in the manioc fields but, as a shading canopy develops, further weeding becomes unnecessary. This contrasts, for example, with areas in lowland Papua New Guinea where soils are more fertile and weed invasion is a persistent problem in more open fields containing yams, sweet potato and taro (Eden, 1974, 1984). A particular problem in that region is *Imperata cylindrica*, which is a widespread and aggressive colonist of disturbed habitats, especially old field areas. In Guyana there does not appear to be as aggressive a counterpart to this species (Nye and Greenland, 1960; Eden, 1984).

In the southern Rupununi, there is also considerable incentive to achieve effective forest regeneration in old field areas, not least because the savanna soils, lacking the nutrient capital available in the forest system, are generally unsuited to cultivation. It is thus desirable for the Wapisiana to achieve forest regeneration in old field areas, so that the supply of accessible cultivable land can be maintained. As indicated, certain features of the agrosystem currently facilitate this objective. At the same time, there are limits to the resilience of the forest system and these preclude any assumption that Wapisiana cultivators will inevitably continue to achieve forest regeneration in abandoned field areas, especially if recent population growth and land pressure persist. In the longer term, a revised relationship may well be required between the ecological and socio-demographic variables in the region. How this will be articulated in terms of resource use and settlement pattern remains to be seen. Some development of the existing road network and transport facilities and the possible relocation of some settlements and service facilities may be helpful. Agricultural innovation, including the development of agroforestry and associated, integrated systems involving livestock, may also contribute, in the longer term, to resolving the growing pressure on local land resources. In the past some government involvement in research and extension schemes has occurred (Eden, 1966), and additional scope currently exists for positive action of this kind.

What is more important in the present context, however, is to stress the role of remote sensing as an aid to land planning and management, both in the southern Rupununi and in other Amazonian environments where indigenous or colonist populations are seeking to sustain extensive agricultural systems in the face of increasing, local population densities. In this respect, remote sensing can be initially employed, as in the present study, to examine and monitor the status of the existing land use system. It can also provide information on the dynamics of the vegetation cover. Such data need to be acquired as part of any formal evaluation of land resources for development. Subsequently, remote sensing can be used to monitor the progress and environmental impact of any resultant agricultural changes or innovations, be they planned or spontaneous. Remote sensing applications of this kind, necessarily undertaken in parallel with ground survey, are no guarantee of effective land management, but they do provide basic information without which such management is less easily achieved.

As is apparent from the present study, the type of imagery used for monitoring purposes is of major importance. The existing aerial photography of the southern Rupununi provides excellent ground information, but the most recent coverage is fifteen years old and a new survey is required. If such is not feasible, alternative use can be made of Landsat imagery. Unfortunately, during the 1970s, very few Landsat images were acquired for the southern Rupununi, and it was not until the present decade that cloud-free MSS coverage became available. Admittedly, the resolution of the existing Landsat imagery is inferior to that of the aerial photographs, creating particular problems for the investigation of localized, indigenous cultivation, but, in the absence of up-to-date photographic cover, Landsat can still provide useful data. In the future, the acquisition of additional MSS imagery will allow digital change detection analysis to be undertaken, while the improved resolution of the Landsat Thematic Mapper (TM) will further increase the scope of such analysis.

In broader terms, the monitoring of indigenous agriculture and its environmental impact is relevant in the search for more sustainable land exploitation in Amazonia. It is possible to introduce more intensive or permanent resource use systems to the region, but hitherto such introductions, whether involving peasant farmers or larger-scale commercial enterprises, have frequently failed to develop into productive, sustained yield systems (Eden, 1978). By understanding the dynamics of existing agrosystems and their environmental impact, it should be possible to design more appropriate and sustainable agrosystems for the future, whether for indigenous groups like the Wapisiana or for recent colonists in the rainforest (Eden, 1982). In either case, sequential remote sensing can contribute significantly to the acquisition of relevant data.

## ACKNOWLEDGEMENTS

Thanks are due to Professor T. L. Hills, McGill University for making available some aerial photographs of southwest Guyana. Acknowledgement is made to the Ministry of Defence, London for permission to reproduce aerial photographs and to the National Oceanic and Atmospheric Administration, USA and the Conselho Nacional de Desenvolvimento Científico e Technológico/Instituto de Pesquisas Espaciais, Brazil for allowing the reproduction of Landsat images.

## REFERENCES

Bishop, J. P., 1978. Development of a sustained yield tropical agrosystem in the upper Amazon. *Agro-Ecosystems*, **4**, 459–461.

Denevan, W. M., 1978. The causes and consequences of shifting cultivation in relation to tropical forest survival. In Denevan, W. M. (Ed.), *The Role of Geographical Research in Latin America*. Muncie: Conference of Latin Americanist Geographers, pp. 67–81.

Denevan, W. M., Treacy, J. M., Alcorn, J. B., Padoch, C., Denslow, J. and Paitan, S. F., 1984. Indigenous agroforestry in the Peruvian Amazon: Bora Indian Management of swidden fallows. *Interciencia*, **9**, 346–357.

Ducke, A. and Black, G. A., 1953. Phytogeographical notes on the Brazilian Amazon. *Anais Acad. bras. Cienc.*, **25**, 1–46.

Dummett, M. J., 1968. Notes on Amerindian agriculture. *Savanna Res. Ser., McGill Univ.*, **9**, 21–28.

Eden, M. J., 1964. The savanna ecosystem—northern Rupununi, British Guiana. *Savanna Res. Ser., McGill Univ.*, **1**, 1–216.

Eden, M. J., 1966. Some problems of Amerindian development in the Rupununi District. *J. Guyana Museum Zoo*, **42**, 36–40.

Eden, M. J., 1971. Some aspects of weathering and landforms in Guyana (British Guiana). *Z. Geomorph.*, N.F. **15**, 181–198.

Eden, M. J., 1974. The origin and status of savanna and grassland in southern Papua. *Trans.Inst.Brit.Geog.*, **63**, 97–110.

Eden, M. J., 1978. Ecology and land development: the case of Amazonian rainforest. *Trans.Inst.Brit.Geog.*, N.S. **3**, 444–463.

Eden, M. J., 1980. A traditional agro-system in the Amazon region of Colombia. In Furtado, J. I. (Ed.), *Tropical Ecology and Development. Proceedings of 5th International Symposium of Tropical Ecology*. Kuala Lumpur: International Society of Tropical Ecology, pp. 509–514.

Eden, M. J., 1982. Silvicultural and agroforestry developments in the Amazon basin of Brazil. *Commonw. For. Rev.*, **61**, 195–202.

Eden, M. J., 1984. Forest cultivation and derived savanna and grassland: comparative studies from southern Papua and south west Guyana. Paper delivered to *International Savanna Symposium*, Brisbane, Australia, May 1984 (in press).

Evans, C. and Meggers, B. J., 1960. *Archaeological Investigations in British Guiana. Bulletin 177.* Washington: Bureau of American Ethnology, Smithsonian Institution.

FAO, 1966. *Report on the Soil Survey Project, British Guiana*. vol. 3, *Reconnaissance Soil Survey*. Rome: Food and Agriculture Organization.

Hecht, S. B., 1982. Agroforestry in the Amazon basin: practice, theory and limits of a promising land use. In Hecht, S. B. (Ed.), *Amazonia. Agriculture and Land Use Research*. Cali: Centro Internacional de Agricultura Tropical, pp. 331–371.

Hills, T. L., 1964. *Progress Report, Savanna Research Project*. Unpublished report, Geography Department, McGill University, Montreal.

Hills, T. L., 1968. Amerindian agriculture. *Savanna Res. Ser., McGill Univ.*, **9**, 7–20.
Hills, T. L., 1973. The savanna biome: a case study of human impact on biotic communities. In Manners, I. R. and Mikesell, M. W., (Eds), *Perspectives on Environment*. Washington: Association of American Geographers, pp. 342–373.
McCann, T., 1972. *Some Recent Processes of Acculturation and of Development among the Wapisiana of Guyana*. Unpublished manuscript, Georgetown.
Meggers, B. J., 1971. *Amazonia. Man and Culture in a Counterfeit Paradise*. New York: Aldine Atherton.
Migliazza, E. C., 1978. The integration of the indigenous peoples of the Territory of Roraima, Brazil. *Int. Wk Grp Indigenous Affairs*, **32**, 1–29.
Myers, J. G., 1936. Savannah and forest vegetation of the interior of the Guiana plateau. *J.Ecol.*, **24**, 162–184.
Myers, N., 1980. *Conversion of Tropical Moist Forests*. Washington: National Academy of Sciences.
Nye, P. H. and Greenland, D. J., 1960. *The Soil under Shifting Cultivation*. Farnham Royal: Commonwealth Agricultural Bureau.
Potter, L., 1978. The Amerindians of Guyana and their environment. Paper presented to *10th Conf. Caribbean Historians*, St. Thomas, Virgin Islands, March–April 1978.
Prance, G. T. and Schubart, H. O. R., 1978. Notes on the vegetation of Amazonia. 1. A preliminary note on the origin of the open white sand campinas of the lower Rio Negro. *Brittonia*, **30**, 60–63.
Salisbury, R. F., 1968. Ethnographic notes on Wapisiana agriculture. *Savanna Res. Ser., McGill Univ.*, **9**, 7–20.
Scott, G. A. J., 1978. Grassland development in the Gran Pajonal of eastern Peru. *Hawaii Monogr. Geog.*, **1**, 1–187.
Sombroek, W. G., 1966. *Amazon Soils*. Wageningen: Centre for Agricultural Publications and Documentation.
Talbot, L. M. and Kesel, R. H., 1975. The tropical savanna ecosystem. *Geosci.Man*, **10**, 15–26.
Taylor, B. W., 1968. Changes in the forest/savanna boundary in the perhumid lowlands of New Guinea. *Savanna Res. Ser., McGill Univ.*, **13**, 62–65.
Waddell, E. W., 1963. *The Anthropic Factor in a Savanna Environment*. M.A. Thesis, McGill University.

Remote Sensing and Tropical Land Management
Edited by M. J. Eden and J. T. Parry


# CHAPTER 15

# *Monitoring Tropical Wetland Rice Production Systems: a Test for Orbital Remote Sensing*

JEAN-PAUL MALINGREAU
*NASA GSFC, Greenbelt, U.S.A.*

Rice is the major food crop of tropical Asia, and is the focus of intensive land and water management practices. The crop is grown across a large and varied territory and under a wide range of cultivation practices. It forms the basis of the economy of many countries and provides the staple food of numerous people. Its production system can be examined from many viewpoints; the bulk of rice-related research has naturally concentrated on the agronomy of the crop, with emphasis on the selection of high-yielding and pest-resistant varieties, and on improvements in cultivation practices. Advances made during the 'green revolution' are due to intensive research of this kind. During the last decade, when the vagaries of the weather created large fluctuations in world food output, crop production modelling became an object of attention. Important advances were made in the formulation and use of yield models for crops such as wheat (Robertson, 1973), grass (Seligman, 1976), sorghum (Arkin *et al.*, 1976), corn (Hanks and Ashcroft, 1977) and, to a lesser extent, rice (Van Ittersum, 1972; Iwaki, 1975; Murata, 1975; Robertson, 1975; Angus and Zandstra, 1979). Most of these models were based on a physiological or statistical understanding of plant responses to the weather. Many of them were predictive models that led to aggregate estimates of crop performance in a given growing unit, i.e. a field, farm or region. Currently, there is a trend towards the renewed examination of intermediate levels of the production hierarchy, i.e. irrigation systems, river basins, etc. This trend is related to the present emphasis placed by development programmes on land and water resource management, environmental conservation, agroecosystem approaches, equity in access to resources and the like.

There is thus a wide spectrum of information being sought about the rice crop. Before assessing the role of crop monitoring technologies, it is important to specify for what purpose the information is being collected and to whom it is addressed. Information required by the central planner or the food emergency coordinator will be different to that required by the irrigation manager, the plant breeder or the agronomist. A careful identification of the various needs is required in order to evaluate the data available from satellite remote sensing. In addition, a better understanding of crop growing conditions and associated data requirements should lead to a more realistic appraisal of the operational value of remote sensing.

Remote sensing has made considerable progress during the last decade. The technology offers new perspectives for the inventory and monitoring of natural resources, but its application to the specific problems of land and water management in tropical agriculture has presented challenges that are far from being resolved. The difficulties are especially acute when the objective is to monitor the fast-changing features of irrigated agriculture. The over-optimistic attitudes which accompanied the launching of the first earth resource satellites in the early 1970s have now been moderated; current emphasis is being placed on the integration of satellite data with existing models of crop growth and production and with conventional ground monitoring activities.

This paper seeks to identify how existing space observation techniques can be used to derive a better understanding of crop production patterns in traditional, wetland rice production areas of southeast Asia, particularly Indonesia. Two data sources will be examined: the Landsat multispectral scanner (MSS) and the Advanced Very High Resolution Radiometer (AVHRR) carried on NOAA satellites. Underlying the presentation is an attempt to define more operational approaches to rice crop monitoring in the tropics.

## THE WETLAND RICE PRODUCTION SYSTEM

'Wetland paddy' refers to a system of rice cultivation where the crop is submerged for the greater part of its life cycle. Favourable topographic positions are sought by rice farmers, and fields are levelled and surrounded by low bunds. Water is accumulated by irrigation or by direct rainfall. Climate, land and water resources, crop management practices, and the plant itself comprise the four major subsystems of the production system. The interconnections of the system are outlined in Fig. 1. Within this general framework, the potential contribution of remote sensing techniques will be examined.

### The climate subsystem

The weather drives the production in the sense that it determines the period favourable for rice growth. Day-length is a limiting factor at higher latitudes; the critical photoperiod is the longest one within which the plant will flower. The photoperiodic sensitivity depends upon the crop variety. Threshold

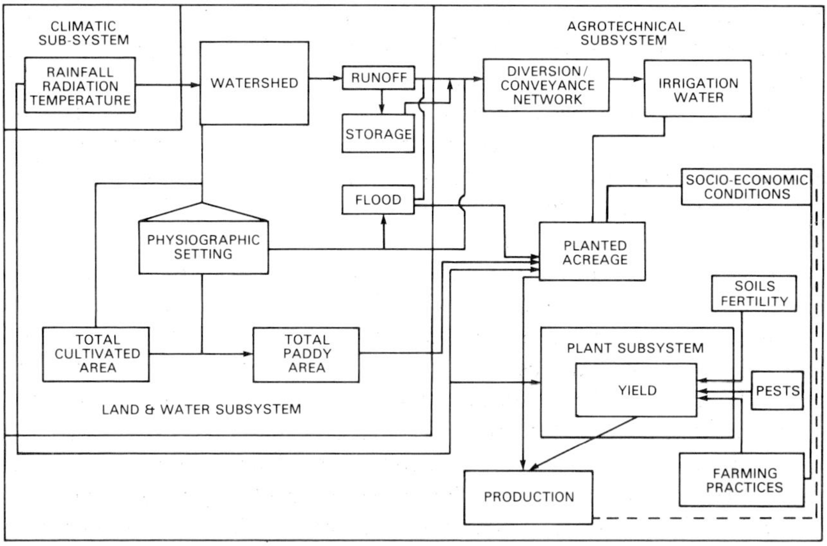


Fig. 1 A wetland rice production model, showing the climatic, land and water, agrotechnical, and plant subsystems.

temperatures determine the length of the growing period (vegetative phase) and floral initiation; low and high temperatures can thus adversely affect the crop performance. Temperature summation varies with varieties but, in general, rice requires between 2000 and 4000 degree-days from sowing to maturity, or a period of 80–160 days when grown at a mean temperature of 25°C. High radiation levels are especially important during the last 45 days of the crop (IRRI, 1970).

The estimation of the mean growing period for a given region is basic to the understanding of crop phenology and the dynamics of production. Further, any assessment of the impact of the weather on crop performance rests upon the determination of the local biological time scale. The verification of crop phenological models is an important application of remote sensing observations.

In the tropical belt, and at elevations where the mean temperature is above 18–20°C, the availability of water is often the major limiting factor in crop growth. In the absence of reservoirs of significant size, it is the rainfall-runoff regime that normally regulates field activities and crop performance. Management practices, including irrigation scheduling, labour allocation, and crop patterns, modulate the details of planting and thus induce spatial and temporal variations in cropping.

An agroclimatic stratification, based upon the occurrence of monthly rainfall of significant value for rice growth, provides important background information for rice monitoring. Such a stratification will reduce the range of possible cropping patterns in a given region and thus guide the analysis of field or remotely sensed data. An agrometeorological stratification based on the number of consecutive wet ($p > 200$ mm) and dry ($p < 200$ mm) months has been found to be effective for this purpose (IRRI, 1974; Oldeman, 1975).

## The land and water subsystem

Land and water resources largely determine the relationship between a production area and crop performance. Cropping patterns and, to some extent, production are related to both broad landscape characteristics and local edaphic and hydrologic conditions (Geiser and Voelke, 1978; Morris and Zandstra, 1978; Malingreau, 1980). For example, the position of a production area on a mountain slope or the distance and relative elevation of a field with respect to water supply influence the scheduling of field activities and water availability, especially if the supply is limiting. Drainage and flood conditions in a coastal plain determine the possibility of planting one, two or even three crops per year.

A land- and water-related physiographic classification is thus another useful tool for understanding production patterns. An example of such a classification is given in Table 1. Physiographic or land stratification requires observations of a synoptic nature, and remote sensing can frequently assist in determining the relationship between site characteristics and production.

Table 1. A physiographic stratification of rice-growing areas in West Java.

| |
|---|
| 1. *Coastal Plain System* |
| A. Alluvio-marine subsystem |
| (marshes, deltaic deposits, ancient seashores, low-lying lands, etc.) |
| B. Alluvial subsystem |
| i. alluvial plain |
| non-flooding area |
| flood-prone area |
| ii. meander belt |
| C. Marine terraces |
| 2. *Alluvio-colluvial System* |
| A. Inter-hill miniplain |
| B. Basin |
| C. Fan |
| D. Valley floor |
| E. Valley head |
| F. Stream-side settings, gorges, gullies, etc. |
| 3. *Hills and Piedmont Transition System* |
| 4. *Volcanic (Mountainous) System* |
| A. Volcanic middle slope |
| B. Volcanic lower slope |
| C. Footslope and piedmont slope |
| D. Volcanic (inter-mountain) plain |

Table 2. Classification of wetland rice production in Java.

1. Irrigated *sawah**: an area that is artificially supplied with water from a surface or underground source.
   1.1 Two or more crops of rice per year: the normal water supply allows at least two crops of irrigated rice per year; in places, five crops in two years may be grown.
   1.2 One or two crops of rice per year: a variable water supply may permit a second crop of irrigated rice in the dry season. If water is short, a secondary (non-rice) crop, known as *palawija*, may be grown.
   1.3 One crop of rice per year: water supply permits only one irrigated rice crop in the rainy season. A secondary crop may be grown in the dry season.
   1.4 Sugar cane cropping: temporarily cultivated, for 16–18 month periods, in lowland rice-growing areas.
2. Rainfed *sawah*: an area in which fields are watered by impounded rainfall, sometimes supplemented by localized runoff collection.
   2.1 One crop of rice per year: the rainy season rice crop is followed by a dry season fallow; the latter is sometimes grazed.
   2.2 One crop of rice per year: the rainy season rice crop is immediately preceded or followed by a secondary crop, depending upon the local rainfall pattern.

* *Sawah*: a cultivated area with level and bunded fields supporting at least one crop of rice in the annual rotation.

## The agrotechnical subsystem of production

Management practices control the system of production within the constraints imposed by climatic and site factors. Technical improvements in irrigation systems, mechanization, use of fertilizers and pesticides, selection of improved varieties, and other factors have drastically affected the rice landscape of many regions of Southeast Asia in the last decade. Cropping practices (including the choice of crop) are in fact in a constant state of adjustment between site potential, technical innovation, and economic conditions.

Wetland rice production systems are commonly classified according to the type of water delivery (i.e. irrigated or rainfed) and the cropping patterns (i.e. one or more crops per year). Several classifications are in use, and comparisons are not always easy (Uhlig, 1969; Chandler, 1979; Malingreau and Christiani, 1981). An example of an hierarchical classification which has been developed in Indonesia is given in Table 2. It is important to note that the availability of an appropriate cropping system classification is a condition for the success of remote sensing observations of the rice crop.

### The plant subsystem

The response of rice plants to environmental factors has been documented through the decades of research and many production models are available in the literature. In phenological models, the developmental stage of the crop is the dependent variable (Jin Zhiqing, 1984). In production models, the crop stage is one of the driving variables. Relative yield and stress models are applied to assess the loss incurred by the plant when exposed to suboptimum environmental conditions. Water availability is an important constraint. Indeed, it has been shown that yield reduction in rice is roughly proportional to the duration of drought (Reyes, 1973; Wickham, 1973). The 'stress-day' concept can be used to quantify drought duration; this is defined as the number of days accumulated over the whole crop period, or critical portions of it, during which the stress intensity is greater than a prescribed value. In the case of wetland rice, stress can be recorded shortly after the soil moisture is below saturation; thus, the presence or absence of surface water in the field can be taken as a general indicator for measuring stress days. This 'yield reduction' approach assumes that, at the time of planting, the crop has an opportunity to achieve a 'potential' yield (Herdt and Bernsten, 1975), which can be determined experimentally for a given environment. 'Maximum possible yield' can also be used; it is the maximum yield value actually observed in a given area. Potential or possible yields are the basic ingredients of agroecological maps.

Regional production models are usually based on generalized relationships which ignore the fact that in traditional agroecosystems many rice varieties may be planted, each responding in its own fashion to environmental influences. Assessing the mix of varieties is a difficult operation which has hitherto been carried out by field sampling (Malingreau, 1980). The use of high yielding varieties, however, has latterly resulted in more uniform planting, especially in better irrigated areas.

More sophisticated models of plant growth are being developed, many of which are very demanding in terms of inputs. For crop modelling over large areas, one is perforce limited to the measurement of a few easily accessible parameters. These may relate directly to the crop itself (i.e. its stress status) or to the environment (i.e. field water, temperature, etc.). At this stage, the sophistication introduced by remote sensing techniques relates more to repetitive measurements over large areas than to any refinement of the choice of parameters. Of necessity, a reliable monitoring system has to focus on simple observations, at least in the first stage of its development.

## GENERAL PATTERNS OF PRODUCTION

As seen in the previous section, various factors influence the patterns of production. Over large areas, the combination of factors can lead to a complex mosaic of rice fields, containing different varieties, submitted to various management practices, and planted or harvested at different times. The heterogeneity of a paddy scene is most striking in areas where perennial water is available and non-periodic varieties can be planted at any time.

These observations, which are familiar to anyone travelling in southeast Asia, lead to an examination of the dynamics of production and a consideration of the level at which they can realistically be assessed. The choice regarding the 'resolution of the dynamic model' (Van Keulen, 1976) is dictated by the amount of information sought in the analysis as well as by practical constraints relating to the system of observation. This section will firstly describe an hierarchy of wetland rice production levels, and then focus on the dynamics of selected levels in the hierarchy.

**An hierarchy of production systems**

It is common practice in crop production studies to divide the plant yield into components and to analyse their individual response to production factors. Experimental results in rice, for example, can be evaluated in terms of average weight per grain, number of grains per spikelet, number of spikelets per panicle or number of productive tillers per plant. Each of these levels, representing increasingly complex organs of production, can be used to analyse the plant performance.

Functional or structural hierarchies can also be examined at higher levels of production, although appropriate criteria are not easily defined. Functional homogeneity, reflecting similar responses by individuals at a given level of the hierarchy, would be an ideal criterion, but remains difficult to measure. Here an hierarchy of production levels is proposed on the basis of increasing spatial variability (Table 3). At each level of the hierarchy above the plant level, a degree of cohesiveness is recognized and the factors which determine intra-level variability are identified. Admittedly, the cohesiveness of a production unit may vary with time, but, barring major changes in the irrigation system or the occurrence of catastrophic events, there are aggregate units of rice production that remain remarkably stable through time. If such stability in cropping patterns, irrigation scheduling, and other production variables can be verified, the monitoring of the rice crop can be undertaken within a pre-established spatial framework whose fundamental production characteristics need only be periodically reviewed (Morris and Zandstra, 1979; Malingreau, 1980).

In many parts of Indonesia, one can recognize, just above the field level, an hierarchical unit called the *golongan* (Table 3) which consists of a large group of fields. This unit exhibits cohesiveness in the sense that it shows continuity in space and uniformity in time. Fields in the unit usually belong to a single water distribution scheme; they develop together and can thus be monitored on the same biological time-scale. Because of the size of the *golongan* (only

Table 3. An hierarchy of wetland rice production levels and their major characteristics.

| Level in the hierarchy | Cohesiveness characteristics | Factors which exhibit intra-level variability | Corresponding administrative units |
|---|---|---|---|
| 1. *Plant* | – | – | – |
| 2. *Field* | input uniformity | biology, soil, water | farm |
| 3. group of fields (*golongan*) | continuity in space; uniformity in time (single biological time scale); climatic uniformity | crop variety; farming practices; crop inputs (fertilizers, pesticides) | hamlet, village |
| 4. Group of *golongan* | continuity in space; non-uniformity in time | crop variety; climatic influences; crop inputs (fertilizers, pesticides); water availability | village, subdistrict |
| 5. Catchment | non-continuity in space; non-uniformity in time | as above, plus more marked climatic differences | district |
| 6. Island | non-continuity in space; non-uniformity in time | as above, plus more marked climatic differences | province |

Fig. 2 Groups of rice fields (*golongan*) at various development stages in the Garut region, Java. 1: Field preparation/transplanting, 2: young crop/tillering, 3: full cover/end of vegetative phase, 4: ripening crop. Black and white, infrared aerial photograph, July 1969 (*By courtesy of the Center for Soil Research and the National Coordination for Surveys and Mapping (BAKOSURTANAL) of the Republic of Indonesia.*)

a few thousand hectares at the most), climatic influences can be considered constant. The main source of variability is introduced by farming practices.

At a higher level in the production hierarchy (level 4 in Table 3), one observes that although fields are contiguous, different cropping schedules are enforced. Across the unit, the crop experiences a different history and is thus subjected to different combinations of climatic variables. If resources are short, different time schedules can make the difference between crop success or failure or between one or two crops per year.

At the catchment level (level 5), the paddies do not present continuity in space, because of topographic variation or competing land uses, nor uniformity in time, because of climatic or hydrologic variations.

In evaluating this hierarchy, it is concluded that level 3 (the group of fields or *golongan*) is a logical unit of observation for rice-producing areas in Indonesia. Because of geographical variations, flexibility must be maintained when identifying such hierarchies in other regions. As illustrated in Fig. 2, aerial photography is an appropriate aid to identifying production hierarchies.

## The dynamics of production

The temporal aspects of production deserve special attention because of their importance in monitoring. The selection of appropriate intervals of observation depends primarily on the dynamics of the system itself. Ideally, the temporal resolution should be such that no significant event occurs between observations, although circumstances at times dictate otherwise in the extensive and fast-changing landscapes of wetland rice.

The dynamics of rice production are here examined, firstly, in relation to the physiological development of the crop and, secondly, in terms of the overall production pattern. While the latter represents the final objective of crop monitoring, it is the former which defines the signals received by a remote sensing instrument.

### *Crop physiological development*

Agronomic observations of grain crops traditionally refer to stages and phases of growth and development which are identified through morphological and, in some instances, physiological observations. Each of the stages (including field preparation and fallowing) has its own dynamic properties; the stage gives a field its particular appearance and defines the rate at which it undergoes change. On a macroscale, observations are related to the reflectance of a mix of land cover features associated with the crop stages within the instantaneous field of view of the sensor. The multispectral signal can be translated into a photophenological stage, and when the various stages are assembled they constitute a 'photocrop calendar' (Hay, 1974), which is an image-related, temporal description of the crop life cycle.

The following five stages are identified in wetland rice cultivation.

*Fallow*. At the end of the dry season the soil in paddy fields is usually bare and dry. A grass cover may appear if soil moisture is available.

*Flooded*. At the onset of the rains or with the arrival of irrigation water the fields are flooded for puddling. The water depth varies from 2–3 to 15 cm. The water may be clear or turbid. Depending upon labour availability, field preparation may take 3–4 weeks. Nurseries are established during this period; nursery time is about 25 days.

*Growing crop*. Tropical rice varieties complete their life cycle within 110–150 days. Vegetative, reproductive, and ripening phases can be recognized. The vegetative phase (tillering, elongation, and booting) determines the total duration of the cycle, since the reproductive phase has a more or less constant duration for all varieties. Heading, panicle initiation, and flowering last about 25–35 days. Panicle initiation and anthesis are critical stages when the plant may be most adversely affected by environmental constraints. Irrigation may be interrupted for a week or ten days for fertilization or weeding purposes.

*Ripening crop*. Leaves change colour in ascending order, although the upper leaves and stems of some varieties may remain green until the end of this phase. Grain filling, milk, and dough stages can be recognized. In some systems, irrigation is stopped during the latter part of this period; in others, water may remain in fields up to harvest.

*Stubble*. When a hand-held knife is used for harvest, a large amount of straw is left in the field. Plant remains are hauled away, burned or later incorporated into the soil.

These stages indicate the temporal variety of land cover types associated with wetland rice cultivation.

### *The production pattern*

Agricultural production ($P$) is the result of crop performance over a given area:

$$P = Y \times A$$

where $Y$ = yield
$A$ = acreage.

This simple equation can be examined in different ways. As previously indicated, the yield component has been the focus of attention by plant physiologists and agronomists. The area component has received less attention in crop modelling, probably because it is, in essence, non-biologically determined. Large-scale monitoring projects are often confronted with serious difficulties in measuring crop area.

Three area-related variables can be identified. The first is the maximum plantable area or paddy ($A_{max}$) which is the total area physically prepared for wetland rice cultivation. The second variable is the planted area ($A_p$), which is a time-dependent variable representing the area where the production process has actually been initiated. The third variable is the harvested area ($A_h$), which is a time-related variable representing the area which has produced an economic return. The variables are related as follows:

$$A_p = A_{max} \times LUC$$

where *LUC* is a Land Utilisation Coefficent ($\leqslant 1$) depending upon season and economic factors

$$A_h = A_p - \text{crop losses}$$

Production is generally calculated on the basis of the harvested area ($A_h$), but it is important from a land and water management point of view to determine the efficiency of the production process. The probability of harvesting an area equal to the planted area depends upon weather conditions, pest attacks, and management practices.

When planted and harvested areas are plotted against time, information is provided on growing periods, the dynamics of field stages, and the overall efficiency of the system. The comparison between planted and harvested areas for three agroecological zones in a large catchment area in Java is shown in Fig. 3. At Bangodua there is a prominent bi-seasonality in the cropping pattern of an irrigated coastal plain, although, in 1978, three crops were planted instead of two because of more favourable rainfall-runoff conditions. A bi-modal pattern of production on a mountain footslope is shown for Sumsel. Irrigation is favourable in such sites and the $A_h$ curve closely follows the $A_p$ curve at a four-month interval. In the high plain area at Garut, planting and harvesting continue throughout the year. Fig. 3 clearly illustrates that crop monitoring requirements vary between lowland tropical environments and temperature-controlled upland zones.

## REMOTE SENSING FOR WETLAND RICE MONITORING

The analysis of a rice production system, which is complex, fractioned, and highly dynamic, provides a severe test for remote sensing techniques. Information on the rice crop, however, is only one of the ingredients for assessing the capabilities of remote sensing techniques. It must also be demonstrated that the crop cannot merely be 'seen' by a given sensor, but that the sensor can do so at appropriate time intervals and with acceptable reliability. In other words, from a resource management point of view, the relevance of remote sensing

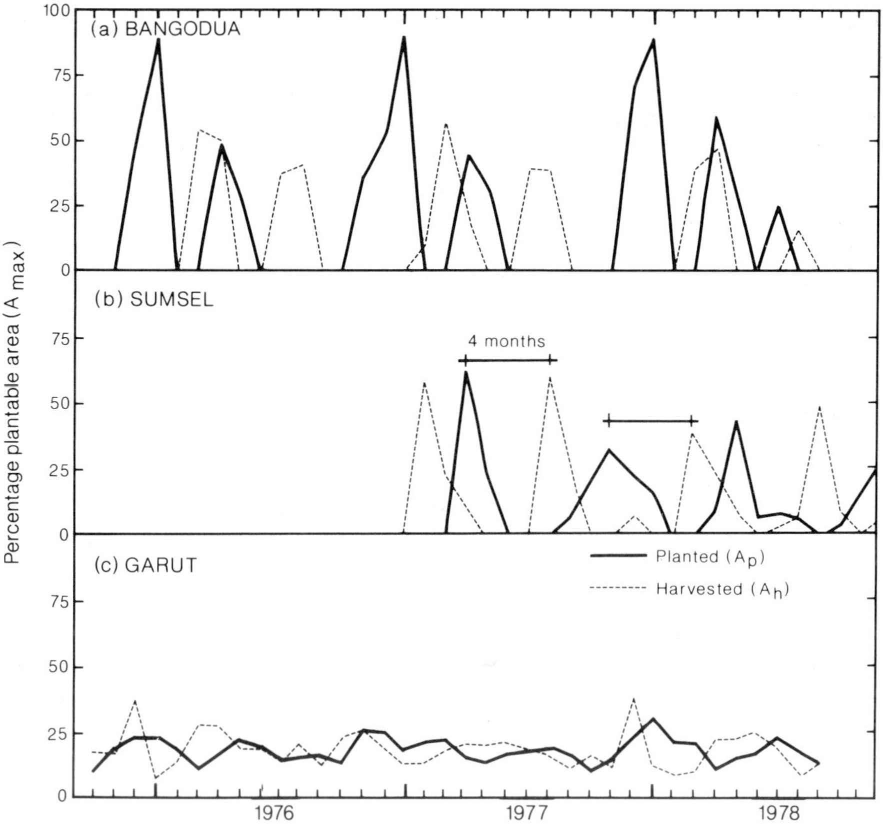


Fig. 3 Time-development curve for planted and harvested rice areas in three agroecological zones of Java: (a) coastal plain, Bangodua, (b) mountain footslope, Sumsel, and (c) high mountain plain, Garut (data from Indonesian Department of Agriculture).

must be demonstrated operationally as well as at the research and development level. This section will examine relevant remote sensing technologies and their operational characteristics.

## Landsat and rice monitoring

The suitability of the Landsat multispectral scanner (MSS) for observing the fragmented landscape of Asia has been a point of some controversy. Initial questions arise about its capability for determining the maximum plantable area for wetland rice and for stratifying the landscape into homogeneous production areas. Further questions relate to the use of multispectral data for reliably identifying and monitoring a rice crop. Of particular concern is the problem of ground resolution with respect to the characteristics of the paddy scene.

The accurate measurement of the maximum plantable area, or paddy, from Landsat data depends upon the availability of images for discriminating between land physically prepared for rice production and for other land uses. Since the main physical characteristics of the wetland rice production system are the levelling of fields, the terracing of the landscape and, in some instances, the construction of irrigation canals, high resolution will commonly be required for accurate identification. However, at certain stages of the crop cycle, the paddy presents characteristics that can be used as identification keys. The most obvious is the relative homogeneity of the paddy during the inter-crop or fallow period by comparison with areas of mixed land use. Again, during the flooding period, the paddy fields can be identified with a high level of confidence, as long as no other crops are under irrigation and the irrigated areas can be separated from natural flooding (overland flow or poor drainage). At maturity, the rice crop may again present some land cover homogeneity which facilitates its delineation. It is usually where irrigation or drainage conditions are less optimal that accurate delimitation of the paddy area becomes more difficult with Landsat imagery.

Aerial photography is an alternative source of data for inventory and stratification purposes. Even with medium- to small-scale photography (1:50 000–1:100 000), a near-exhaustive identification and delineation of wetland paddy is possible with minimum fieldwork (Malingreau, 1978). While this operation is somewhat tedious when carried out over large areas, it would only need to be performed every decade or so under prevailing rates of land use change. Stereoscopic examination of aerial photographs also permits a stratification of the rice landscape at the landform and cropping system levels. Infrared photographs are particularly useful for this purpose. Once the general framework is established, Landsat imagery can then be used for monitoring specific production strata whose basic spatial and temporal characteristics are known. A multistage approach of this kind is the most efficient, and often the only way to extract useful information on the rice crop from an irregular sequence of Landsat images.

The Landsat spectral characteristics are well known and need not be reviewed here (Freden and Gordon, 1983). Transformations taking place in a rice field during crop growth have two distinct components which variously affect the signal received by the sensor. The most important component is the change in the hemispherical leaf reflectance and transmittance. Alterations in the tissue structure lead to the build-up and subsequent disappearance of the photosynthetically active chloroplasts (Suits, 1972). Equally important are the changes occurring to the geometry of the canopy (Gaussman *et al.*, 1976), which lead to shifts in the relative significance of green matter, senescent leaves, litter, and the soil and water background. These changes can be expressed on a time-reflectance curve (Fig. 4).

Vegetation indices can be used to enhance crop features through a manipulation of reflectance values. Indices have the advantage of standardizing the representation of crop spectral responses. Since infrared and red reflectance

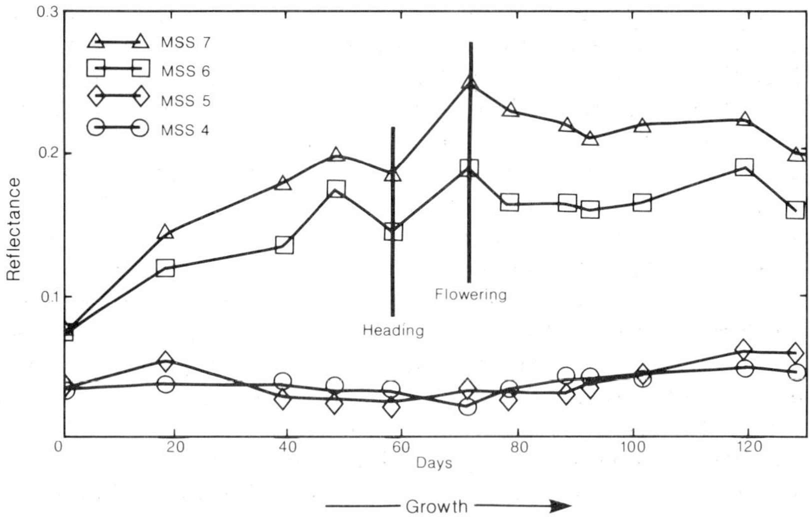


Fig. 4 Changes of reflectance with time for growing rice in four Landsat bands (after CCE, 1981).

exhibit inverse behaviour during plant growth, it is useful to examine their ratio. Fig. 5 illustrates a time sequence for Landsat MSS band 7:5 ratio values, as rice fields progress from flooding to harvest. Although the effect of the soil-water background is clearly detected, it is noteworthy that the band 7:5 ratio remains essentially stable up to the tillering-elongation period. This means that rice plants cannot be detected earlier than four to five weeks after transplant. When this period is added to the nursery time, a detection threshold period of almost seven weeks (out of a total crop duration of 17 weeks) is obtained. From an orbiting satellite, with a 16–18 day revisiting period, a rice crop can therefore be rather elusive. Spectral measurements performed with field radiometers in the Niger basin in West Africa (CCE 1981) yield a similar range of values to those in Fig. 5.

Another approach, which has not yet been applied to rice in the tropics, is the calculation of a 'Greenness transform' (Kauth and Thomas, 1976). In this case, the rice crop follows a trajectory which can be fitted to an exponential function of the type $Y = At^{B}e^{-Ct^{D}}$ (Fig. 6). This is similar to that derived by Badhwar (1980) from Landsat reflectance measurements of wheat in the United States. The coefficients of the greenness profile can be calculated on a regional basis, and the curve provides a reference for estimating the date of individual crop stages. Other vegetation indices are available, but few have been applied to rice monitoring.

Temporal trajectories of spectral change are useful in the classification of remote sensing data, especially where several crop types need to be separated on limited, temporal series of images. 'Biowindows' can be identified by

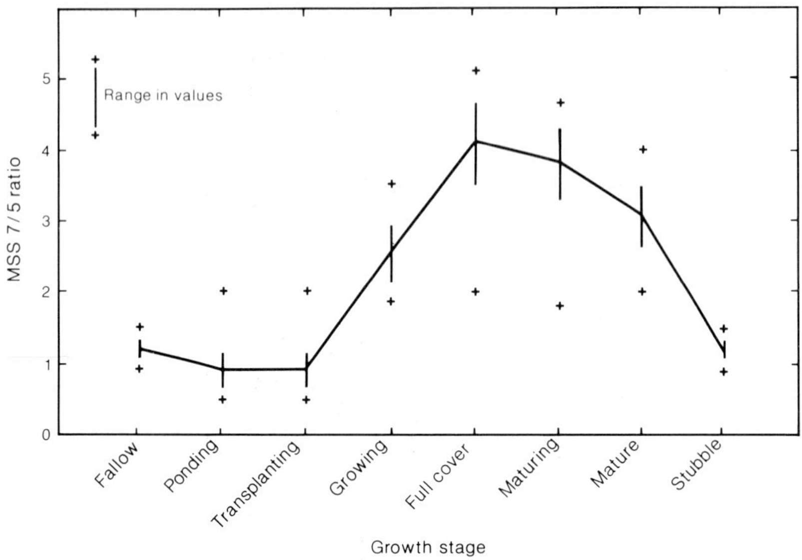


Fig. 5 Changes in the Landsat MSS band 7:5 ratio during rice crop development in West Java (after Malingreau, 1980).

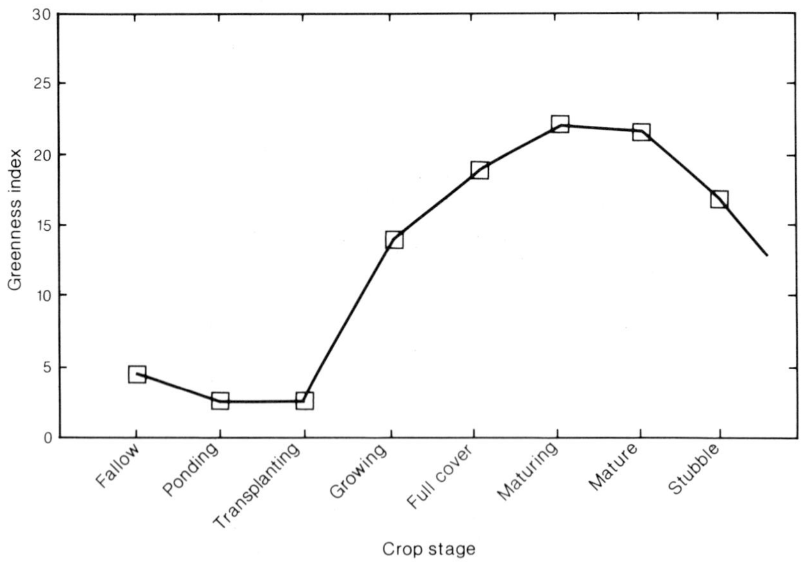


Fig. 6 Changes in the Landsat-derived 'greenness index' during rice crop development in West Java (after Malingreau, 1980).

Table 4. Photophenological calendar for wetland rice based on data from Java.

| Crop stage | Enhanced colour composite image (bands 4, 5, and 7) | Single-band MSS images 4 | 5 | 6 | 7 |
|---|---|---|---|---|---|
| 1. Fallow | light grey, light yellow; brown and green patches | white-light grey | white-light grey | white-light grey | white-light grey |
| 2. Field preparation | dark blue, blue, dark to light green | medium grey | medium grey | dark grey | dark grey-black |
| 3. Transplanting | dark blue, blue, dark to light green | medium grey | medium grey | dark grey | dark grey-black |
| 4. Growing crop (tillering) | pink, blue | medium grey | dark grey | light grey | light grey |
| 5. Full cover; end of vegetative period | red (bluish tint) | dark grey | dark grey-black | light grey | light grey-white |
| 6. Maturing | reddish-orange | medium grey | medium grey | light grey | light grey |
| 7. Maturing | orange, yellow, some pink | medium grey | medium grey | light grey | light grey |
| 8. Stubble/fallow mix | yellow, pink, blue, green | mix of different tones | | | |

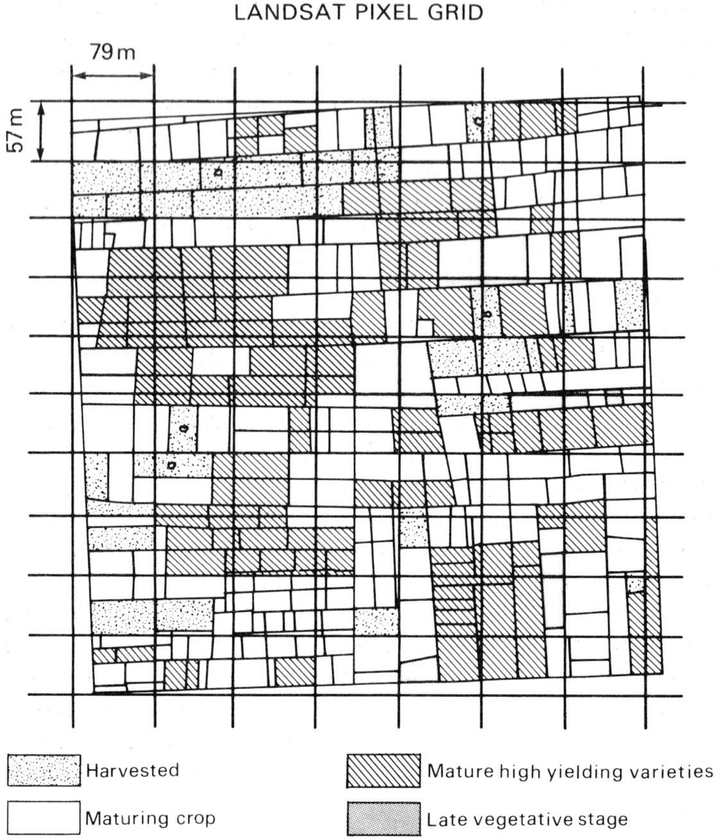


Fig. 7 Paddy fields in the coastal plain of Java, mapped from an aerial photograph (scale 1:5000), with the Landsat MSS pixel grid superimposed.

comparing the spectral profiles of the crops to be separated. In addition, the curves offer points of reference for the assessment of crop stress. Thus, if normal crop behaviour is known, shifts in the trajectory can be interpreted in terms of abnormal behaviour.

Since most Landsat interpretation is still carried out by visual analysis of individual MSS bands or their colour composites, a rice photophenological calendar is presented in Table 4. The information is derived from analysis of single-band images (MSS bands 4, 5, 6, and 7) and an enhanced colour composite (Plates 15 and 16). When using this approach, it is important to remember that processing techniques may influence the colour balance of individual Landsat prints.

The ground resolution of the Landsat multispectral scanner (about 0.5 ha) is often considered a major limitation in the examination of the fragmented landscape of South and Southeast Asia. As indicated in Fig. 7, the average field size is considerably smaller than the MSS pixel size. The Landsat colour composite image (bands 4, 5, and 7) of the same area taken at the end of the vegetative stage shows that the pixel pattern does not coincide with the individual field layout (Plate 15). However, the image does confirm the relatively homogeneous crop conditions that prevail in the area. Similarly, a colour composite of a rice production area in montane Java (Plate 16) reveals clear groups of fields at different stages of crop development. For monitoring purposes, it is more useful to look at these general patterns of production than at the individual fields. Even if higher resolution data were available for this complex landscape, they could place an inordinate burden on data processing systems and would probably reduce interpretation accuracy because of noise in the data set (Irons *et al.*, 1985). High resolution data are even less important in a multistage procedure, where aerial photography and fieldwork establish the framework within which the Landsat data are analysed.

The success of monitoring depends on the ability of the sensor to measure crop-related parameters and to do so at appropriate intervals. As previously indicated, wet rice cultivation essentially requires year-round monitoring. For the individual rice crop, at least monthly observations are required; and since important, yield-determining changes may occur over shorter periods, a frequency of twice a month is preferable. Unfortunately, the Landsat acquisition record over the past 12 years has not been encouraging, particularly in rice-growing areas of tropical Asia. In the northern Cimanuk basin of Java, for example, only 11 images (<30% cloud cover) have been acquired during the period 1971–83, of which no more than six were of good quality. Under these circumstances, one is tempted to dismiss Landsat for agricultural monitoring.

Cloudiness is of course partly to blame but some caution is required since recent analysis suggests that, in rice areas affected by monsoon/orographic weather conditions, the period 0900–1000 hours, i.e. the time of the Landsat overpass, does not exhibit levels of cloudiness significantly higher than in some temperate areas (Malingreau, 1980). The lack of ground receiving stations in Asia prevented direct data acquisition until the early 1980s. Receiving stations have now been established in India, Thailand, Bangladesh, Indonesia, and Australia, but improved observation is still required for effective rice monitoring.

## The Advanced Very High Resolution Radiometer (AVHRR) and rice monitoring

The AVHRR carried on the NOAA series of weather satellites is a less familiar instrument for vegetation, land, and water resource analysis, but recent studies have shown that the sensor is well suited to vegetation monitoring (Townshend and Tucker, 1981; Schneider and McGinnis, 1982). As yet, wetland rice has not been investigated, but preliminary results indicate that this high frequency/low resolution sensor could be ideally suited for the purpose.

The NOAA-7 satellite is a polar-orbiting, sun-synchronous satellite of the TIROS-N series (Kidwell, 1983). It operates at an altitude of 850 km, with local equatorial crossing times at 0230 and 1430 hours. Its orbital period is 102 minutes, and its repeat cycle is 9.2 days. Because of its wide scanning swath (2700 km), the satellite can effectively revisit every area of the globe on a daily basis, thereby increasing the probability of obtaining cloud-free data. The AVHRR has the following spectral channels:

Channel 1: 0.58–0.68 $\mu$m
Channel 2: 0.72–1.10 $\mu$m
Channel 3: 3.55–3.93 $\mu$m
Channel 4: 10.5–11.5 $\mu$m
Channel 5: 11.5–12.5 $\mu$m

The visible and near-infrared channels contain information relating to vegetation, while the thermal channels allow the observation of surface temperature patterns. The instantaneous field of view is 1.4 mrad, which represents a ground resolution of 1.1 km at nadir. Because of angular effects, this resolution decreases towards the edge of the scanning swath.

The NOAA data can be directly received in the so-called LAC (Local Area Coverage) format which offers full resolution. Receiving stations are now operational in India, Bangladesh, Thailand, Indonesia, Philippines, and the People's Republic of China. A resampling of the data is performed to produce the GAC (Global Area Coverage) data, which are available on a daily basis at NOAA-NESDIS, Washington.

Although the AVHRR is characterized by low resolution and by broad band width, its data have the following advantages for crop monitoring:

1. high temporal resolution and thus a high probability of obtaining cloud-free images;
2. appropriate channels and band widths for the calculation of vegetation indices;
3. direct read-out by stations in range and thus a reduction in data delivery time;
4. a lower data load than higher resolution sensors;
5. low cost of data acquisition and processing for repetitive coverage.

As already mentioned, indices derived from the visible and near-infrared reflectance data are highly correlated with vegetation conditions. A common index is the Normalized Difference Vegetation Index (NDVI), derived as follows:

$$\text{NDVI} = \frac{(\text{Channel 2} - \text{Channel 1})}{(\text{Channel 2} + \text{Channel 1})}$$

From this, a Global Vegetation Index (GVI) has been made available at NOAA on a weekly basis since April 1982 (NCDC-NOAA, 1984). The index is based on the GAC data, resampled to a 15 km resolution at the equator, and it displays the maximum NDVI for the week. NDVI values can also be calculated from the 1.1 km LAC data.

Vegetation index analysis has been applied to various natural ecosystems and the index variations have been shown to follow changes in the above-ground biomass (Tucker *et al.*, 1983). Since the technique provides incomparable operational advantages, a preliminary attempt has been made to assess the vegetation index response to the dynamics of rice production. Examples of temporal variations in the NDVI are presented from the north coastal plain of Java and from a wheat/rice-growing area near Shanghai in the Jiangsu Province of China. To remove disturbances caused by cloudiness and haze, the weekly GVI data have been maximized over tri-weekly periods (Holben, 1985).

Admittedly, the coarse spatial resolution offered by the GVI product is not optimum for monitoring this kind of landscape. The response represents an aggregate of influences on the red and infrared reflectances, but, as indicated in Figs 8 and 9, the seasonality of the agroecosystems is well marked. The NDVI for the Java scene (Fig. 8) shows the rainy and dry season crops and the two inter-crop periods. It also illustrated differences in the cropping schedule and the cohesiveness of operations from year to year. Because of an extended dry season in 1982, new planting did not take place until December to January, while in 1983, after a normal dry season, planting began in late October to early November. The low NDVI values for the 1982 dry season also reflect

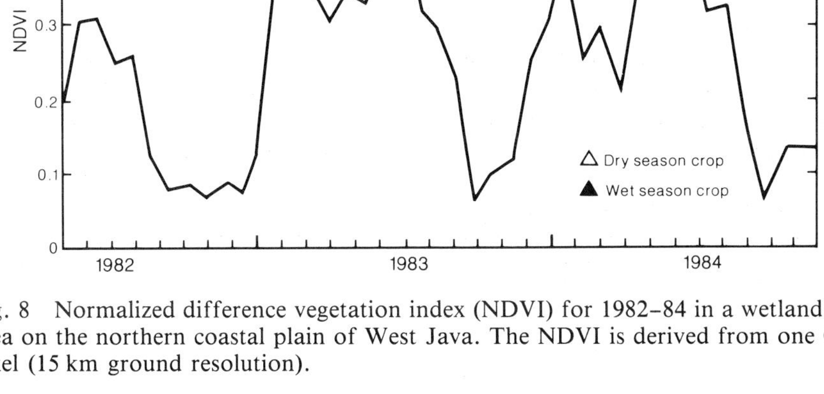


Fig. 8 Normalized difference vegetation index (NDVI) for 1982–84 in a wetland rice area on the northern coastal plain of West Java. The NDVI is derived from one GVI pixel (15 km ground resolution).

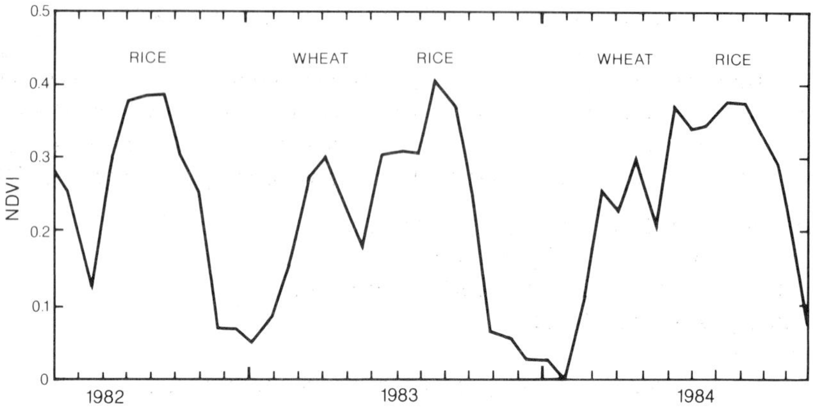


Fig. 9 Normalized difference vegetation index (NDVI) for 1982–84 in a wheat/rice area near Shanghai, Jiangsu Province, China. The NDVI is derived from one GVI pixel (15 km ground resolution).

poor growing conditions at that time. Research is now under way to relate the amplitude, periodicity and smoothness of the NDVI curves to cropping patterns and growing conditions. The implications for crop forecasting, disaster monitoring, production assessment, irrigation scheduling, and so on are only beginning to emerge (Malingreau, 1984).

The NDVI data for Jiangsu Province, in the subtropical belt of China, shows a bimodal cropping system characterized by a wheat/rice alternation (Fig. 9). The wheat is grown in the winter and harvested in the spring; the rice is then planted. The maximum NDVI value for the wheat is lower than for the rice, reflecting possible differences in green biomass at full growth. The NDVI summer peak is relatively wide, and does not immediately show the two successive crops of rice that are sometimes planted (Jin Zhiqing, 1984). Because of the rapid succession of these rice crops, the land cover during the transition period is probably too heterogeneous to be detailed by the data. In Plate 17 a NOAA AVHRR image of Jiangsu Province shows the spatial distribution of the NDVI during the wheat-growing period.

Despite the obvious limitations of the AVHRR in terms of ground resolution, it is evident that a sequence of data, as in Figs 8 and 9, can provide important information for the verification of phenological models and the assessment of crop growing conditions. The full interpretation of the data requires a better understanding of the mechanisms which produce the visible and infrared signals at such coarse resolution. However, the NOAA weather satellites are the only platforms that have so far demonstrated a real operational value for monitoring rice agroecosystems.

## AN OPERATIONAL REMOTE SENSING SYSTEM FOR CROP MONITORING IN THE TROPICS

An operational remote sensing system has different requirements compared with a research or development system. Relevant considerations include the user-preparedness to handle the data as well as the remote sensing technology itself. The following characteristics are sought in an operational system (Benton, 1982):

1. the system must be capable of answering the user's needs;
2. it must provide continuity of data;
3. it must be reliable within acceptable cost limits;
4. it must provide timely data;
5. it must be consistent;
6. it must be amenable to changes as the user's needs develop.

Operational criteria also depend on the type of application envisaged. In Asia, spacecraft remote sensing can be judged operational for geologic and cartographic applications, for national forest inventories and for long-term land use and land cover monitoring. For short-term monitoring of crop growth, the system still has serious limitations. Current uncertainties due to cloud cover or technical failure drastically reduce the value of the Landsat platform. Moreover, even if the system were able to produce the required data, it is arguable that no country in Asia is ready to exploit fully the resultant data loads. Problems of data processing, archiving, and distribution are formidable considering the size, complexity, and dynamics of rice-growing areas over the continent. Improving the ground resolution, and thus the data load, is likely to compound the problems encountered in attempting to make the technology more operational.

Table 5 summarizes the resolution requirements and the current status of remote sensing platforms for the collection of data related to wetland rice in the tropics. A rating of the system reliability, also included in the table, integrates technical as well as non-technical factors that affect overall performance. These data indicate that the operational reliability of the various remote sensing systems is highest in the following cases:

1. the measurement of surface water conditions at low resolution with AVHRR data;
2. bi-monthly vegetation index assessment at low resolution with AVHRR data;
3. the assessment of long-term land use change with Landsat MSS data;
4. the stratification of rice-growing areas by a multistage procedure based on aerial photographs and Landsat.

Land-related information is thus best obtained using the Landsat platform, while data on vegetation and water dynamics are best collected at low resolution by daily weather satellite.

Table 5. Status of orbital remote sensing platforms for wetland rice monitoring in the tropics.

| Parameter | Platform | Operational reliability* |
|---|---|---|
| I. Surface water conditions | | |
| a. Weekly assessment of water/no water status over large area | | |
| at 4 km resolution | AVHRR GAC | high |
| at 1 km resolution | AVHRR LAC | medium |
| at 1 ha (or less) resolution | LANDSAT | low |
| b. Daily assessment of surface water status (in critical conditions only) | | |
| at 4 km resolution | AVHRR GAC | medium |
| at 1 km resolution | AVHRR LAC | medium |
| II. Crop performance | | |
| a. Bi-monthly to monthly crop stage assessment at 'group of fields' level | LANDSAT MSS, TM† | low |
| b. Bi-monthly to monthly vegetation index (green biomass) assessment at catchment level | | |
| at 15 km resolution | AVHRR GVI | high |
| at 4 km resolution | AVHRR GAC | high |
| at 1 km resolution | AVHRR LAC | medium |
| c. Bi-monthly to monthly crop status assessment (stage, stress) at 'group of fields' level | LANDSAT MSS, TM | low |
| d. Bi-monthly to monthly crop status assessment (stage, stress, variety) at field level | LANDSAT TM, SPOT | (low) |
| III. Land use changes | | |
| a. Monitoring land-use changes, including expansion/reduction of agricultural area | | |
| at 4 km resolution | AVHRR | medium |
| at 1 km resolution | AVHRR | medium |
| b. Detailed land use change assessment | LANDSAT MSS | high |
| c. Stratification of rice-growing area into homogeneous units for modelling purpose | LANDSAT MSS | high |
| IV. Seasonal assessment of crop conditions: varietal mix, crop stage, stress or pest attacks | LANDSAT<br>aircraft | medium<br>high |
| V. Decennial 'wall to wall' measurement of paddy land. Sampling for average field size | aircraft | high |

* This is the 'system reliability' which includes satellite characteristics, weather conditions, and constraints on data processing and analysis at the selected spatial/temporal resolution.

† The Landsat Thematic Mapper and SPOT data are expected to be available over most of Southeast Asia by 1986.

It is also apparent from the study that, although sensor-related research is still needed, more attention should be given to other segments of the technology. Of particular importance are the processing, archiving, and analysis of high frequency data and their integration with current models of rice production. The trade-offs between spectral, spatial, and temporal resolution need to be reassessed, and the current bias towards the acquisition of more data at higher ground resolution should be questioned.

Many Asian countries have acquired or are in the process of acquiring sophisticated equipment for receiving and analysing satellite remote sensing data. Such systems can collect very large amounts of information about the rice crop. The challenge is now to show that the data can be appropriately interpreted and can contribute to the better management of the productive resources of the region.

## REFERENCES

Angus, T. F. and Zandstra, H. G., 1979. Climatic factors and the modeling of rice growth and yield. In IRRI, *Agrometeorology of the Rice Crop*. Los Banos, Philippines: International Rice Research Institute.

Arkin, G. F., Vanderlip, R. L. and Richie, J. T., 1976. A dynamic grain sorghum growth model. *Trans.Am.Soc.agric.Eng.*, **19**, 622–626.

Badhwar, G. D., 1980. Crop emergence determination from spectral data. *Photogramm.Engng Remote Sensing*, **46**, 369–377.

Benton, G. S., 1982. Future operational remote sensing programs. In Johannsen, C. J. and Sanders, J. L. (Eds), *Remote Sensing for Resources Management*. Ankeny: Soil Conservation Society of America, pp. 589–598.

CCE, 1981. *Prévision des Productions Rizicoles par Télédétection dans le Bassin du Niger en amont du Delta Central Nigerien. Rapport d'Activité*. Ispra: Centre Commun de Recherche.

Chandler, R. F., 1979. *Rice in the Tropics: a Guide to the Development of National Programs*. Boulder: Westview Press.

Freden, S. C. and Gordon Jr, F., 1983. Landsat satellites. In Colwell, R. N. (Ed.), *Manual of Remote Sensing*. Falls Church: American Society of Photogrammetry, pp. 517–570.

Gaussman, H. W., Rodriguez, R. A. and Richardson, A. J., 1976. Infinite reflectance of dead compared with live vegetation. *Agron. J.*, **68**, 295–296.

Geiser, V. and Voelke, H., 1978. *Rice Yield Forecasting for Landsat Imagery. Report on a Pilot Project to Evaluate the Use of Landsat Imagery for the Collection of Agricultural Data in Sri Lanka*. Zurich: University of Zurich.

Hanks, R. J. and Ashcroft, G. L., 1977. Crop production as influenced by climate, soil, water and plant factors. *Proc. 13th Amer.Met.Soc. Conf. on Agriculture and Forest Meteorology*, Purdue University, Lafayette, 256–260.

Hay, C. M., 1974. Agricultural inventory techniques with orbital and high altitude imagery. *Photogramm. Engng Remote Sensing*, **60**, 1283–1293.

Herdt, R. W. and Bernsten, R. H., 1975. *Methodology for Assessing Rice Yield Constraints*. International Rice Agroeconomic Network, Working Paper No. 2, International Rice Research Institute, Los Banos, Philippines.

Holben, B. N., 1985. Characteristics of maximum value composition of temporal AVHRR data. *Remote Sensing Environ.* (in press).

Irons, J. R., Markham, B. L., Nelson, R. F., Latty, R. S. and Stauffer, M. L., 1985. The effects of spatial resolution on the classification of Thematic Mapper data. *Int.J.Remote Sensing*, **6**, 1385–1404.

IRRI, 1970. *Rice Production Manual.* Los Banos, Philippines: International Rice Research Institute.

IRRI, 1974. *An Agroclimatic Classification for Evaluating Cropping Systems Potentials in Southeast Asian Rice Growing Areas.* Los Banos, Philippines: International Rice Research Institute.

Iwaki, H., 1975. Computer simulation of vegetative growth of rice plants. In Murata, Y. (Ed.), *Japanese International Biological Program. Synthesis*, vol. 11. Tokyo: J.I.B.P., pp. 105–121.

Jin Zhiqing, 1984. *Assessing Climatic Suitability for Rice Production in China.* M.Sc. Thesis, University of the Philippines, Los Banos.

Kauth, R. J. and Thomas, G. S., 1976. The tasselled cap—a graphic description of the spectral-temporal development of agricultural crops as seen by Landsat. In LARS, *Proceedings of Symposium on Machine Processing of Remote Sensing Data*, Lafayette, pp. 41–51.

Kidwell, K. B., 1983. *NOAA Polar Orbiter Data Users Guide.* Washington: Satellite Data Service Division, National Oceanographic and Atmospheric Administration.

Malingreau, J. P., 1978. Remote sensing inputs into a wetland rice regional production model. *Proc.12th int. Symp. Remote Sensing Environ.*, **3**, 1617–1642.

Malingreau, J. P., 1980. *The Wetland Rice Production System and its Monitoring by Remote Sensing in Indonesia.* Ph.D. Dissertation, University of California, Davis.

Malingreau, J. P., 1984. Remote sensing and disaster monitoring. A review of applications in Indonesia. Paper delivered at *18th int. Symp. Remote Sensing Environ.*, Paris.

Malingreau, J. P. and Christiani, R., 1981. A land cover/land use classification for Indonesia. *Indonesian J. Geog.*, **11**, 41, 13–47.

Morris, R. A. and Zandstra, H. G., 1978. Soil and climatic determinants in relation to cropping patterns. In IRRI, *International Rice Research Conference.* Los Banos, Philippines: International Rice Research Institute, pp. 25–32.

Morris, R. A. and Zandstra, H. G., 1979. Land and climate in relation to cropping patterns. In IRRI, *Rainfed Lowland Rice Selected Papers from the 1978 IRRI Conference.* Los Banos, Philippines: International Rice Research Institute, pp. 255–274.

Murata, Y., 1975. Estimation and simulation of rice yield from climatic factors. *Agric. Met.*, **15**, 117–131.

NCDC-NOAA, 1984. *Global Vegetation Index User's Guide.* Washington: Satellite Data Services Division, National Climatic Data Center-National Oceanographic and Atmospheric Administration.

Oldeman, L. R., 1975. *An Agroclimatic Map of Java.* Centrai Kesearch Institute of Agriculture No. 13, Bogor, Indonesia.

Reyes, A. J., 1973. An analysis of some factors affecting rice yield response to water. In IRRI, *Water Management in Philippines Irrigation Systems: Research and Operations.* Los Banos, Philippines: International Rice Research Institute, pp. 37–52.

Robertson, G. W., 1973. *A Factorial Yield Weather Model for Wheat.* World Meteorological Organization No. 396. Proc. WMO Symp. on the Agrometeorology of the Wheat Crop, Geneva.

Robertson, G. W., 1975. *Rice and Weather.* World Meteorological Organization No. 243, Tech. Note No. 144, Geneva.

Schneider, S. R. and McGinnis, D. F., 1982. The NOAA-AVHRR: a new satellite sensor for monitoring crop growth. In LARS, *1982 Machine Processing of Remote Sensing Data Symposium.* Lafayette: Laboratory for Applied Remote Sensing, pp. 281–290.

Seligman, M. G., 1976. A critical appraisal of some grassland models. In Arnold, G. W. and deWitt, C. T. (Eds), *Critical Evaluation of Systems Analysis in Ecosystems Research and Management*. Wageningen: Centre for Agricultural Publishing and Documentation, pp. 60–97.

Suits, G. H., 1972. The calculation of the directional reflectance of a vegetative canopy. *Remote Sensing Environ.*, **2**, 117–125.

Townshend, J. R. G. and Tucker, C. J., 1981. Utility of AVHRR of NOAA-6 and 7 for vegetation mapping. In Remote Sensing Society, *Matching Remote Sensing Technologies and their Applications Proceedings*. London: Remote Sensing Society, pp. 97–109.

Tucker, C. J., Vanpraet, C., Boerwinkel, E. and Gaston, A., 1983. Satellite remote sensing of total dry matter production in the Senegalese Sahel. *Remote Sensing Environ.*, **13**, 461–474.

Uhlig, H., 1969. Draft presentation of a framework for the geographical terminology of rice cultivation. *Int.Congr.Meeting of the Standing Committee on Geography, Pacific Science Association*, Singapore.

Van Keulen, H., 1976. Evaluation of models. In Arnold, G. W. and deWitt, C. T. (Eds), *Critical Evaluation of Systems Analyses in Ecosystems Research and Mangement*. Wageningen: Centre for Agricultural Publishing and Documentation, pp. 22–29.

Van Ittersum, K., 1972. A calculation of potential rice yield. *Neth. J.agric.Sci.*, **20**, 20–21.

Wickham, T., 1973. Predicting yield benefits in lowland rice through a water balance model. In IRRI, *Water Management in Philippines Irrigation Systems: Research and Operations*. Los Banos, Philippines: International Rice Research Institute, pp. 151–181.

Remote Sensing and Tropical Land Management
Edited by M. J. Eden and J. T. Parry


# CHAPTER 16

# *Monitoring the Vegetation of Semi-arid Tropical Rangelands with the NOAA-7 Advanced Very High Resolution Radiometer*

STEPHEN D. PRINCE
*School of Biological Sciences, Queen Mary College, University of London, UK*

Rangelands are areas of low biological production which are mainly utilized for the grazing of livestock. The areas occupy in excess of 40% of the earth's surface. In tropical regions, they support sparse human populations whose principal means of livelihood is livestock production, although small areas may also be planted with crops. It has always been difficult to obtain accurate and timely information about conditions in rangelands because of the vast areas involved and the low economic value of the land. The techniques used to survey and monitor agricultural land in more productive regions are often unsuitable for use in rangelands, because they are costly and therefore inappropriate in areas of low capital intensity. In these circumstances, remote sensing from earth resource satellites can make an important contribution which not only complements the data obtained by means of ground and aerial surveys but also enhances them by increasing the amount of information that can be gained from more intensive methods. Remote sensing from earth-orbiting spacecraft cannot replace ground and aerial measurement but the synoptic and repetitive coverage of large areas offers an entirely new source of information for the management of rangelands.

Rangeland monitoring is desirable for short-term management and for longer-term planning. The recent droughts and famines in the semi-arid regions of Africa have given a new urgency to the development of rapid means of collecting information for use by government and international agencies. The livestock production of semi-arid rangelands on a year-to-year basis is largely determined by the rainfall, which controls the amount of primary production

Table 1. Sensors and platforms for measuring red and near-infrared reflectance.

| Platform | Sensor | Approximate pixel size | Approximate area covered (9 pixels— 1 scene) |
|---|---|---|---|
| Aircraft | e.g. Daedalus multispectral scanner | variable | local coverage only |
| SPOT | High Resolution Visible (HRV) | 400 $m^2$ | 0.004–10 000 $km^2$ |
| Landsat 4 and 5 | Thematic Mapper (TM) | 900 $m^2$ | 0.01–10 000 $km^2$ (¼ MSS scene) |
| Landsat 1 to 5 | Multispectral Scanner (MSS) | 6400 $m^2$ | 0.1–50 000 $km^2$ |
| NOAA-6, 7, and 8 | Advanced Very High Resolution Radiometer (AVHRR) | LAC 1.2 $km^2$<br>GAC 16 $km^2$ | 10-3 000 000 $km^2$ |

and therefore the fodder available to the herbivores. Equally, rainfall is the primary determinant of crop production in the cultivated areas bordering the rangelands and occurring in pockets within them. As yet, there is no means of monitoring rainfall or soil moisture over a large enough area and with sufficient accuracy to predict annual primary production directly and, in any case, it is likely that more accurate prediction of livestock production will be possible by monitoring the vegetation, which is the proximate factor controlling secondary production.

Remote sensing of vegetation amount depends on the differential reflection of red and near-infrared radiation by green vegetation. Sensors which are capable of resolving the red and near-infrared spectral bands are available on a number of platforms at the present time and more will be available in the near future. The platforms, the sensors, their resolution and scene coverage are listed in Table 1. The first satellite to carry a multispectral linescanner which gave near-global coverage for a considerable period of time was Landsat 1, and the acquisition of data has been maintained by successive satellites in the series. For this reason, most of the techniques developed for the remote sensing of vegetation have employed Landsat MSS data.

Early studies were directed towards agricultural crops (e.g. LACIE, 1979) but more recently the possibility of monitoring natural vegetation has become important. This is particularly significant in rangelands where it is the natural vegetation that is the main source of fodder for livestock. It has been established that natural vegetation types can be distinguished using Landsat MSS data, that vegetation cover and standing crop can be estimated, and that, by careful stratification of scenes with calibration against ground data, mean values for regions can be obtained (Poulton *et al.*, 1975; Hutchinson, 1982). Primary production has been determined by monitoring both the amount and duration of 'greenness'. The temporal resolution required depends upon the rate of change of 'greenness' in the study area but, in semi-arid rangelands, where 'greenness' is often transient even during the growing season, at least monthly acquisitions of satellite data are desirable.

The need for increased detail in remotely sensed data has been provided for recently by the Thematic Mapper (TM), carried on Landsat 4 and 5, which has a ground resolution of 30 m in the visible and near-infrared bands compared with 80 m for the MSS. For local monitoring, in areas up to about $100 \times 100$ km, this represents a major improvement although it greatly increases the quantity of data and thus the costs of acquisition and processing. One effect of increased ground resolution, therefore, is to reduce the total area that it is practicable to include in a particular study. This can be detrimental to monitoring at the regional or national scale. However, there are other factors affecting the selection of a particular ground resolution. Certain phenomena may actually be obscured by information at higher levels of resolution (Allen and Starr, 1982), so that failure to select an appropriate scale can reduce the ability to discriminate certain ground features by remote sensing (Townshend, 1981a).

The only current sources of lower resolution satellite data suitable for vegetation monitoring are the National Oceanic and Atmospheric Administration (NOAA) meteorological satellites NOAA-6, 7, and 8. All of these carry Advanced Very High Resolution Radiometers (AVHRR) which operate in the 0.58–0.68 $\mu$m and 0.725-1.10 $\mu$m bands and are therefore able to detect green vegetation (Schneider *et al.*, 1981). Since the launch of NOAA-6 in 1979, a number of studies have reported on the use of AVHRR data for monitoring vegetation (Gray and McCrary, 1981; Greegor and Norwine, 1981; Townshend and Tucker, 1981, 1984; Brown *et al.*, 1982; Cicone and Metzler, 1982; Ormsby, 1982; Schneider and McGinnis, 1982; Yates and Tarpley, 1982; Tucker *et al.*, 1983(a) and (b); Tucker *et al.*, 1984). The main features of AVHRR data compared with Landsat MSS data are identified in Table 2. NOAA-7 is more useful for vegetation monitoring than its predecessors because it collects data at 1430 hours local time, compared with 0730 hours for NOAA-6 (and 8). Although cloud cover can be more extensive at 1430 hours, shadows are much less troublesome.

Global Area Coverage (GAC) data are generated on-board the satellites by performing a running average of the digital values of four pixels from a block of fifteen. The GAC data are recorded on-board and transmitted to ground stations when convenient. The primary data are called Local Area Coverage (LAC) and are only available to ground receiving stations in range of the satellite, although limited quantities can be stored on-board for later transmission to a ground station.

Table 2. Characteristics of NOAA AVHRR and Landsat MSS (after Townshend and Tucker, 1984).

| | NOAA AVHRR | Landsat MSS |
|---|---|---|
| Inclination of orbit | 99.092° | 98.91°–99.1° |
| Height above surface | 833–870 km | 916.6 km |
| Number of orbits day$^{-1}$ | 14.25 | 14 |
| Times of equatorial crossing (h) | NOAA-6 and 8 07.30 and 1930<br>NOAA-7 02.30 and 14.30 | 09.30 |
| Orbital period (min) | 102 | 103.3 |
| Latitudinal coverage | 90° N to 90° S | 82° N to 82° S |
| Cycle duration | circa 1 day | 18 days |
| Ground coverage (1 scene) | 2700 km × 1100 km | 185 km × 185 km |
| Field of view | ±56° | ±5.78° |
| Instantaneous field of view (IFOV) | 1.39–1.51 mrad | 0.086 mrad |
| Ground resolution | | |
| nadir | 1.1 km | 79 m |
| maximum off-nadir | 2.4 km along track<br>6.9 km across track | 79.5 m along track<br>80 m across track |
| Samples per IFOV | 1.362 | 1.411 |
| Number of channels | 5 | 4 |
| Channels—spectral bandwidth ($\mu$m) | ch. 1 0.58–0.68<br>ch. 2 0.725–1.1<br>ch. 3 3.55–3.93<br>ch. 4 10.5–11.5<br>ch. 5 11.5–12.5* | ch. 4 0.5–0.6<br>ch. 5 0.6–0.7<br>ch. 6 0.7–0.8<br>ch. 7 0.8–1.1 |
| Data precision | 10 bits | 6 or 7 bits |

*Not on NOAA-6.

An important feature of AVHRR data is their high frequency of acquisition. Most points on the earth's surface are recorded approximately once a day compared with the 9- or 18-day gap between successive images of Landsat MSS scenes. This high temporal resolution permits the detection of transient green vegetation and the problem of cloud cover can be overcome by substitution of data from preceding or succeeding orbits. However, the high temporal frequency is achieved at the expense of a wide scan angle (±56°), which causes variation of air mass between the satellite and the earth's surface from 1 at nadir to 2.2 at extreme viewing angles (Holben and Fraser, 1984). Fortunately, the temporal frequency of acquisition is sufficiently high to introduce some redundancy into the data and this feature has been exploited in software developed at the United States National Aeronautics and Space Administration Goddard Space Flight Center (NASA/GSFC), Maryland, in which composite images are derived from a sequence of individual AVHRR images.

The NASA/GSFC software undertakes three main tasks. A normalized difference vegetation index (NDVI) is calculated for each pixel using channels 1 (red) and 2 (near-infrared) of the AVHRR data. The use of NDVI (channel 2 − channel 1)/(channel 2 + channel 1) is discussed by Tucker (1979). A ratio index of this type is useful since it reduces the effects of variation in surface topography on reflectance and also the effects of variation in sun elevation in different parts of a scene. Perry and Lautenschlager (1984) have shown that NDVI is functionally equivalent to many other widely used vegetation indices. The second step is to mask out pixels in which there is a high proportion of cloud cover. This is made possible by using the channel 5 (thermal infrared) data. A cloud screen is created by setting a threshold temperature and the NDVI of all pixels which are below the threshold are set to zero, thereby masking out the pixel (Gatlin *et al.*, 1983; Tucker *et al.*, 1984). The third stage involves registration of each NDVI image to a standard projection, followed by selection of the maximum NDVI value for each pixel out of a temporal sequence of images. Normally ten images are used which are calculated from successive daily acquisition of AVHRR data. Thus, the final image is a temporal composite of approximately ten separate images. This procedure permits the generation of largely cloud-free images for most of the world. As well as reducing cloud screening, the procedure is effective in selecting those pixels closest to the satellite's nadir, thereby eliminating the off-nadir pixels which are most affected by atmospheric scattering and tend to give significantly lower NDVI values (Holben and Fraser, 1984).

The NDVI is not sensitive to the quantity of dead leaves, wood, and non-green stems or to other non-green material in the vegetation canopy, or, of course, to the root system. But it is the annual primary production, including both green and non-green material, and its nutritional quality which are functionally most closely connected to livestock production. Fortunately, there is evidence that integration of NDVI with respect to time gives a greenness duration index which is related to primary production (Markham *et al.*, 1981; Tucker *et al.*, 1981; Tucker *et al.*, 1983a). The NASA/GSFC compositing procedure for AVHRR data reduces the temporal frequency of the processed NDVI images to approximately 10-day intervals, but this has been found adequate to detect the transient green flushes that occur in desert regions (Tucker *et al.*, 1983b) and so it is unlikely to miss similar events in rangelands.

The low temporal frequency of Landsat data, particularly in areas having higher cloud cover than deserts, limits the role of Landsat in monitoring vegetation to features which do not undergo significant changes with a higher frequency than 1–2 months. Nevertheless, Landsat has an important role in rangeland monitoring in stratification of terrain into land classes in order to delimit similar vegetation types within which a particular calibration of NDVI, with respect to ground conditions, can be expected to hold.

## GENERAL FEATURES OF NOAA-7 AVHRR GAC MULTITEMPORAL COMPOSITE NDVI IMAGES

Examples of 4 km resolution NDVI composites of southern Africa, obtained between August 1983 and April 1984, are shown in Plates 18–24. The colours assigned are browns for very low NDVI, followed by green, red, and blue for increasing values. The summer images, from November onwards, have some residual cloud cover, particularly in the northern and eastern areas, and this is shown in black. For completely cloud-free images in the wetter areas, longer compositing periods would be required, but this is not necessary in the semi-arid regions. National boundaries have been superimposed on the images for convenience in interpretation.

The pattern of high and low values of NDVI in the images corresponds to the spatial distribution of green vegetation cover in the field. Permanently high values occur in several areas along the Drakensberg mountains in South Africa and extend to the Soutpansberg in northern Transvaal (Fig. 1). Similarly, both the eastern highlands of Zimbabwe and the northern escarpment of the Zambezi valley in Zambia have permanently high NDVI values, corresponding to green vegetation throughout the year. Larger areas of permanent green vegetation are indicated in southern Angola and between the Zambezi and Kafue rivers

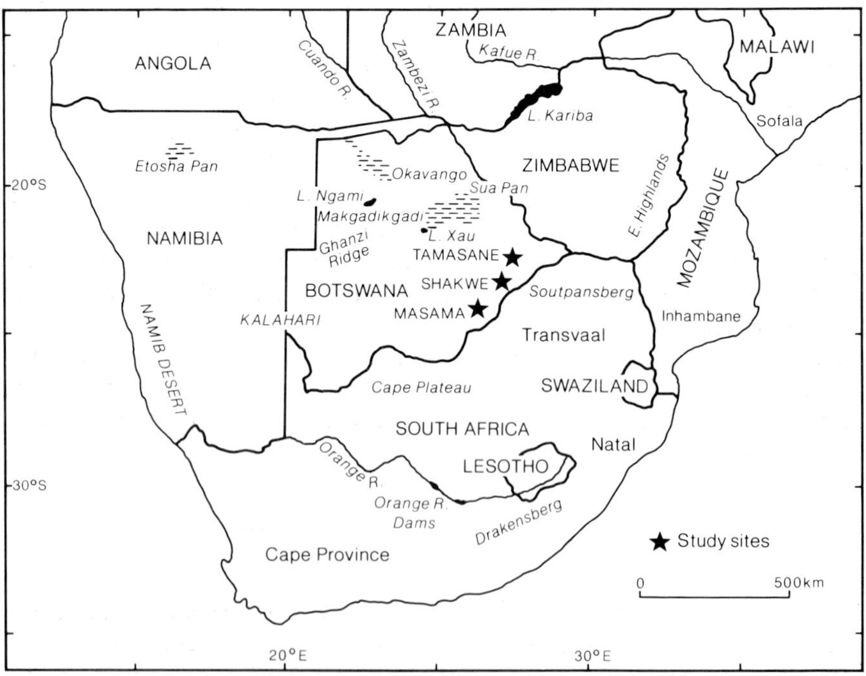


Fig. 1 Map of southern Africa showing location of places referred to in the text.

in Zambia. The last two areas are separated by low NDVI values in the flood-plain grasslands between the Cuando and Zambezi rivers. Coastal Inhambani and Sofala in Mozambique and the coastal strip of Natal in South Africa also have permanent green vegetation. The Cape Plateau in the north of Cape Province is greener than the surrounding north Cape, although with rather low NDVI values, and the Orange river valley is picked out by higher values than in the very arid country through which the river flows. In the winter dry season (August–October), the Okavango delta in Botswana is green when the surrounding semi-arid savannas are completely brown. The deserts and arid regions of southern Africa are clearly shown, notably the Namib desert, the Kalahari, and north Cape, the last two being distinguished from the first by intermittently green vegetation. Some local features are well illustrated in the images, including the Makgadikgadi pans in Botswana and Etosha pan in Namibia.

The pattern and sequence of green vegetation on the images, indicated by the NDVI values, reflect the three major rainfall regimes of southern Africa, viz. rainfall associated with the intertropical and the subtropical convergence zones and the winter rainfall of the southern Cape. The intertropical convergence zone extended as far south as northern Namibia, northern Botswana, and central Zimbabwe in January 1984. The subtropical low pressure over southern Africa during the summer also attracted rain, mainly over eastern and central South Africa. In contrast, the Cape region received mainly winter rainfall.

These seasonal patterns of rainfall and vegetation are important indicators of differing types of vegetation. A land cover classification for the whole of Africa has been derived from a classification of two NOAA images acquired in different seasons (Tucker *et al.*, 1983c), and there is much scope for study of the actual phenology of vegetation over large areas using this technique (Justice *et al.*, 1985). As longer sequences of data are accumulated, fundamental biogeographical processes will be amenable to investigation on a much broader scale than has hitherto been possible. The ability to make observations for all points over most of the earth's surface, instead of having to interpolate between more or less limited ground and aerial observations, is an important advance. An attractive feature of the NOAA imagery is the way in which some interpretations can be undertaken visually and without the need for detailed knowledge of digital techniques.

## INTERPRETATION OF NOAA-7 AVHRR GAC COMPOSITE NDVI DIGITAL DATA

The images derived from NDVI data are generally in accord with the expected distribution of 'greenness' through the year but, if the data are to be used for monitoring the vegetation, a more detailed analysis of the relationship with ground conditions is needed. Investigations into the relationship between vegetation indices calculated from satellite radiometer data and the cover, biomass, and production of vegetation have mostly been carried out over extensive agricultural crops which are monospecific, have a simple physiognomy, and are at the same stage of growth. In rangelands, however, the natural vegetation is the primary concern. Much less is known about the behaviour of vegetation indices in these complex canopies, with only one study reporting the use of AVHRR NDVI data in semi-arid tropical rangelands (Tucker *et al.*, 1983a).

In this study, six acquisitions of LAC data for the Ferlo area of the Senegalese Sahel were examined at intervals of between 7 and 18 days throughout a 52-day period during the wet season. The first image was acquired about three weeks after the start of the rains. Average precipitation for the region varies between 200 mm in the north and 500 mm in the south. During the latter 14 days of the study, 18 sites were visited in each of which the herbaceous vegetation in nine 1 $m^2$ plots was clipped and weighed. These data were used to establish the relationship between the dry biomass in each site and an integrated NDVI for the six AVHRR images. Although the authors emphasize the limitations of the data—the variable tree cover (2–15% and occasionally up to 100%), the underestimation of the end-of-season biomass by sampling before the end of the growing season, and the small area of ground which was sampled—their results do show a linear relationship between integrated NDVI and the herbaceous biomass (Fig. 2).

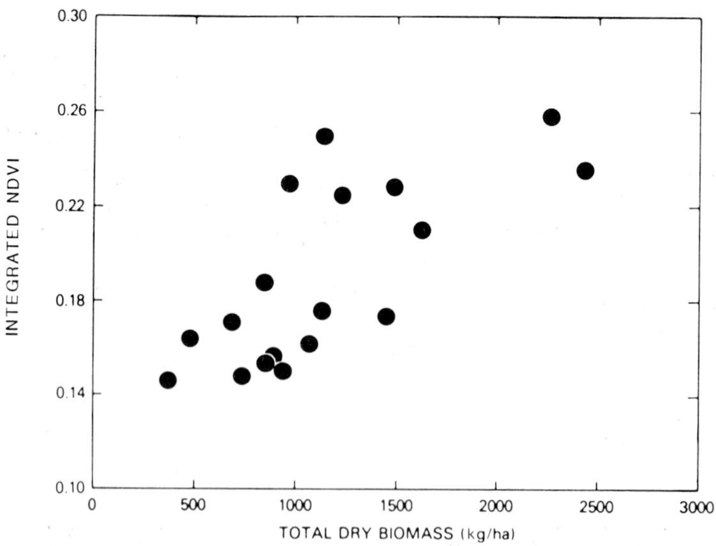


Fig. 2 Relationship between integrated NOAA AVHRR normalized difference vegetation index (NDVI) and the total end-of-season dry biomass of herbaceous vegetation in the Ferlo area of Senegal (after Tucker *et al.*, 1983a).

The relationship between the NDVI and vegetation biomass is likely to vary in different vegetation types because of differences in species, structure, and phenology. In order to investigate this and further to develop the techniques for calibrating AVHRR NDVI data, a study was undertaken by the Food and Agriculture Organization of the United Nations for the United Nations Development Programme (FAO/UNDP) Range Management Project (BOT/82/001), in conjunction with the Botswana Government (Animal Production Research Unit, Ministry of Agriculture). The results of the study, undertaken during the 1983–84 growing season, are described below. The processing of the NOAA-7 AVHRR GAC data was carried out by NASA/GSFC according to the standard compositing procedures already described. The analysis is based on the NDVI images and digital data supplied by NASA/GSFC.

## CORRELATION OF NOAA-7 AVHRR GAC COMPOSITE NDVI WITH PRECIPITATION AND VEGETATION

The NASA/GSFC software permits output of the NDVI values of individual pixels selected from the composite images. The temporal sequence of these values permits visual comparisons of the NDVI with qualitative observations of the vegetation in known areas and with rainfall records for meteorological stations. Fig. 3 shows such sequences for three study sites in Botswana and also for part

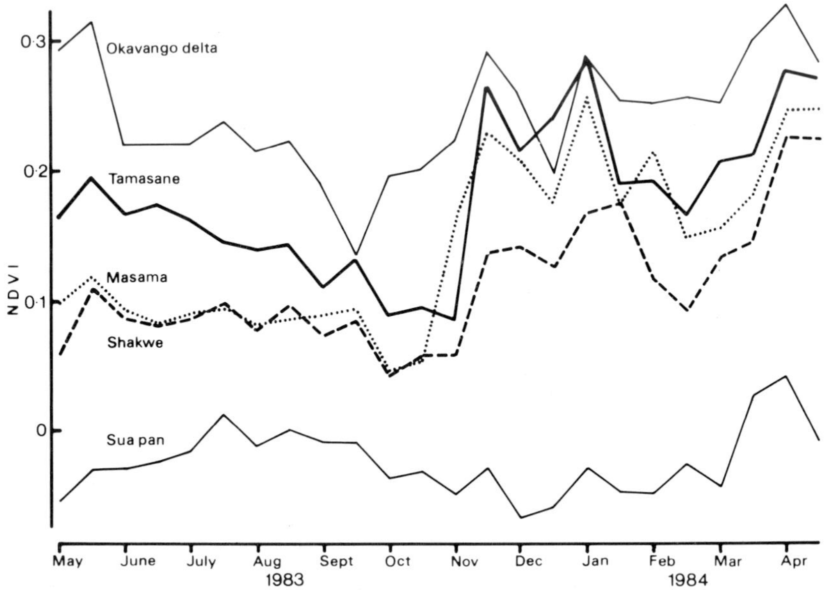


Fig. 3 Temporal sequence of NDVI for the three study sites (Tamasane, Masama, and Shakwe) and two control sites (Okavango delta and Sua pan).

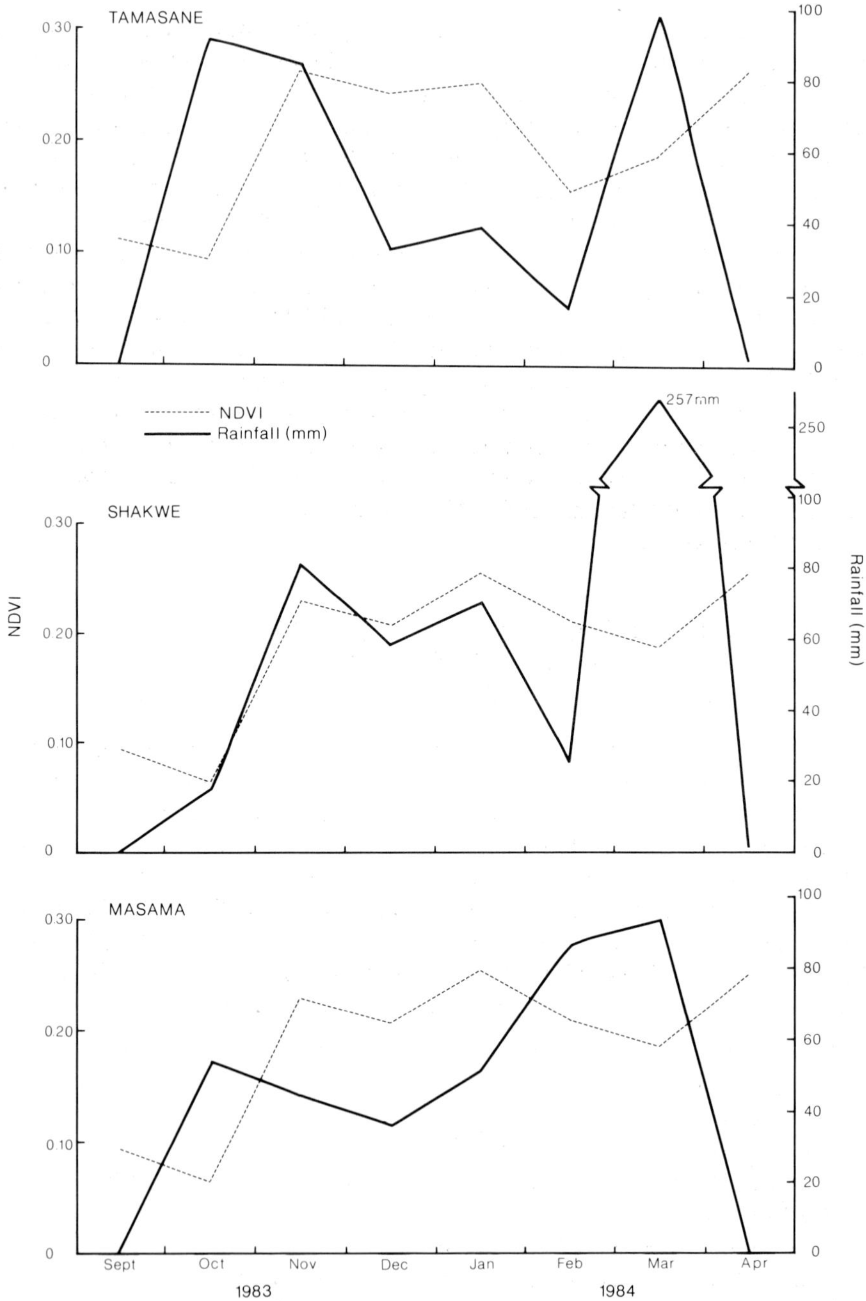


Fig. 4 Monthly data for NDVI in the three study sites and precipitation at the nearest available meteorological stations, during the 1983–84 wet season. Rainfall data from Botswana Meteorological Services.

of the permanent swamp within the Okavango delta and for Sua pan in the Makgadikgadi basin. The Okavango pixel can be expected to be permanently occupied by green vegetation and open water and much of Sua pan can be expected to be permanently bare of vegetation, the surface consisting of sand encrusted with sodium and potassium salts. The NDVI for the Okavango pixel declined in the late dry season which suggests that the quantity of green vegetation decreased at this time, and in the Sua pan pixel there was a slight decline in the NDVI in the mid-rainy season which may have resulted from differences in the spectral reflectance of the pan surface when wet and dry.

In the study sites at Tamasane, Masama, and Shakwe, the rise in NDVI in November, following the low dry season values, correlates well with precipitation records for the nearest meteorological stations (Fig. 4) and with general observations on the progress of the wet season. After early rains at Tamasane and Masama, there was a mid-season drought followed by renewed rainfall in late March 1984. At Shakwe the wet season started very late and green vegetation did not appear until January 1984. All these changes are paralleled by changes in the NDVI values.

## QUANTITATIVE COMPARISON OF NOAA-7 AVHRR GAC COMPOSITE NDVI VALUES WITH GROUND MEASUREMENTS IN BOTSWANA

It has been established that AVHRR NDVI values are correlated with the occurrence of green vegetation, but it is necessary to examine the details of the relationship if the data are to be used for quantitative monitoring of vegetation. Experience with Landsat MSS data suggests that quite different relationships between vegetation indices derived from satellite radiometers and the actual vegetation may be found in different vegetation types. The field sites used in Botswana are described below in order to specify the extent and limitations of the investigation of the AVHRR composite NDVI data reported here.

Three study sites were selected in eastern Botswana (Fig. 1), as follows:

1. Tamasane (lat. 22° 24′S, long. 27° 29′E) 30 km NE of Palapye, 15 km E of Dikabeya.
   Mean annual precipitation 387 mm.
   *Colophospermum mopane* woodland.
   Communal land with widely scattered boreholes and little cultivation.
2. Shakwe (lat. 23° 8′S, long. 27° 6′E) 30 km E of Mahalapye.
   Mean annual precipitation 444 mm.
   *Combretum apiculatum–Acacia nigrescens* woodland.
   Communal livestock area near large village with some cultivation.
3. Masama (lat. 23° 54′S, long. 26° 20′E) 10 km S of Dibete quarantine gate.
   Mean annual precipitation 526 mm.

*Terminalia sericea* woodland.
Fenced ranch and neighbouring communal livestock area with little cultivation.

Study sites were selected after careful examination of recent Landsat images (MSS band 5 at 1:1 000 000 scale and colour composites of bands 4, 5 and 7 at 1:250 000), panchromatic aerial photographs at 1:50 000 scale, and appropriate geology, soil, and vegetation maps. Reconnaissance on the ground and by light aircraft was undertaken to confirm that the sites were reasonably uniform and representative of the region. The aerial and ground photographs in Figs 5, 6, and 7 give a general impression of each site. Apart from some isolated small hills outside the study area to the east of Tamasane, there is little change of relief in any of the sites.

Each site was approximately the size of one NOAA AVHRR GAC pixel (4 × 4 km). Within this area, twenty 200 × 200 m plots were located subjectively in order to represent the major variation in the vegetation present. Measurements were made of the cover and biomass of live and dead vegetation separately, and the cover of smaller shrubs and the amount of bare ground were recorded. In each plot, thirty 0.5 × 0.5 m quadrats were arranged in a semi-random pattern. Cover was estimated visually and biomass was measured by clipping the vegetation, followed by drying and weighing. These measurements were made in September, November, January, and March, 1983–84. The canopy cover of trees and larger shrubs in the plots was measured using aerial photographs obtained during the same period. A motor-driven, automatic 35 mm camera (Nikon F3) was mounted in the fuselage of a high-wing aeroplane (Cessna 150) fitted with a radar altimeter. The plots were marked with 1 $m^2$ sheets of galvanized iron, which were painted a different colour for each corner. The best results were obtained at a flying height of 300 m, using a 28 mm lens set to an aperture of about f 3.5 in order to ensure a shutter speed of 1/200 s. Standard colour film was used. The canopy cover was obtained using a dot-grid on to which the transparencies were projected at × 10 magnification. A projector with a variable focal-length lens enabled the plot markers to be registered with the corners of the grid on every photograph, thereby adjusting for minor variations in the aircraft's altitude and attitude.

The ground data are summarized in Tables 3, 4, and 5. Maximum and minimum values for cover and biomass are given in order to indicate the range of conditions encountered. In Fig. 8, the ground data are assembled for each date and site, and the mean values are plotted against the mean AVHRR GAC composite NDVI value for the block of nine pixels whose nominal centre lies closest to the centre of each study site. The mean of nine pixels is used in order to reduce the effects of drift in the locations of pixels between successive images.

The relationship between the mean cover of green herbaceous vegetation in the study sites in each observation period and the AVHRR GAC composite NDVI values is reasonably linear and holds down to cover values of less than

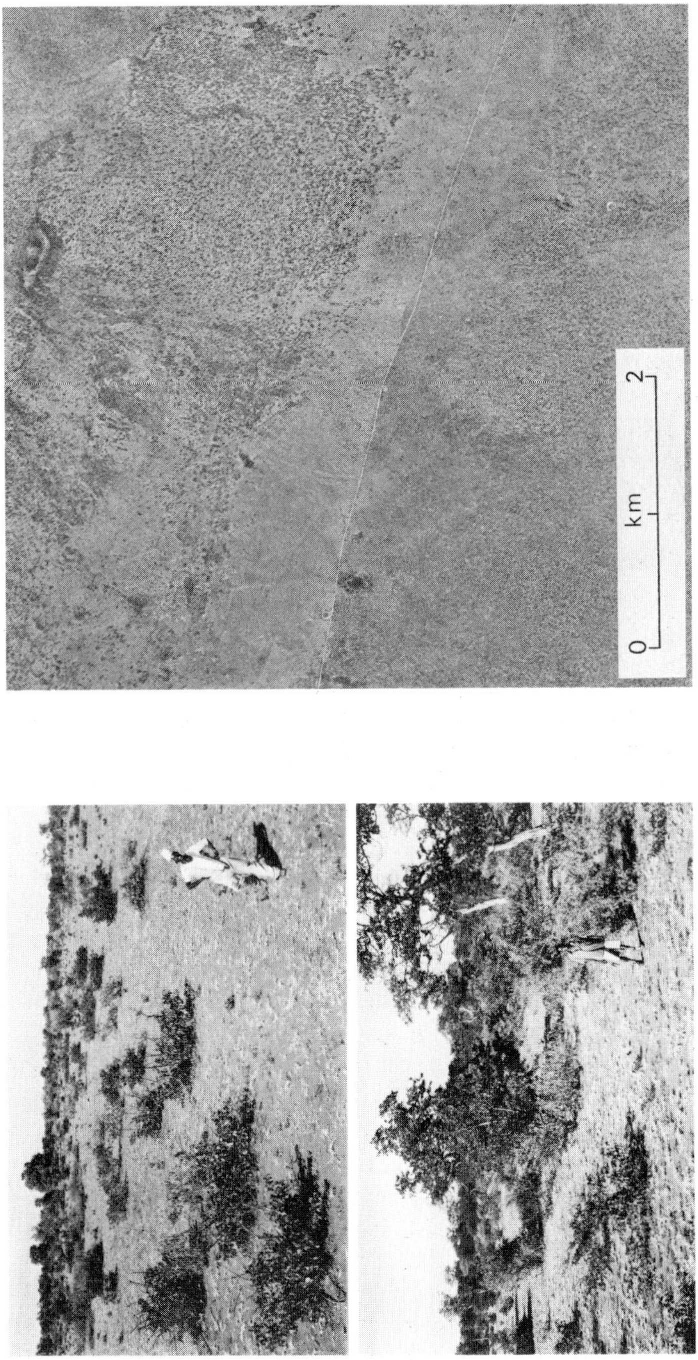


Fig. 5 Tamasane study site. Panchromatic aerial photograph and ground photographs of sample plots. (*Reproduced by permission of Department of Surveys and Lands, Gaborone, Botswana.*)

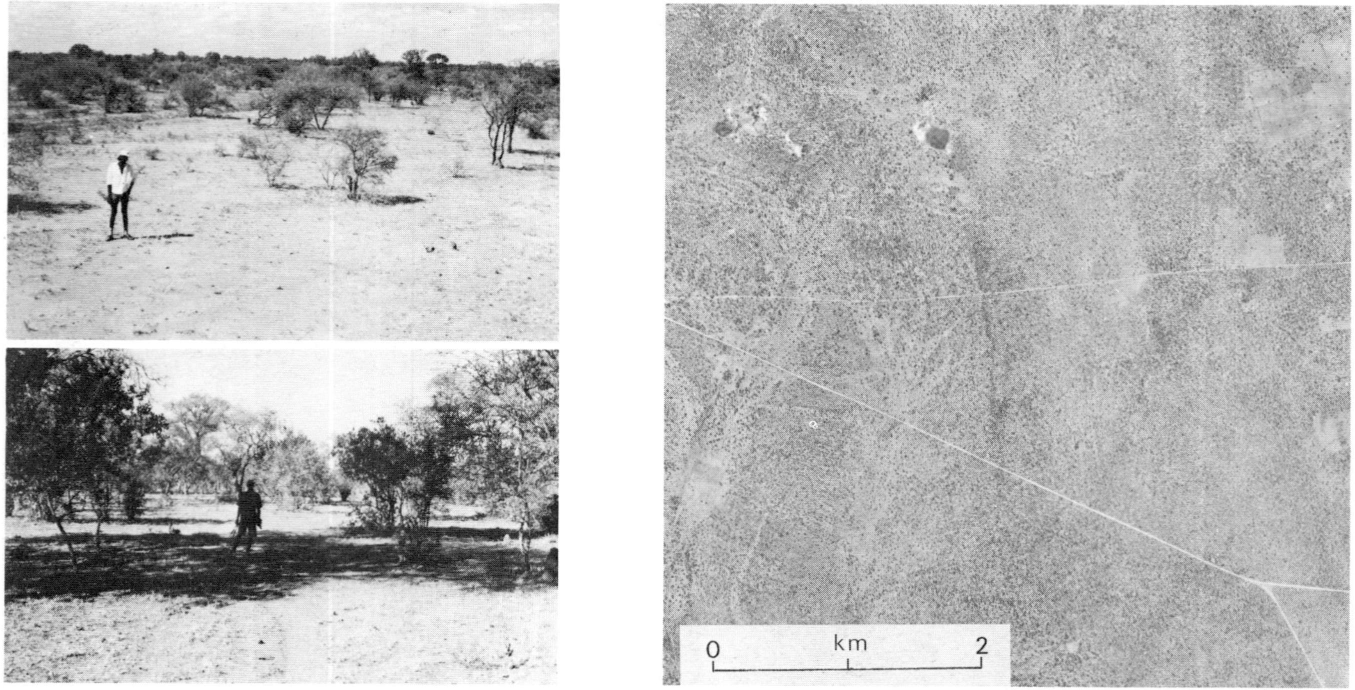


Fig. 6 Shakwe study side. Panchromatic aerial photograph and ground photographs of sample plots in the natural vegetation and cultivated land bordering a village. (*Reproduced by permission of Department of Surveys and Lands, Gaborone, Botswana.*)

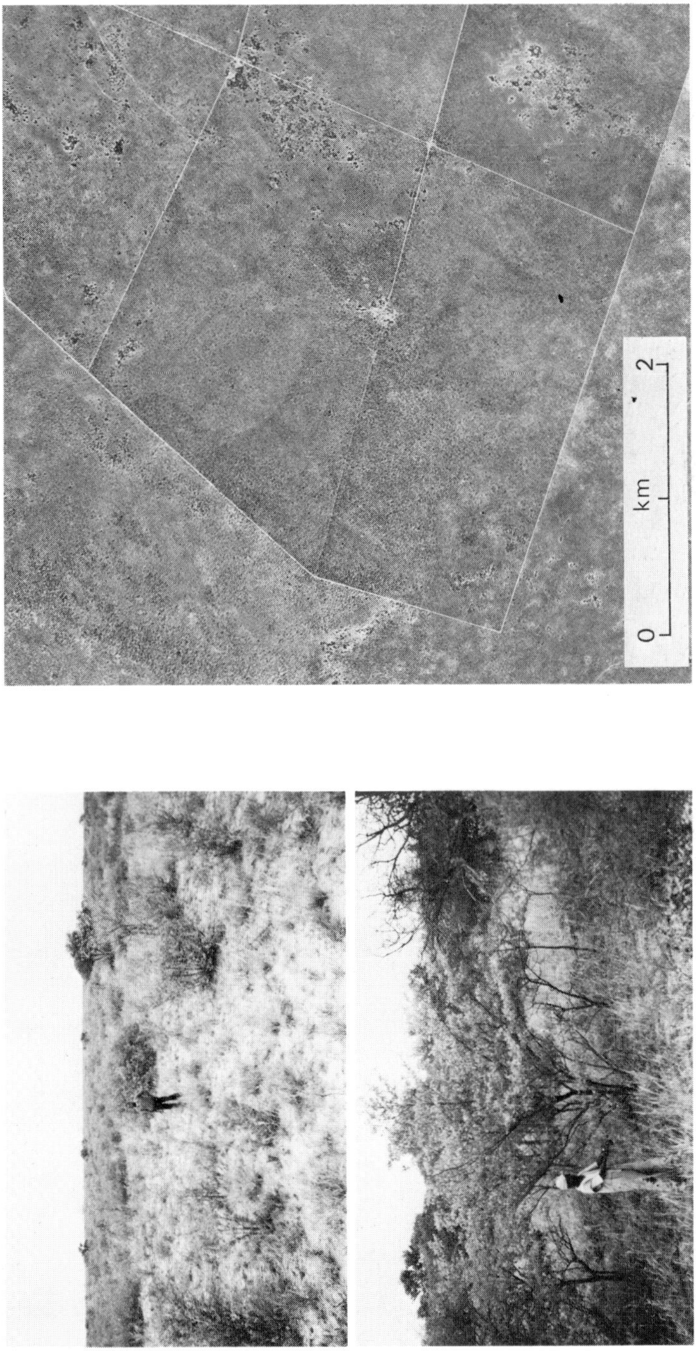


Fig. 7 Masama study site. Panchromatic aerial photograph and ground photographs of sample plots in the paddock of Masama ranch and in the communal grazing area to the south. (*Reproduced by permission of Department of Surveys and Lands, Gaborone, Botswana.*)

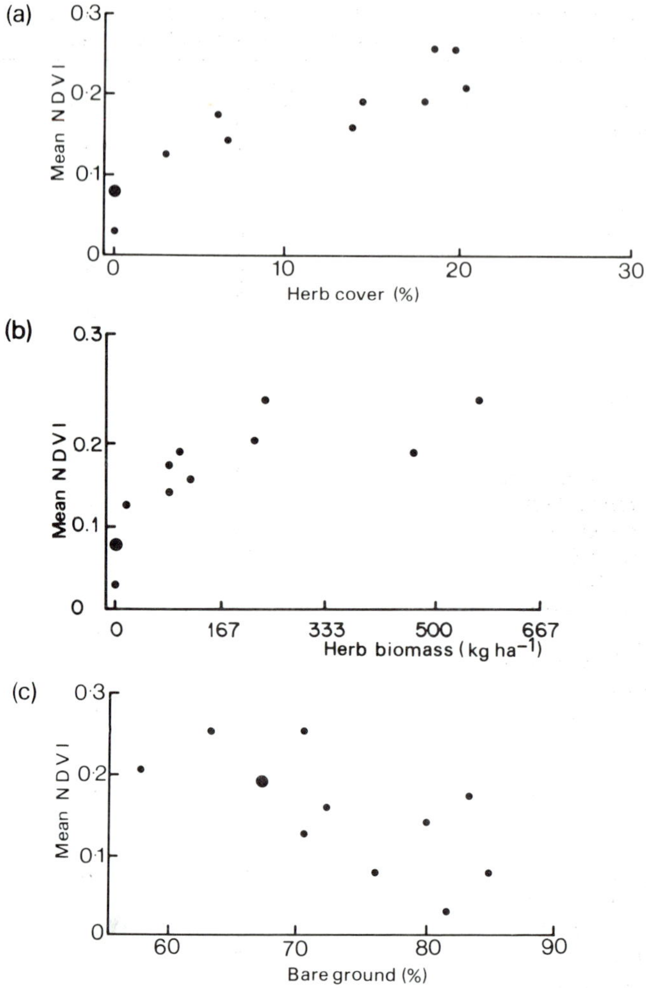


Fig. 8 Scatter diagrams of data for the three study sites showing the relationship between NDVI values and (a) % herb cover, (b) herb biomass, and (c) % bare ground. Each point represents the mean of 9 AVHRR pixels and 20 plots, for one site and date. Data are presented for four dates during the 1983–84 wet season. Larger dots indicate coincident points.

10% (Fig. 8(a)). This is remarkable in view of the difficulties experienced in reaching as low as 10% cover with Landsat MSS data (Allan and Richards, 1983). Biomass values for green herbage have a curvilinear relationship with AVHRR NDVI and values down to about 100 kg $ha^{-1}$ can be detected (Fig. 8(b)). Bare ground is effectively the complement of cover of green herbaceous vegetation and shows the expected negative relationship with NDVI (Fig. 8(c)).

Table 3 Abundance of woody species at the three study sites. Subjective estimates of cover-abundance based on observations in twenty 4 ha plots at each site.

| Species | Tamasane | Shakwe | Masama |
|---|---|---|---|
| *Acacia erioloba* | 10 | 3 | |
| *Acacia erubescens* | 1 | 4 | |
| *Acacia fleckii* | | | 22 |
| *Acacia mellifera* | | 1 | 7 |
| *Acacia nigrescens* | 29 | 52 | |
| *Acacia nilotica* | 12 | 26 | |
| *Acacia robusta* | 2 | 21 | |
| *Acacia tortilis* | 18 | 35 | |
| *Acacia sp.* | | | 5 |
| *Albizia harveyi* | 21 | 28 | |
| *Burkea africana* | | | 42 |
| *Bauhinia petersiana* | | | 62 |
| *Boscia foetida* | | | 20 |
| *Colophospermum mopane* | 83 | | |
| *Combretum apiculatum* | 36 | 68 | 4 |
| *Combretum heroense* | 13 | 23 | 3 |
| *Combretum imberbe* | 26 | 33 | |
| *Combretum zeyheri* | | | 30 |
| *Commiphora angolensis* | 15 | 13 | 4 |
| *Commiphora mollis* | | 17 | |
| *Commiphora pyracanthoides* | 47 | 41 | 14 |
| *Croton gratissimus* | | | 8 |
| *Dichrostachys cinerea* | 31 | 45 | 67 |
| *Diospyros lycoides* | | | 4 |
| *Ehretia rigida* | 11 | 8 | |
| *Elephantorrhiza burkei* | | 22 | |
| *Euclea undulata* | 3 | 1 | 5 |
| *Grewia bicolor* | 49 | 39 | 7 |
| *Grewia flava* | 34 | 20 | 63 |
| *Grewia retinervis* | | | 45 |
| *Lonchocarpus capassa* | 6 | | |
| *Maytenus senegalensis* | 8 | 5 | |
| *Maytenus tenuispina* | | | 55 |
| *Mundulea sericea* | | | 1 |
| *Ochna pulchra* | | | 47 |
| *Ozoroa paniculosa* | 26 | 20 | 18 |
| *Peltophorum africanum* | | 9 | 13 |
| *Rhigozum brevispinosum* | 8 | | 12 |
| *Rhus leptodictya* | 2 | 1 | 13 |
| *Sclerocarya caffra* | 8 | 23 | |
| *Strychnos sp.* | | | 1 |
| *Tarchonanthus camphoratus* | | | 3 |
| *Terminalia sericea* | 3 | 3 | 74 |
| *Tylosema fassoglensis* | | | 50 |
| *Ximenia caffra* | 22 | 4 | |
| *Ziziphus mucronata* | 1 | 6 | 13 |

Table 4. Summary of field data for the three study sites. Mean, maximum, and minimum values for each site and date.

| | | Percentage cover | | | | Biomass (kg ha$^{-1}$) | |
|---|---|---|---|---|---|---|---|
| | | Bare ground | Green shrubs <1 m | Litter | Live herbs | Litter | Live herbs |
| *September 1983* | | | | | | | |
| Tamasane | Mean | 84.86 | 0 | 14.74 | 0 | 616 | 0 |
| | Max | 94.4 | 0 | 25.9 | 0 | 1645 | 0 |
| | Min | 73.1 | 0 | 5.6 | 0 | 231 | 0 |
| Shakwe | Mean | 81.58 | 0 | 18.01 | 0 | 792 | 0 |
| | Max | 97.0 | 0 | 42.5 | 0 | 2037 | 0 |
| | Min | 57.5 | 0 | 2.9 | 0 | 31 | 0 |
| Masama | Mean | 76.31 | 0 | 20.43 | 0 | 997 | 0 |
| | Max | 94.8 | 0 | 31.7 | 0 | 1867 | 0 |
| | Min | 64.1 | 0 | 4.3 | 0 | 279 | 0 |
| *November 1983* | | | | | | | |
| Tamasane | Mean | 67.35 | 6.96 | 6.86 | 18.01 | 283 | 94 |
| | Max | 78.3 | 17.2 | 16.3 | 32.8 | 776 | 191 |
| | Min | 53.2 | 2.0 | 1.8 | 8.5 | 76 | 33 |
| Shakwe | Mean | 70.55 | 2.69 | 23.35 | 2.78 | 414 | 17 |
| | Max | 90.2 | 8.3 | 56.4 | 8.8 | 1103 | 63 |
| | Min | 34.2 | 0 | 0.5 | 0.5 | 5 | 1 |
| Masama | Mean | 57.43 | 8.78 | 12.95 | 20.55 | 473 | 215 |
| | Max | 70.3 | 16.2 | 22.8 | 33.0 | 857 | 393 |
| | Min | 47.8 | 3.4 | 1.0 | 12.7 | 0 | 97 |
| *January 1984* | | | | | | | |
| Tamasane | Mean | 70.52 | 3.19 | 6.47 | 18.84 | 185 | 229 |
| | Max | 78.7 | 12.3 | 19.1 | 30.3 | 450 | 359 |
| | Min | 58.3 | 0 | 1.2 | 7.4 | 53 | 99 |
| Shakwe | Mean | 72.03 | 1.75 | 11.98 | 13.54 | 296 | 118 |
| | Max | 87.0 | 6.7 | 41.1 | 31.3 | 1209 | 255 |
| | Min | 43.4 | 0 | 1.3 | 7.6 | 69 | 54 |
| Masama | Mean | 63.24 | 5.06 | 11.47 | 20.08 | 374 | 563 |
| | Max | 75.0 | 15.5 | 21.3 | 26.1 | 1041 | 797 |
| | Min | 48.7 | 0 | 1.8 | 13.8 | 0 | 340 |
| *March 1984* | | | | | | | |
| Tamasane | Mean | 83.43 | 4.77 | 5.42 | 6.18 | 136 | 86 |
| | Max | 90.7 | 9.0 | 17.7 | 14.9 | 236 | 160 |
| | Min | 74.6 | 1.0 | 0.8 | 1.6 | 55 | 37 |
| Shakwe | Mean | 79.77 | 3.65 | 10.29 | 6.31 | 204 | 79 |
| | Max | 94.9 | 18.2 | 31.2 | 24.8 | 780 | 129 |
| | Min | 62.6 | 0 | 0.8 | 2.1 | 13 | 31 |
| Masama | Mean | 67.2 | 8.07 | 9.13 | 14.42 | 417 | 468 |
| | Max | 80.1 | 20.8 | 25.0 | 27.7 | 1117 | 1069 |
| | Min | 32.8 | 0.3 | 2.1 | 7.2 | 161 | 265 |

Table 5. Canopy cover of trees and shrubs at the three study sites measured from 1:1000 scale colour aerial photographs.

| | Tamasane | Shakwe | Masama |
|---|---|---|---|
| Mean canopy cover (%) | 28.5 | 29.3 | 31.75 |
| Standard error ($n=20$) | 2.24 | 2.27 | 2.43 |
| Maximum cover (%) | 43.0 | 50.0 | 45.0 |
| Minimum cover (%) | 9.0 | 15.0 | 0 |

These results are encouraging but not conclusive. They indicate that cover and even biomass of herbaceous vegetation can be resolved into at least three classes, and they support the conclusions reached in the general discussion of the AVHRR NDVI images above. The relationship of the variation of NDVI to green herbaceous cover, biomass and bare ground was examined in separate analyses, using generalized linear interactive modelling (GLIM) (NAG, 1978). Green herbaceous cover accounted for 82% of the variation ($p<0.001$), biomass for 54% ($p=0.01$), and bare ground for 46% ($p=0.05$). In the analyses of biomass and bare ground, the date of the measurements significantly reduced the residual variation in NDVI ($p=0.05$), which suggests that the predictive value of NDVI could be increased if separate calibrations were used for each date. However, it is important to bear in mind that the 20 plots represent only 5% of the area of one GAC pixel and that the area which was actually harvested and estimated was only 0.001% of a pixel. This low sampling rate is inevitable with AVHRR GAC composite data because of the large instantaneous fields of view. Furthermore, conclusions based on these results must be treated with caution, since the sample plots within each site were not intended to represent quantitatively the vegetation in each site, but rather the range of vegetation conditions present. For this reason, simple averaging of the cover and biomass is unlikely to give an accurate mean value for the GAC pixel as a whole. Nevertheless, the results do indicate that AVHRR composite NDVI data can be used to determine quantitatively the distribution of herbaceous cover and biomass which is a valuable capability for certain types of rangeland monitoring.

## INTERPOLATION OF GROUND MEASUREMENTS USING LANDSAT MSS DATA

A large number of ground measurements would be needed in order to sample adequately an AVHRR GAC pixel of 16 $km^2$ by traditional methods. Some vegetation is sufficiently uniform for a single, small sample to be adequate without any more sophisticated design being needed, but this is unusual and would need to be demonstrated rather than assumed. The normal method of sampling heterogeneous vegetation is to create uniform areas by subdivision and to sample separately within each one (Greig-Smith, 1983). Stratification of this sort depends either on prior knowledge of the pattern of variation, in this case the cover and biomass of herbaceous vegetation, or on an initial survey that describes the variation. In the Botswana study, the three sites were each selected to be internally uniform on the basis of soils and broad vegetation types and so no obvious additional stratification was available. But it is not possible to use aerial photographs quantitatively to assess the range and distribution of herbaceous vegetation in the same manner in which they are used in this study to assess the cover of woody vegetation. It is thus unlikely that the 20 plots were an adequate sample of the variation present. Short of a much larger sample, which would be impractical, the only means of reducing the bias is by remote sensing.

As a part of the Range Management Project, Landsat MSS data were acquired for each of the ground sampling dates (September, November, January, March, 1983–84) and also for May 1983. Nine MSS pixels were registered with each of the plots in the three sites and regressions were derived for the band 7/5 ratios (near-infrared/red) on cover and biomass of herbaceous vegetation and on the amount of bare ground. These regressions were then applied to all of the MSS pixels within the NOAA AVHRR GAC pixel centred on each study site, so that an overall mean value for cover and biomass of herbaceous vegetation and amount of bare ground could be quantitatively interpolated from the 20 sample plots to the whole AVHRR pixel.

Cloud obscured most of the study areas at Tamasane (scene 171–76) and Masama (scene 172–76) in January 1984 and Shakwe (scene 171–76) was affected by haze. Cloud cover and technical problems with Landsat delayed the acquisition of scene 172–76 (Masama) for March 1984 until the following month. Apart from some line-slip in scene 172–76, the MSS data were otherwise of good quality. The low-level 35 mm colour aerial photographs which showed the plot markers were used to transfer the plot corner locations to the appropriate 1:50 000 panchromatic aerial photographs (Figs 5, 6, and 7). Transparent overlays prepared from these photographs were attached to the monitor screen of a Dipix Aries II image analysis computer and the X and Y linear zoom facility was used to register the MSS sub-scene to a selection of landmarks transferred from the photographs. The procedure ensured that no errors in position of more than three pixels occurred in either X or Y directions. Since most of the plots were placed in larger, uniform patches of vegetation, even a position error of three pixels would not be particularly serious. Having located the centre pixel, the band 5 and band 7 values for the nine pixels centred on the plot were extracted. The ratios of the mean band 7 values and the mean band 5 values for the nine pixels were used in the regression analysis.

The variation in the three dependent variables, viz. bare ground, green herbaceous cover, and green herbaceous biomass, was examined using GLIM (NAG, 1978). The principal independent variable in each case was the band 7/5 ratio derived from the Landsat MSS data for the pixels registered with each sample plot. Site and date factors and their interactions with the vegetation index were also included.

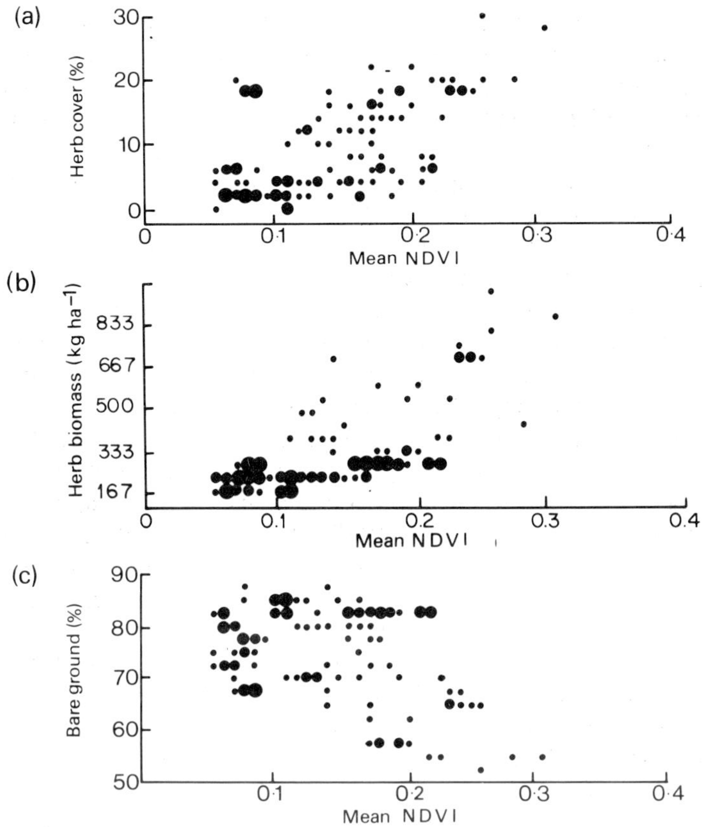


Fig. 9 Scatter diagrams showing the relationship between NDVI values and Landsat MSS interpolations of (a) % herb cover, (b) herb biomass, and (c) % bare ground. Larger dots indicate coincident points.

The results indicated that relatively little of the variation in the dependent variables was explained by the MSS band 7/5 ratio alone, viz. green herbaceous cover 4% ($p = 0.01$), green herbaceous biomass 35% ($p < 0.001$), and bare ground 3% ($p = 0.01$). When date and site factors were included as well as their interactions with the band ratio (when significant), the percentage of the variation accounted for increased to 74%, 84%, and 54%, respectively (all significant at $p < 0.001$% level). Clearly, just one regression would give a poor prediction of the dependent variables and so separate equations for each site and date were used. The signs of the regression coefficients emphasize this point. The negative relationship of bare ground to the band 7/5 ratio is confirmed by negative coefficients, except for Tamasane in September and November 1983. The coefficient values for Tamasane for green herbaceous cover throughout the period and for green herbaceous biomass in September and November 1983 were all negative. This unexpected result, discussed in more detail elsewhere, is clearly an object lesson on the dangers of uncritical extrapolation of calibrations derived in one vegetation type to others.

Landsat MSS subscenes corresponding to the NOAA-7 AVHRR GAC pixels were extracted using the nominal AVHRR pixel centres. Approximately 3600 MSS pixels were needed for each AVHRR GAC pixel. Since the study sites had been located in large areas of uniform vegetation, the MSS data were extracted for all nine AVHRR GAC pixels centred over the study sites. Some of these additional pixels were cloud-free in January 1984 and were therefore included even though the pixel over the study site itself was obscured. Data for May 1983 were extracted and processed using the regressions derived for March/April 1984. In this way the bare ground, green herbaceous cover, and biomass for a total of 123 AVHRR GAC pixels were obtained using Landsat MSS as an intermediary.

The results are shown in Fig. 9. The linear relationships of NDVI with the interpolated values for green herbaceous cover, biomass, and bare ground were analysed using GLIM (NAG, 1978). Green herbaceous cover accounted for 30% of the variation, herbaceous biomass for 47%, and bare ground for 15% (all significant at $p<0.001$). The residuals were correlated with site and date to varying extents and the maximal models using cover, biomass, and bare ground, with date and site, accounted for 88%, 83% and 91% of the variation in NDVI, respectively (all significant at $p<0.001$). Improvement of the ground data used to calibrate the AVHRR data by interpolation with Landsat MSS data thus confirms the ability of the AVHRR GAC composite NDVI to measure herbaceous cover, herbaceous biomass, and bare ground in rangelands.

## INTEGRATED AVHRR NDVI AND ANNUAL PRIMARY PRODUCTION

A procedure for regular monitoring of the herbaceous vegetation in rangelands would permit the annual above-ground net primary production of fodder to be estimated for specific vegetation types. Once again obtaining adequate field data for comparison with the vegetation indices derived from satellite-mounted radiometers is the main constraint. There are errors not only in interpolation of the ground measurements to the whole field of view of the satellite radiometer, but also in the estimates of net primary production due to consumption by herbivores during the harvest intervals. Exclosure plots are normally used to estimate the consumption of vegetation by the larger herbivores but this technique requires a very intensive field sampling programme.

In their study of the Ferlo area of Senegal, Tucker *et al.* (1983a) measured standing biomass towards the end of the growing season, in order to provide an estimate of total dry matter accumulation for comparison with the multi-temporal integration of AVHRR NDVI (Fig. 2). In order to use Landsat MSS interpolation to apply ground data to the complete AVHRR fields of view, it is necessary to obtain the end-of-season biomass from satellite measurements, but satellite radiometers can only detect green vegetation. With the Botswana data, the possibility of using a multitemporal integration of green herbaceous biomass to estimate annual production was investigated in the study sites by regressing green biomass duration against the sum of litter and green herbaceous biomass for March 1984. Insofar as the March figures represent the end-of-season biomass, this is a test of the validity of using the AVHRR NDVI estimates of green herbaceous biomass to calculate annual production. In spite of the errors due to herbivore consumption, 61% of the variation in biomass duration was accounted for by end-of-season biomass and the relationship was linear, passing through the origin. It seems reasonable to conclude, therefore, that Landsat MSS interpolations can be used to obtain the green herbaceous biomass for AVHRR pixels and that these values can be integrated to estimate total dry matter accumulation.

The composite NDVI data for the nine AVHRR GAC pixels in each site were integrated from 1 October 1983 to 31 March 1984 and plotted against interpolated green herbaceous biomass integrated for the same period (Fig. 10). There are very marked effects of site on the relationship. The Masama site contributes all of the points with higher biomass values, but they seem to exhibit the same general relationship as for Shakwe and for the Tamasane points which have the lowest integrated NDVI values. However, the Tamasane data taken as a whole seem to be distinct, having an overall higher integrated NDVI in relation to biomass than the other two sites. It is probable that the higher values at Tamasane are related to the presence of *Colophospermum mopane*, which is the dominant tree. *Colophospermum mopane* differs from the trees which dominate the other two sites in its canopy architecture and leaf morphology. Also its leaves generally appear earlier than those of the other species and fall later. Which, if any, of these features is responsible for the upward displacement of the Tamasane data is a matter for

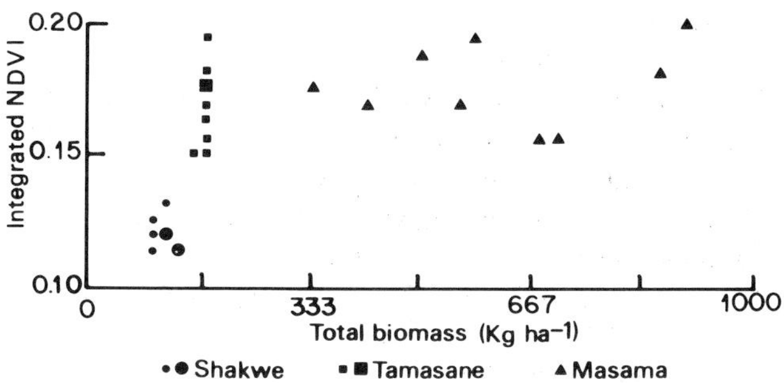


Fig. 10 The relationship between NDVI values integrated between October 1983 and March 1984 and estimated annual production of herbaceous biomass at each study site. Larger symbols indicate coincident points.

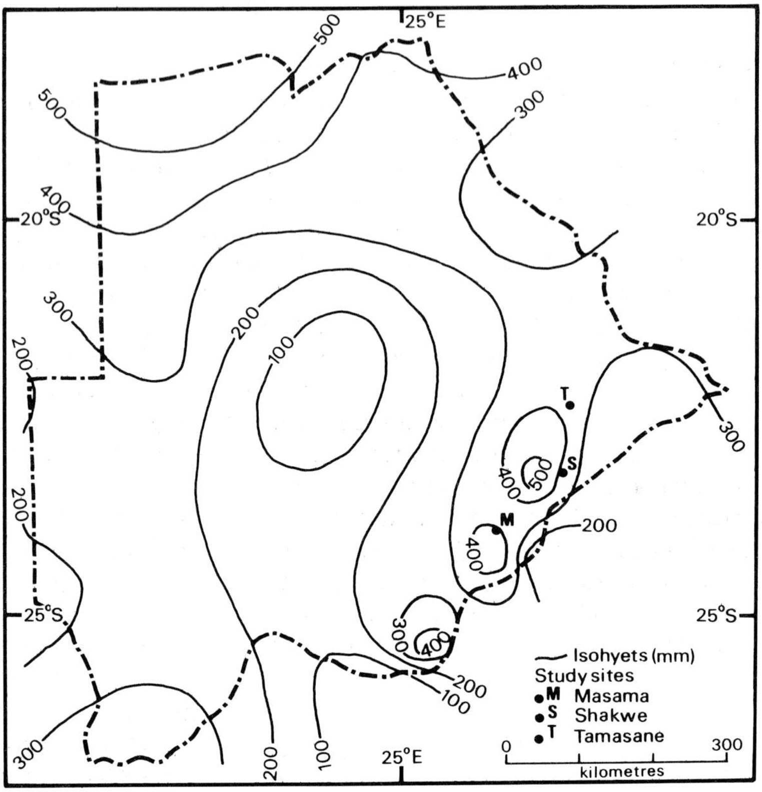


Fig. 11 Total precipitation (mm) for the period July 1983 to April 1984. Data from Botswana Meteorological Services.

conjecture, but it is at least clear that vegetation types should be treated separately.

The image created from an integration of 24 bimonthly NOAA-7 AVHRR GAC composites (Plate 25) must be interpreted with caution, bearing in mind the effect of vegetation type on the interpretation of integrated AVHRR NDVI values. It is not justified to assume that the higher values are simply indicative of rainfall (Fig. 11). For example, on the Ghanzi ridge (Fig. 1), bedrock near the surface provides access to ground water which is denied to the vegetation on the deeper Kalahari sand to the south (Cole, 1976). It is not certain, therefore, whether the higher integrated NDVI values on the Ghanzi ridge indicate higher rainfall, higher soil moisture or simply a different vegetation type. Nevertheless, the integrated image does allow comparisons within areas having similar vegetation and it is now essential to delimit these vegetation units if more reliable interpretation of NOAA AVHRR NDVI imagery is to be made.

## FUTURE DEVELOPMENTS

NOAA AVHRR data, processed using the compositing procedure pioneered by NOAA and NASA/GSFC, are an important new source of information for rangeland monitoring. This type of information has not been available previously and so it is necessary to develop means of utilizing the data as well as refine the technique itself. At a superficial level, the imagery produced by the compositing and integrating software is easier to interpret than any previous form of remote sensing product for rangeland monitoring, since little technical skill is needed to draw broad conclusions from the distribution of green vegetation indicated by high NDVI values. The chief innovation is the high temporal frequency and large spatial coverage. For the first time, national and regional planners can monitor vegetation at a high frequency and at the broad scale most appropriate to their needs. Already these products are proving themselves in routine monitoring of desert locust habitats by FAO (J. U. Hielkema, personal communication).

The present analysis of AVHRR NDVI data from Botswana indicates that more detailed information can be derived from the data than the false colour images alone can provide. However, it will not be possible to exploit the data fully unless progress is made in the design of field data collection programmes. Reflectance models for heterogeneous vegetation at the scale of AVHRR pixels are needed which deal with the problem of matching the large size of the pixels with what can actually be measured on the ground. Higher-resolution satellite imagery, airborne radiometers, and aerial photography will enable interpolation procedures, such as those used in Botswana, to be investigated. Vegetation measurements can also be improved, particularly by the development of stratifications at a scale and of a type appropriate to AVHRR GAC composite NDVI imagery. Attention to the phenology of greenness in the various physiognomic components of the vegetation may allow considerable improvement in precision. Investigation of the use of multitemporal NDVI data for estimation of annual production requires measurements in large enclosure plots, such as fenced ranches (Fig. 7) or vacant quarantine areas in Botswana. The distribution of dry matter production between above and below ground parts of the vegetation may need to be taken into account if multitemporal integration of greenness is to be used to estimate only above ground production. The AVHRR data would be improved if the GAC data were calculated by averaging all 15 instantaneous fields of view on the satellite. Improvements in the registration of the multiple images used to create the composite images are under consideration.

The results of the analysis of data collected in 1983–84 in Botswana show that the amount of bare ground, the cover, and the biomass of green herbaceous vegetation can all be measured using NOAA-7 AVHRR GAC composite NDVI data. The precision of the relationship between current satellite and ground data is likely to be acceptable at a regional or national scale and to be improved as a result of current developments at the satellite and field data levels. Annual production of herbaceous biomass can be estimated by integration of NDVI data, although some vegetation types appear to have a different relationship of production to integrated NDVI and so it may be that no single relationship will be acceptable for use in vegetationally heterogeneous areas. It is important to emphasize that, at the present time, these characteristics of vegetation, measured over such large areas, cannot be obtained either by any other remote sensing technique or by direct measurement. Thus, the question is not whether AVHRR data are better for monitoring the vegetation at this scale than other methods, but whether there are practical uses for the image products and the estimates of forage which can be achieved with the system. Even without expected technical improvements, the AVHRR system remains the only practical means of routine rangeland monitoring for areas greater than 10 000 $km^2$.

## ACKNOWLEDGEMENTS

Mr W. L. Astle, Project Leader, FAO/UNDP Range Management Project BOT/82/001 collected the field and aerial data in Botswana. The field programme was devised and initiated jointly by the author and Mr Astle. The NOAA-7 AVHRR data were processed by Drs C. J. Tucker and C. O. Justice, Earth Resources Branch, NASA/GSFC, Maryland. Landsat MSS data extraction was carried out by Dr A. Singh (Reading University) and Mr T. Elsey (University College London Remote Sensing Unit). The Food and Agriculture Organization of the United Nations is gratefully acknowledged for permission to publish data collected under the auspices of BOT/82/001, also the Government of the Republic of Botswana for permission to use the aerial photographs in Figs 5, 6, and 7 and Dr C. J. Tucker for permission to reproduce Fig. 2. Dr R. N. Carter assisted with the data analysis at Queen Mary College, University of London.

## REFERENCES

Allan, J. A. and Richards, T. S., 1983. Remote sensing for identifying low density vegetation cover in semi-arid coastal north-west Egypt. In *Remote Sensing for Rangeland Monitoring and Management*. Reading: Remote Sensing Society, pp. 69–79.

Allen, T. F. H. and Starr, T. B., 1982. *Hierarchy: Perspectives for Ecological Complexity*. Chicago: University of Chicago Press.

Brown, R. J., Bernier, M. and Fedosejevs, G., 1982. Geometric and radiometric considerations of NOAA AVHRR imagery. *Proc. 8th int. Symp. on Machine Processing of Remotely Sensed Data*. Purdue University, Indiana, 374–381.

Cicone, R. C. and Metzler, M. D., 1982. Comparisons of Landsat MSS, Nimbus 7 CZCS and NOAA 6/7 AVHRR sensors for land use analysis. *Proc. 8th int. Symp. on Machine Processing of Remotely Sensed Data*, Purdue University, Indiana, 291–297.

Cole, M. M., 1976. The vegetation of the Ghanzi area of western Botswana. *J. Biogeog.*, **3**, 169–196.

Gatlin, J. A., Sullivan, R. J. and Tucker, C. J., 1983. Monitoring global vegetation using NOAA-7 AVHRR data. *Proc. IGAARS Symp.*, San Francisco, I.PF2, 7.1–7.5.

Gray, T. I. and McCrary, D. G., 1981. *Meteorological Satellite Data—a Tool to Describe the Health of the World's Agriculture.* AgRISTARS Rept. EW-N1-04042, NASA/JSC, Houston, Texas.

Greegor, D. H. and Norwine, J., 1981. *A Gradient Model of Vegetation and Climate Utilizing NOAA Satellite Imagery—Phase I: Texas Transect.* AgRISTARS Rept. JSC-17435, FC-J1-04176, NASA/JSC, Houston, Texas.

Greig-Smith, P., 1983. *Quantitative Plant Ecology.* Oxford: Blackwell.

Holben, B. and Fraser, R. S., 1984. Red and near-infrared sensor response to off-nadir viewing. *Int. J. Remote Sensing*, **5**, 145–160.

Hutchinson, C. F., 1982. Remote sensing of arid and semi-arid rangeland. *Int. Geoscience and Remote Sensing Symp.*, Munich, 1–6.

Justice, C. O., Townshend, J. R. G., Holben, B. N. and Tucker, C. J., 1985. Analysis of the phenology of global vegetation using meteorological satellite data. *Int. J. Remote Sensing*, **6**, 1271–1318.

LACIE, 1979. *Proceedings of Technical Sessions. The LACIE Symposium.* JSC-16015, NASA, Houston.

Markham, B. L., Kimes, D. S., Tucker, C. J. and McMurtrey, J. E., 1981. Temporal spectral response of a corn canopy. *Photogramm. Engng Remote Sensing*, **48**, 1599–1605.

NAG, 1978. *The GLIM System.* Oxford: Numerical Algorithms Group.

Ormsby, J. P., 1982. Classification of simulated and actual NOAA-6 AVHRR data for hydrological land-surface feature definition. *IEEE Trans. Geoscience and Remote Sensing* GE-20, 262–268.

Perry, C. R. and Lautenschlager, L. F., 1984. Functional equivalence of spectral vegetation indices. *Remote Sensing Environ.*, **14**, 169–182.

Poulton, C. E. *et al.*, 1975. Range resources: inventory, evaluation and monitoring. In Reeves, R. G. (Ed.), *Manual of Remote Sensing.* Washington: American Society of Photogrammetry, pp. 1427–1478.

Schneider, S. R. and McGinnis, D. F., 1982. The NOAA/AVHRR: a new satellite sensor for monitoring crop growth. *Proc. 8th int. Symp. on Machine Processing of Remotely Sensed Data*, Purdue University, Indiana, 281–290.

Schneider, S. R., McGinnis, D. F. and Gatlin, J. A., 1981. *Use of NOAA/AVHRR Visible and Near Infrared Data for Land Remote Sensing.* NOAA Tech. Rept., NESS 84, Washington.

Townshend, J. R. G., 1981. Effects of spatial resolution on the classification of land cover type. In Fuller, R. M. (Ed.), *Ecological Mapping from Ground, Air and Space.* Cambridge: Institute of Terrestrial Ecology, pp. 101–112.

Townshend, J. R. G. and Tucker, C. J., 1981. Utility of AVHRR of NOAA-6 and 7 for vegetation mapping. In *Matching Remote Sensing Technologies and their Applications Proceedings.* Reading: Remote Sensing Society, pp. 97–109.

Townshend, J. R. G. and Tucker, C. J., 1984. Objective assessment of Advanced Very High Resolution Radiometer data for land cover mapping. *Int. J. Remote Sensing*, **5**, 497–504.

Tucker, C. J., 1979. Red and photographic infrared linear combinations for monitoring vegetation. *Remote Sensing Environ.*, **8**, 127–150.

Tucker, C. J., Holben, B. N., Elgin, J. H. and McMurtrey, J. E., 1981. Remote sensing of total dry matter accumulation in winter wheat. *Remote Sensing Environ.*, **11**, 171–189.

Tucker, C. J., Vanpraet, C., Boerwinkel, E. and Gaston, A., 1983a. Satellite remote sensing of total dry matter production in the Senegalese Sahel. *Remote Sensing Environ.*, **13**, 461–474.

Tucker, C. J., Hielkema, J. U. and Roffey, J., 1983b. *Remote Sensing of Ecological Conditions for Desert Locust Control.* NASA/GSFC Tech. Mem., TM84978, Goddard Space Flight Center, Greenbelt.

Tucker, C. J., Townshend, J. R. G. and Goff, T. E., 1983c. *Continental land cover classification using meteorological satellite data.* NASA/GSFC Tech. Mem., TM 85060, Goddard Space Flight Center, Greenbelt.

Tucker, C. J., Gatlin, J. A. and Schneider, S. R., 1984. Monitoring the vegetation in the Nile Delta with NOAA-6 and NOAA-7 AVHRR imagery. *Photogramm. Engng Remote Sensing*, **50**, 53–61.

Yates, H. W. and Tarpley, J. D., 1982. The role of meteorological satellites in agricultural remote sensing. *Proc. 8th int. Symp. on Machine Processing of Remotely Sensed Data*, Purdue University, Indiana, 23–32.

# PART V

# The Prospect for Remote Sensing

Remote Sensing and Tropical Land Management
Edited by M. J. Eden and J. T. Parry


# CHAPTER 17

# *Background, Perspective, and Issues for Remote Sensing in the Tropics*

JOHN T. PARRY
*Department of Geography, McGill University, Canada*

One of the results of the revolution in communications which began with the launch of the first Telstar communication satellite in 1962 was the realization that, in a rapidly changing world, regional, and even local, problems cannot be ignored. The global scenario of the 1980s is disquieting. The steadily increasing world population, the escalating costs of obtaining scarcer non-renewable resources, the increasing demands on renewable resources and the spectre of global pollution suggest that mankind will need every benefit that technology can provide in order to survive. Abject poverty is the common denominator of life in many developing countries and it is likely that the poorest people in the world will be the ones most directly affected by regional traumas such as soil exhaustion and crop failure, deforestation and flooding, urban congestion and pollution, and continental desertification.

Raw statistics project a precarious future. The current rate of population growth at approximately 1.8% per annum will result in a global community in excess of 6.35 billion by the year 2000 (CEQ and DOS, 1982). In the same period (1985–2000), the average annual growth of the per capita gross national product will be only 1.38% for the countries of Africa, 1.84% for Latin America, and 2.27% for Asia and Oceania (CEQ and DOS, 1982). As indicated in Fig. 1, the great majority of developing countries in the tropical zone have a per capita gross national product or gross domestic product of less than US$400 per annum (World Bank, 1981). The developing world is desperately poor today, yet the projections indicate that its people will become even poorer during the last 15 years of the century. The World Bank has estimated that the number of malnourished people in the less developed countries will increase from 400–600 million in the 1980s to 1.3 billion by the year 2000, and that the hunger belt will expand well beyond the tropics (Reutlinger and Selowsky, 1976).

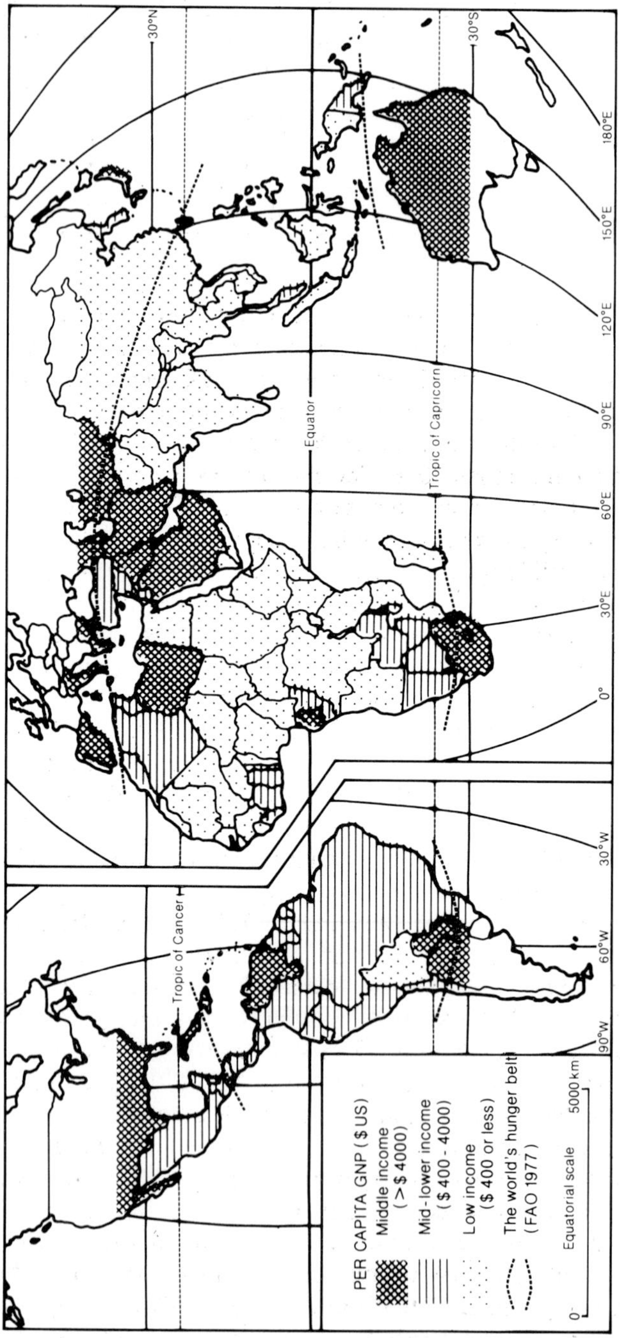


Fig. 1 The developing countries of the tropical and mid-latitude zones: per capita gross national or gross domestic product (data from World Bank, 1981).

Table 1. Estimate of world forest resources by the year 2000 (CEQ and DOS, 1982).

| | Closed forest (million ha) | | | Growing stock (billion m$^3$) | | |
|---|---|---|---|---|---|---|
| | 1978 | 2000 | % loss | 1978 | 2000 | % loss |
| Latin America | 550 | 329 | 40.2 | 94 | 54 | 42.5 |
| Africa | 188 | 150 | 20.2 | 39 | 31 | 20.5 |
| Asia and Pacific LDCs | 361 | 181 | 49.9 | 38 | 19 | 50.0 |
| Subtotal (LDCs) | 1099 | 660 | 39.9 | 171 | 104 | 39.2 |

The earth-orbiting satellites of the last decade have been silent witnesses to several regional and continental traumas, and their sensors have provided a graphic, sequential record of deteriorating conditions in many parts of the tropics. One of the most readily observed phenomena is the destruction of tropical forest. Since the beginning of the century, the area of wet tropical forest has decreased by more than 50%. The current forest stock of the less developed countries in the tropics, approximately 1.1 billion ha of mature closed forest, is being consumed at a rate of 20 million ha per year (Table 1). On this basis, most of the accessible forest will have been removed by the year 2000 (CEQ and DOS, 1982). In Java, an island with an acute population problem, an examination of aerial photographs shows that forest resources are being depleted at the rate of 13% per year (Soemerwato, 1974). Less than 10% of the island is now forest-covered. In Thailand, illegal logging has become so lucrative, especially in the teak forests of the Tha Sala District, that the Royal Thai army has been given wide-ranging powers of patrol, search and seizure (ESCAP 1984b). In northwest Thailand, the Landsat record shows that the major problem is not illegal logging but shifting cultivation and cutting for fuel, which result in the clearance of 250 000 ha per year, representing nearly 2% of the total area (Cheosakul *et al.*, 1973).

In Africa, agriculture is the main cause of deforestation (Persson, 1977). The moist tropical forests of the continent now cover less than half the area for which they are considered the natural climax. The greatest changes are occurring in West Africa where large populations exist. For example, in the Côte d'Ivoire, the forest area has been reduced by 30% in a ten-year period (Howard and Lanly, 1975). However, the most serious effects are being experienced in the dry subtropical woodlands of the Sahel zone. All the countries of this zone depend on firewood as a basic fuel: the average annual per capita consumption where wood is abundant is about 450 kg, decreasing to 225 kg where wood is scarce (Eckholm, 1976). Local scarcities of fuelwood are evidenced by the bare, treeless landscapes, clearly visible on Landsat, which extend to distances of 50–100 km around the larger settlements (Bardinet *et al.*, 1978). However, the problem is no longer local. The Club du Sahel (1978) has reported that total deforestation is inevitable in the Sahel zone wherever population densities exceed

25 persons $km^{-2}$. Even in Kenya, which is better endowed with forest resources than the countries of the Sahel, the indigenous forest now covers only 1.9% of the national territory and remote sensing shows that approximately 16% of that forest is being lost in each ten-year period (Ochanda and Epp, 1982). For Africa as a whole, only 6% of the land area is forested and, if clearance continues at the present rate, the forest cover will have been reduced to 5% by the year 2000.

Less dramatic than the loss of forest cover, but also of fundamental significance is the loss of soil fertility and the decreasing ability of the land surface to resist erosion. In many areas where shifting cultivation is practised, the fallow interval has been significantly diminished. For example, in Guyana where population pressure is far less severe than in Africa, the bush fallow period has decreased from 10–20 years to 2–9 years in the period 1968–1978 (Potter, 1978). In Africa, where there is a critical need for increased output, food production is actually declining: Myers (1980) reported a 5% decrease during the 1960s and 1970s, and there seems little prospect of reversing this trend. The cause of declining productivity, according to Voss (1975), is long continued environmental mismanagement. In many parts of Africa, soils tend to be highly weathered, deficient in humus and liable to lose fertility under cultivation. In the absence of mineral fertilizers, which are now prohibitively expensive for Third World farmers, successive crops show declining yields, afford less nutrition, and provide less vegetative cover for the soil, with a consequent increase in soil loss from the field. Stocking (1983) has noted an important relationship between crop cover and soil erosion in Zambia: well-fertilized crops planted at the beginning of the rains provided not only higher yields but also more than twice the effective protection against soil erosion, compared with poorer crops planted a month after the onset of the rains. The erosive potential of tropical rain is immense, and the trigger mechanisms for soil erosion are provided by decreased surface cover resulting from poor crops, the inappropriate timing of agricultural practices such as planting, tillage and burning, or the removal of protective bush and forest cover (Ooi, 1983). From the Upper Volta, Voss (1975) has provided a classic example of these linkages. Over a period of 20 years he found that millet yields had declined from 729 to 352 kg $ha^{-1}$, rates of runoff had tripled, and soil losses increased from 143 to 1318 t $ha^{-1}$ $year^{-1}$. From Vietnam, soil losses of 2000 t $ha^{-1}$ $year^{-1}$ were reported from state farms specializing in industrial crops in 1975. For the 30-year period 1945–75, the forest area in Vietnam has decreased from 48% to 29%, resulting in catastrophic gully erosion in mountain areas (ESCAP, 1984a). The incidence of flooding has approximately doubled, and massive sediment loading of major rivers has resulted in the death of fish and the disruption of aquatic food chains.

The global trauma that has become most apparent from the vantage point of the satellite is desertification (Walker and Robinove, 1981). Along the sensitive interface of the desert and the subtropical grasslands, the possibility of collapse of the ecosystem chains has been a cause of concern to environmental scientists and government advisors for more than two decades (Andrawis and Khidir, 1980). The great variability of rainfall in the Sahel zone is well-documented for both prehistoric and historic periods (Adefolalu, 1983). Rainfall records beginning in the late 19th century show periods with significantly less than 'normal' rainfall, followed by a slow return to longer periods of wet conditions. The cyclicity of wet and dry periods in this region is clearly apparent. For example, in Fig. 2, it can be seen from the Banjul area of the Gambia that drought maxima occur approximately every 30 years. Although the present drought has been more severe than its predecessors, extrapolation suggests a return to pluvial conditions in the early 1990s as long as the ecosystem chains remain intact. Continuance of the pattern will result in an even more severe drought in the first decade of the 21st century.

The meteorological causes of the periodic aridity are not well understood. However, the dangers of injudicious arable practices, firewood gathering and overgrazing have been stressed for more than a quarter century. The real tragedy of the Sahel, beyond the acute shortages of water, the severe decline in agricultural production, and the massive mortality of human beings and livestock, has been the unpreparedness of governments and international agencies and the failure to obtain information that could be used to mitigate the worst effects of the disruption. Currently, about 35% of the world's land surface (45 million km$^2$) is at risk (Fig. 3). Of this area, 75% is already affected (Reining, 1978) and the amount of land reduced to zero net economic productivity is increasing at a rate of 21 million ha annually. The population of the dryland zone was estimated at about 850 million in 1983, with 120 million living in areas experiencing severe or very severe desertification (ESCAP, 1984c). Overexploitation of these fragile marginal lands is the root cause of desertification. Underlying this is bad management, rural neglect and a mesh of political and economic forces, many of which originate outside the areas of impact. It is the failure to perceive these linkages that explains the failure to arrest desertification (Tolba, 1984).

The preceding paragraphs are by way of a prologue. They provide some indication of the magnitude of problems facing tropical countries. What is the role of remote sensing in this scenario? How can optimum benefits be obtained from space technology? In order to attempt an answer to these questions, it is necessary to examine some basic features of remote sensing technology in the context of development and some of the fundamental problems of the development process itself.

## REMOTE SENSING TECHNOLOGY

The role of technology tends to be overestimated in the developed countries because it is such a routine ingredient of daily life and there is a large supporting infrastructure, including skilled scientists and technicians, that makes its use

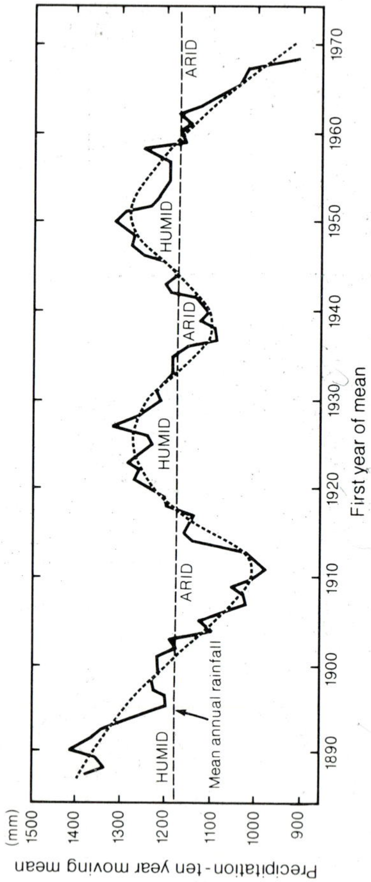


Fig. 2 Ten-year moving means of rainfall 1886–1970, Banjul, the Gambia, West Africa (adapted from Jawara, 1982).

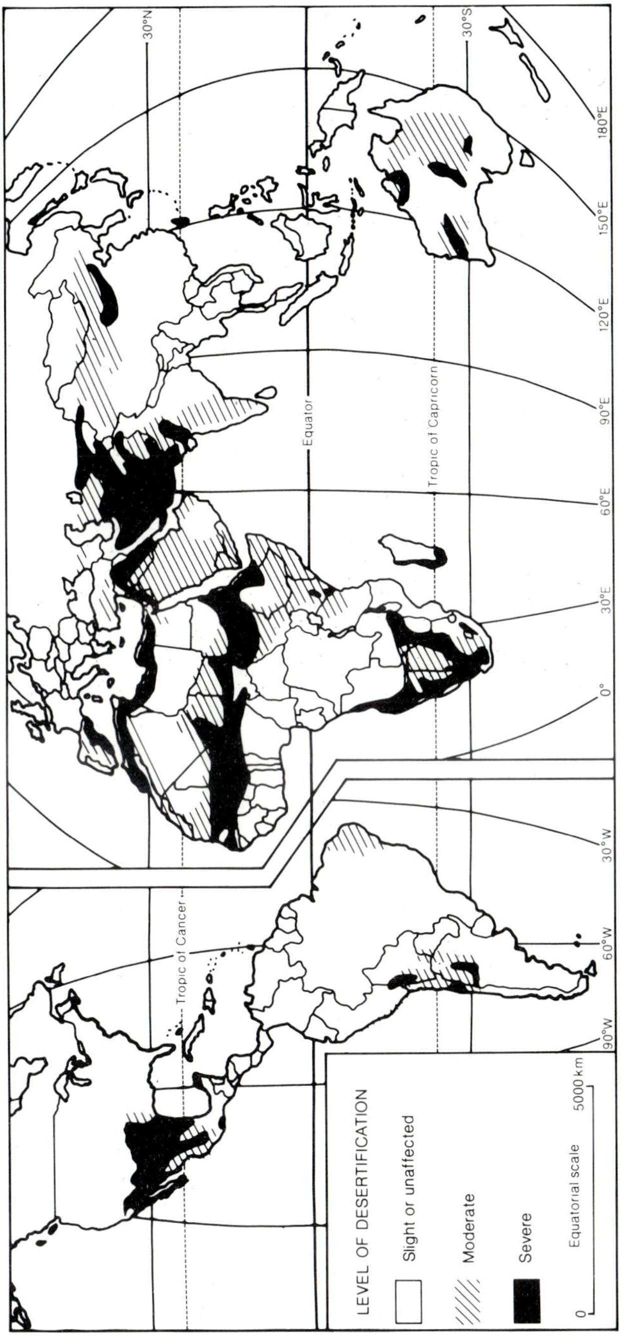


Fig. 3 Desertification in the developing countries of the tropical and mid-latitude zones (after UNO, 1977).

feasible. This abundance of capability and resources allows a substantial amount of high-risk activity. The question that has to be asked is whether it is possible for a technology designed in the advanced world to respond adequately to the very basic needs of the developing world (Malingreau, 1981). In the latter areas, the most urgent problems are those related to primary human needs: food, water, shelter, and health. Basic problems are more urgent and less tractable, and their solution nearly always depends on local conditions and the degree of acceptance by local cultures. The scarcity of material resources and trained manpower does not allow the luxury of misguided efforts or high risk activities. The formula for integrating advanced technology in the development process has been spelled out very effectively by the United States Agency for International Development (Arnold, 1978). Such integration is achieved in the following ways.

1. Systematically identifying, defining, and analysing primary environmental problems and their secondary effects such as desertification and the despoliation of tropical forests.
2. Devising plans for the prudent management of the environment and natural resources, with provision for the monitoring of unintended impacts such as soil erosion, salination, and the pollution of rivers and coastal waters.
3. Giving priority to urgent problems.
4. Perceiving and developing generic solutions that are broadly applicable. In this context the importance of land use planning cannot be overestimated.
5. Providing solutions that enhance the quality of life and are socially acceptable. This means that large-scale and long-term issues must be addressed as well as local and short-term problems.

When set in this context, there is clearly a major role for remote sensing from aircraft and spacecraft. Remote sensing offers a powerful battery of procedures for identifying spatial patterns and monitoring changes in those aspects of the biophysical environment that have a physical expression in terms of electromagnetic energy (Colwell, 1973). Patterns of electromagnetic radiation form only a part of the environment system with which a developing country is concerned, and remote sensing can provide only a part of the information needed for effective planning. Nevertheless, remote sensing techniques have specific attributes that lend themselves directly to applications in the developing world. With the Tiros and Nimbus programmes in the 1960s and the Landsat and NOAA/ITOS programmes in the 1970s, remote sensing became global and the potential for increasing knowledge about the environment of the developing world was vastly increased. The unique qualities of aircraft and spacecraft remote sensing and the innovative consequences of space technology for developing countries were addressed at length in a symposium held in Tehran in 1971 (Afshar, 1971), and these can be summarized as follows:

1. synoptic, large area coverage that is essentially global;
2. accessibility to areas which are difficult to gain access to on the ground;
3. rapidity of information collection;
4. flexibility of information aggregation into different areal units;
5. compatibility of data sets from different dates, permitting verification and serialization suitable for records and data banks.

The main contribution of remote sensing technology in meeting the needs of the developing world is probably best summarized as integrative data gathering, and, as noted by Steiner (1972), its particular function can be categorized in one of three modes:

1. the addition of completely new categories of information not previously collected or even measurable;
2. the extension of existing data gathering systems by improving their accuracy or refining their results;
3. the replacement of part or all of existing data gathering systems, where there is specific advantage in so doing.

## DEVELOPMENT OBJECTIVES AND REMOTE SENSING

Remote sensing is of vital importance for resource management and socio-economic growth in tropical countries because it supplies information, a fundamental resource in itself. However, the significance of the information and the manner in which it is applied depend very much on an individual country's view of development. It is common for the term development to be used to describe a particular process, i.e. some form of qualitative change through time, or the results of that process, i.e. a condition of being developed or underdeveloped according to identifiable characteristics. Development, however, is not a particular set of events or achievements. The simplistic models of development, based on incremental steps from the poverty of the low resource consumption stage (underdeveloped) to the wealth and sophistication of high levels of resource consumption (developed), have been challenged (Brookfield, 1978; Mabogunje, 1980; Unwin, 1983). It is increasingly realized that development involves social and institutional changes as well as economic ones, and the goal of even greater resource consumption is viewed with suspicion. The particular attitudes to development, whether they be traditional, broadly Marxist, or what may be called rejectionist (Maitra, 1980), have had a direct effect on the manner in which remote sensing technology is viewed.

In those countries where the development strategy stresses economic growth, emphasis has been given to specific, short-term applications in the search for minerals or fossil fuels. This has been the focus of remote sensing in the Middle East and the Andean countries of South America (Sillitoe, 1974;

Bentz and Gutman, 1975; Rowan, 1975). Countries with Marxist development models have tended to be more cautious in relation to remote sensing, and the areas of application have been social in nature with emphasis on integrated rural development and the understanding of spatial patterns of land use and settlement. Interesting examples of the use of remote sensing for assessing the basic needs of poor regions can by found in recent studies from Afghanistan (Dayal and Khairzada, 1976), Ethiopia (Allan and Alemayehu, 1975), China (Lister, 1984), and Tanzania (Paul and Mascarenhas, 1981).

The debate concerning imported technology is linked in many respects to the particular view of development to which one subscribes. From the 'rejectionist' standpoint the technology is inappropriate because it has been created in wealthy countries to serve their requirements and its transfer merely increases the power of big business in exploitive development, without contributing to the real needs of the poor country. The rejectionist would argue that implanted technology can have even more deleterious effects for the poor country by creating enclaves of sophistication and networks which have stronger links with the affluent 'outside' world than with the host country. Within the enclaves there is considerable wealth and power, but this benefits only an elite and is a source of dissatisfaction for the many who see technology transfer as a mechanism by which transnational capital and rich countries continue to subjugate the poor countries in the post-colonial period.

Even when one takes a less cynical view of development, it is clear that for a developing country the assessment of the suitability, benefits, and costs of a particular technology is often extremely difficult (Hossain, 1978). Remote sensing from aircraft and spacecraft is one of the most advanced of modern technologies. It requires considerable initial investment and is not labour-intensive; yet in a developing country there is labour in abundance and capital is at a premium. The main product of remote sensing is information about resources, environmental conditions, and economic conditions; yet what can be done with this knowledge if the infrastructure for the understanding and application of remotely sensed data is lacking? For a developing country to harness the powers of this latter-day Pegasus it has to face the problems of mastering a highly complex technology, assimilating the information it provides and utilizing it effectively within the frame of the country's needs and priorities. At the recent ESA-sponsored conference on satellite remote sensing for developing countries (Longdon and Lévy, 1982), it was emphasized repeatedly that the success of remote sensing in the tropics would depend on the effectiveness of technology transfer and on the extent to which the technology itself is adapted or modified to meet the environmental and cultural conditions of the zone.

## APPROPRIATE TECHNOLOGY—BUT WHAT IS APPROPRIATE?

In the following pages some of the specific concerns of developing countries with regard to remote sensing technology and its applications will be addressed, with emphasis on the question of appropriate technology. Appropriate technology is an approach and not a package. It involves the development of tools adapted to actual problems and existing capabilities, and is obviously relevant in many human activities ranging from crop production to small-scale industry. Remote sensing technology is basically a decisional technology, which includes the know-how for diagnosing complex problems and formulating choice strategies at a variety of levels (Goulet, 1977). Its major components are data collection, processing and analysis techniques, alternative selection and planning processes. These components are at the upper level of technology and it might be argued that it is thus irrelevant to discuss appropriate technology. However, there is no good reason why the touchstone should not be applied. Indeed, failure to do so will encourage a blind acceptance of high technology and reinforce a planning paradox that is already apparent in some developing countries: the situation in which increasingly sophisticated methods of analysis are used to devise plans which involve relatively simple technology. Although each developing country has its individual concerns, the issues can be grouped in six interrelated sets.

**Systems limitations of the present generation of satellite sensors**

The basic requirements of all tropical countries are the same: food, water, health, and housing. Many countries still lack the capacity to collect primary data for planning long-term development to meet these basic needs. Remote sensors, particularly satellite systems, can provide such an information base. However, the pertinent question facing the government of a developing country considering an investment in remote sensing technology is whether the systems are suitable for their particular needs. Studies in the United States (Committee on Remote Sensing for Development, 1977) and Europe (Fraysse, 1980) show mixed results for the utility of remote sensing data from spacecraft, and the concerns of the tropical developing countries become very clear if one considers specific system parameters of the Landsat satellites (Pacheco, 1980; Hood, 1982; Voûte, 1982).

Landsat has an 18-day revisit cycle, an equatorial crossing time on the descending node of approximately 0930 hours and a spatial resolution of only 79 m (NASA, 1976). The system parameters were selected to suit land use studies and crop monitoring in the mid-latitudes—a zone with lengthy anticyclonic periods of clear weather and a homogeneous agricultural system characterized by large, single-crop fields. Such systems are not suitable for many tropical countries where fields are small, crops vary from field to field, and planting, growth, ripening and harvesting do not necessarily conform to a specific crop calendar. There is an even more fundamental problem for countries in the equatorial and monsoon regions because of the persistent or seasonal cloud cover. As indicated in Fig. 4, this affects an area in excess of 40.5 million km$^2$ (Winston, 1969).

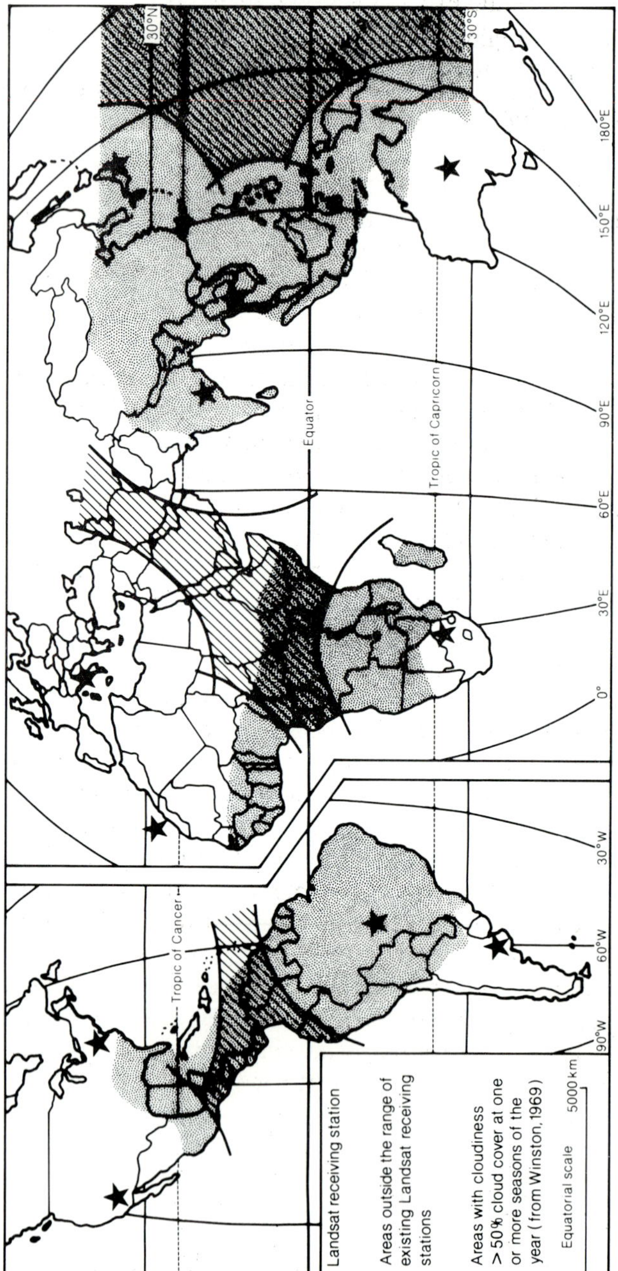


Fig. 4 The problems of cloudiness and satellite data reception for the developing countries of the tropical and mid-latitude zones.

The typical instrumentation of the first generation of resource satellites (apart from the ill-fated Seasat) consists of sensors operating in the visible and near-infrared portion of the spectrum. These are poorly suited to the task of obtaining good quality imagery in regions where the atmosphere is hazy, humid or persistently cloudy (CNRS, 1976). There are two solutions to the problems posed by the tropical atmosphere. One is to change the satellite orbit so that the equatorial crossing occurs in the early morning before the cumulus build-up, the other is to replace the current battery of cameras and multispectral scanners with all-weather sensors, such as radars. Needless to say, the funds required to place such a satellite in orbit are beyond the means of the developing countries most in need of the information it could supply.

The same problem of funding is highlighted by the fact that there are still significant gaps in the network of ground receiving stations in the tropics, which means that considerable areas are outside the limits of direct satellite coverage (Fig. 3). These areas have to rely entirely on on-board recording, which remains one of the weakest elements in the present satellite systems (Rao and Kasturirangan, 1985). The problem is illustrated in East Africa where Landsat coverage has been irregular and of variable quality. Until recently the region lay beyond the range of Landsat receiving stations, and was thus entirely dependent on the on-board tape recording system. Both Landsats 1 and 2 experienced tape recorder problems. The complete absence of images for most of 1974, for example, was the result of the near-total failure of the recording system on Landsat 1. The limited life of the wide-band video tape used in the recorders caused other problems, including deterioration in the quality of the images and also inadequate coverage because of NASA's reluctance to expend the remaining hours of recording time on other than major disasters. These factors explain the limited number of images available for many parts of the tropics and the irregularity of the sequence which rarely provides an 18-day series.

There is no easy solution to the problem of selecting appropriate sensors and sensor parameters for the developing world. However, it is ironic that at a time when there is intense rivalry among technologically advanced nations to produce more sophisticated sensors capable of finer resolution — 30 m for the thematic mapper on Landsat 4, 10 m for the panchromatic mode in SPOT IMAGE — most of the countries of the developing world have no means of using precision digital data, and in any case really need more general resource information than pixel by pixel detail.

### Information autonomy — whose bits and whose bytes?

A second issue that causes continuing concern for developing countries is what may be termed information autonomy. Understandably, most developing countries would like to exercise complete control over the information that is acquired about their resources by earth-orbiting satellites. The problem is at the centre of the issue of development.

A developing nation ought to have effective control of its own resources, although information on those resources is most beneficial to those with the capital and technical ability for exploitation. Historically, these have been the large transnational corporations and enterprises, and information in this context has meant the power to develop the resources and to exploit the economic vulnerability of the developing country. With the new technology of remote sensing, there is the fear that the developing countries will be further disadvantaged vis-à-vis the multinational corporations, whose experts are more proficient in extracting usable information from the satellite data.

The parameter that has caused the greatest controversy in the United Nations and elsewhere is spatial resolution. This was already a problem with the 79-metre resolution of Landsats 1 and 2, and has become more so with the improved precision of the thematic mapper (TM) on Landsat 4. The spatial resolution, or number of pixels per image, is increased from 10.6 million (57 m pixels) for MSS data to 41.6 million (28.5 m pixels) for TM data. In addition, the spectral sensitivity has improved from a 6-bit MSS pixel with 64 grey levels in four bands to an 8-bit TM pixel with 256 grey levels in seven bands (NOAA, 1983). As resolution became finer, the number of developing countries objecting to the United States' policy of open dissemination of earth resources data obtained from non-military space programmes has increased. The French commercial satellite company SPOT IMAGE, which plans to launch a series of earth resources satellites starting in October 1985, has adopted a prior consent policy, by which it will not sell 10 m resolution data or imagery of any country without that country's agreement (SPOT IMAGE, 1984).

As developing countries become more familiar with satellite remote sensing, the concerns relating to security and economic vulnerability may diminish. In the 1980s, the United States will cease to be the sole operator of a civil earth resources satellite, and so any fears of an American monopoly of information from this source can be dismissed. The question remains whether access to information obtained by earth-orbiting spacecraft will be limited by the ability of a country to pay the price charged by purely commercial enterprises, such as SPOT IMAGE and its Japanese, Indian, and other competitors. From open dissemination of earth resources data, which was the goal of the signatories to the Outer Space Treaty of 1967, there is a shift towards increased competition and demand-controlled costing for satellite data (Parry, 1981). It remains to be seen whether this will lead to increased cooperation between those countries with and those without satellites.

## An international framework— false hope or real prospect?

For global environmental monitoring to be generally acceptable in the developing world some type of international framework would seem to be required. The arguments for internationalizing this global technology have been effectively presented by the Committee on Remote Sensing for Development (1977), convened by the United States Board on Science and Technology for International Development. Remote sensing from space should serve transnational objectives and the desirability of a cooperative participatory approach, as opposed to one characterized by dependence or rivalry, is obvious. Activities in the common domain of humanity, such as the problems of identifying, using and managing the world's resources and maintaining the earth's environment, call for collaborative action on an international scale. The complex challenges that were outlined in the introduction and the interdependent responses required to deal with them demand that orbital remote sensing be managed as a cooperative enterprise under some type of international control.

The United Nations Committee on the Peaceful Uses of Outer Space (COPUOS) and its legal subcommittee have debated the problem of satellite sensing since their foundation in 1959 (Christol, 1981). The increased international concern about the future availability of natural resources has focused political attention on the implications of remote sensing programmes. It is now generally accepted that remote sensing is not restricted by existing international law. However, the question as to how data and information should be disseminated remains controversial. During its 1976 session, the COPUOS legal subcommittee addressed the fundamental differences between raw data, processed data, and information derived from the analysis of data. A general understanding seems to be emerging that major restrictions on data dissemination could significantly reduce the benefits available to non-sensing countries, and that prior consent restriction is not necessarily aiding a country concerned about the foreign exploitation of its resources (Stowe, 1977).

A more constructive response to this problem lies in expanding data dissemination and information extraction capabilities throughout the world rather than restricting them. The ability to analyse remotely sensed data is integral and the technologically advanced countries can contribute by committing themselves to broad-based development assistance in remote sensing. There are numerous ways in which this could be accomplished, regional cooperation appearing to be the most attractive in many parts of the world.

As an illustration of the form that regional cooperation should take, one can cite the Regional Remote Sensing Training and User Assistance Facilities at Ougadougou, Burkina Faso and Nairobi, Kenya, established in 1977 and funded through the United States Agency for International Development (Conitz and Lowe, 1977). The centres are staffed by four or five specialists, representing different disciplines, who conduct demonstration projects and are available for consultation. Each centre contains analytical equipment, imagery files, image reproduction facilities, a technical library and equipment for field survey. In addition to providing workshops and training, the centres are expected to become focal points for contact between government officials and planners from the region and scientists from the developed countries. Staff are encouraged to become familiar with the problems of the region in order to provide the technical assistance that is relevant. The East African centre has organized training programmes and special seminars on rangeland, forest conservation, water resources, soil mapping, and topographic map revision (Conitz *et al.*, undated).

Clearly, no set of procedures can guarantee that every country has the same level of access to remote sensing data, but much can be done to ensure that the potential benefits of the technology are widely shared. This should go some way in reducing the concerns of developing countries that the rapid acceleration of remote sensing activities might disadvantage them. As developing countries improve their capabilities in interpreting and making use of remote sensing data, the case for a policy of open data dissemination becomes progressively stronger because the countries know that they are able to obtain exactly the same data about their territory as any third party, using remote sensing methods.

## Continuity—the uncertainty about operational satellites

Another concern which arises directly out of the above is that of continuity. One of the factors that has caused some developing nations to hesitate before committing themselves to major investments in satellite remote sensing—US$5 million for a satellite receiving station or US$1 million for a sophisticated digital image processor—is the fact that there is no assurance of continuity. Landsat was conceived as an experimental satellite, and it is not certain that NOAA will follow the current Landsats 4 and 5 with truly operational satellites, as originally planned by NASA. Satellite technology is still in an experimental phase and, in the absence of any firm commitment to a fully-operational, long-term system, it is hardly surprising that many governments are adopting a wait-and-see attitude. Even a large agency like the United States Department of Agriculture is on record as being unwilling to replace its conventional data collection with a Landsat-based system, because there is no assurance of a continuous data flow over a minimum period of ten years or so (Hart, 1977). The SPOT IMAGE satellite programme appears to offer some continuity to the Landsat programme and some countries, such as Canada, have already modified their receiving stations to accommodate SPOT data.

Many countries would be reassured of the continuity of satellite remote sensing, if some form of international governance was arranged, especially as the uncertainty about the future includes such basic considerations as the cost of data products, quality control, access fees for receiving stations, and the compatibility of software analysis systems. Even if some regular programme of earth resource satellites is agreed upon, there remains the problem of compatibility with present systems. Some governments and agencies fear that future systems will differ so greatly from the current ones that efforts to develop or integrate Landsat information into their data systems would be wasted. Agencies and government departments prefer standard information products that are available on a continuing basis so that existing analytical models can be maintained. Perhaps this uneasiness is unwarranted in view of the flexibility of digital methods and the relative ease with which software translation can be made.

**The user gap—bridging or leaping?**

Possibly the most difficult and complex issue for remote sensing technology is the user gap. This has several dimensions, existing at the technical and application levels. In the former case, there is always a gap between a company or agency which has pioneered a particular technique or product and another agency which has decided to adopt it. The situation is well-illustrated by a comparison between the United States, Canada, European countries, and the advanced countries in Asia and the Pacific. Canada acted with foresight by becoming involved in the Landsat programme from the start and today has an international reputation in the field. For Canada, the learning process was facilitated by the close technological ties between Canada and American industry. European countries were slow to become involved in space technology and the learning gap compared to North America is two to three years, while Asian and Pacific countries follow some three years or more behind Europe (Remote Sensing Working Group, 1978).

The user gap is even more profound at the application level. Even in the United States, there is still a sizeable gap between the technical experimenter and the resource manager. In less developed countries the gap between satellite data in its raw state and information usable by planners and government agencies is much greater. The degree to which new types of data can be incorporated into existing analytical and planning programmes is limited by the underdeveloped state of these procedures. The obstacles to realizing the benefits of remote sensing for economic and social development are interlinked and, although it is useful to isolate particular obstacles for comment, it is the linkages that are often the most significant elements in perpetuating the user gap.

The lack of trained personnel is the primary factor limiting the ability of developing countries to assimilate remote sensing technology. The learning gap exists at all levels: technicians and resource specialists who need training to enable them to analyse remotely sensed data; policy-makers and planners who need to be aware of the potentials of the technology; and scientists and university teachers who require training in the more technically demanding aspects. Some idea of the magnitude of this problem can be gained from the estimate made by the Committee on Remote Sensing for Development (1977) of the needs of a medium-level developing country for trained manpower over a ten-year period: for visual analysis of remote sensing data 180–240 non-degree and 20–50 degree personnel and for quantitative analysis 90–150 non-degree and 10–40 degree personnel. The scale of the problem is realized when these figures are translated into the aggregate needs of more than 100 developing countries.

Directly linked to this is the problem of assimilating the information provided by remote sensing and utilizing it productively. Traditionally, in many former colonial territories, resource information was gathered by different means and often maintained in separate agencies. For these and other reasons, national planners have rarely incorporated current resource or environmental information in their planning scenarios. When used at all, resource information has been limited to particular development projects for which data requirements were more clearly defined and often the regional effects of these projects were ignored. Increasingly, resource data are required over the large area, for environmental monitoring or for integrated development projects involving river basins or regions, but all too often the infrastructure for a comprehensive analysis of the resource base and its incorporation into national planning is completely absent (Odingo, 1983).

The reasons for this are easy to identify, but more difficult to correct. There is a lack of coordination between the physical planners, whose concern is the location and suitability of resources, and the economic planners, who are concerned with fiscal resources and revenue generation. The two groups lack a common vocabulary and frame of reference, and the influential economic planners are often unaware of the quantity and quality of spatial and thematic resource information available through remote sensing (Howard, 1977). The user gap is here most powerfully illustrated. Because central planners make no demand for this data and sectorial ministries and technical services are only concerned with subsets of the data, there is little incentive or effort to use comprehensive, integrated resource information. Remote sensing can provide such information, but the planning system does not make effective use of it. This is ironic because in the developing world the synoptic view is often more important than the regional or local one.

## Transfer: the art of grafting

The final issue to be discussed is that of appropriateness. For many reasons no single organizational or operational formula for a national remote sensing effort can be said to have gained widespread acceptance. This is not surprising because the requirements for resource information gathering and environmental monitoring vary greatly from country to country and also because what is technically desirable may not be politically acceptable or economically justifiable. It is being slowly realized that the establishment of appropriate infrastructures or the adjustment of existing ones is at least as important as training and the actual introduction of hardware or software (World Bank, 1979). Equally, it is becoming apparent that the wholesale transfer of technology, as a total package, is not necessarily an appropriate strategy. 'Depackaging' a technology has the advantage of strengthening the technological and industrial capabilities of developing countries. It can provide an opportunity for local production of system components, for adaptation and development of system modules that suit local requirements, and for the integration of indigenous techniques and products.

An appropriate technology cannot be achieved without a proper assessment by the user. Such an assessment should consider the efficiency and cost-effectiveness of operations, the type and volume of work to be performed, the required adjustments of the infrastructure, the use of the information and the economic and social benefits to be derived. In effect, the type of science and technology policy that is to be adopted must mesh with a country's socio-economic and socio-cultural development policy. In this context it should be remembered that, although remote sensing systems are high technology, the extraction and use of information from them can be carried out at many levels. Significant information can be obtained with simple, visual inspection, and monochrome images can be effectively used as map substitutes (Larin-Alabi, 1978). With intermediate instrumentation, such as a colour additive viewer, a densitometer or density slicer, and a zoom transferscope, much more information can be derived. At a higher technical level, integrated remote sensing and information management procedures allow automatic extraction of data based on spectral signatures and change with time (Schultink, 1983). At the most sophisticated level such data can be entered directly into geographical information systems and used for development planning over large areas (Tuyahov *et al.*, 1982). An instructive example of this is provided by the United States AID programme in Central America and the Caribbean (Shelton and Tilmann, 1978). It is clear from the experience that the level of remote sensing technology can be scaled to a country's abilities and needs.

The full benefit of remote sensing is realized when the data acquired by aircraft or spacecraft are interpreted and integrated with other information to provide resource or geographical data bases that can be used by resource managers and planners (Tomlinson, 1972). The end-to-end concept of an experiment or project is directly relevant here. At the outset specific goals are identified. This is followed by an assessment of the multiple data requirements from aircraft and spacecraft and also from fieldwork. This part of the project offers a good opportunity for multilateralization and cost sharing. A regional approach—as distinct from the present practice in which a particular country services its neighbours—would encourage tropical countries to cooperate in establishing and operating receiving stations. This could be a key step in the internationalization of remote sensing, because the 3000 km radius coverage of a receiving station would inevitably include many countries with common problems and concerns. At the next stage, suitable processing and analysis techniques are selected and again the regional centre can provide an overview in monitoring key environmental variables, developing multisectoral outlooks, and accumulating the range and continuity of information necessary for an effective geographical data base usable by the national planner and policy maker (de Haas and Rijnberg, 1973). An interactive linkage between data use and data generation is essential to ensure that the product continues to meet the user's need and that the user's understanding of the technology keeps pace with the advances in sensor systems and data processing.

## CONCLUSION

If remote sensing is to have a major role in addressing some of the critical issues facing mankind in the final decade of the century, it must be able to prove that it helps solve basic problems, that is, a technology with a human face (Schumacher, 1974). It is not easy to reconcile an advanced technology like remote sensing with the more mundane, but urgent, problems of a growing world population and an environment under stress of forest clearance, desertification, and soil degradation. However, this reconciliation must be achieved if remote sensing is to become a credible and appropriate technology for resource management. In attempting this, four features of the development-technology interface should be kept in mind.

1. Despite their magnitude the problems of development have critical interrelationships. Environment and food are directly related, poverty and health, population growth and economic welfare, and so on. By identifying and clarifying these linkages remote sensing can contribute not only to the understanding of a specific problem under review but also to the linked issues.
2. From the spectrum of available technology, it is important to select those elements that are appropriate for the local conditions and infrastructure. Selectivity is the key to appropriateness in terms of relative economic costs, acceptance by local cultures and the incorporation of local skills and resources.
3. Development problems are not merely technological. Social, political, legal, and economic elements are involved, and solution of the problems requires a multidisciplinary, integrative approach to which remote sensing is well suited.
4. The development process requires an indigenous base of science and technology. The solutions to the problems must be consistent with the aims of the society and not superimposed. An indigenous base is essential for effective technical assistance and technology transfer. It must include the skills to manage the technology and direct it towards practical solutions in both the long and short term.

## REFERENCES


Adefolalu, D. O., 1983. Desertification of the Sahel. In Ooi, J. B. (Ed.), *Natural Resources in Tropical Countries*. Singapore: Singapore University Press, pp. 402–438.

Afshar, H. K. (Ed.), 1971. *The Innovative Consequences of Space Technology and the Problems of the Developing Countries*. Tehran: Tehran University Press.

Allan, J. A. and Alemayehu, T., 1975. Rural population estimates from air photographs—an example from Ethiopia. *ITC Jl*, **5**-1, 85–100.

Andrawis, A. S. and Khidir, M. O. (Eds), 1980. *Proceedings of the Sudan Symposium and Workshop on Remote Sensing*, vols. 1 and 2. Brookings: South Dakota State University.

Arnold, H. A., 1978. The role of science and technology in development. *Proc. 12th int. Symp. Remote Sensing Environ.*, **1**, 175–178.

Bardinet, C., Monget, J. M. and Poisson, M., 1978. Land use mapping in the Sahel zone using Landsat. In Collins, W. G. and van Genderen, J. L. (Eds), *Remote Sensing Applications in Developing Countries*. Birmingham: Remote Sensing Society, pp. 85–90.

Bentz, F. P. and Gutman, S. I., 1975. Landsat data contributions to hydro carbon exploration in foreign regions. *U.S. Geol. Surv. Prof. Paper*, **105**, 83–92.

Brookfield, H., 1978. Third world development. *Prog. hum. Geog.*, **2**, 121–132.

CEQ and DOS (Council on Environmental Quality and the Department of State), 1982. *The Global 2000 Report to the President*. New York: Penguin Books.

Cheosakul, P., Indrambarya, B. and Morgan J. A., 1973. ERTS-1 data utilisation in the field of agriculture in Thailand. *3rd Earth Resources Technol. Satellite Symp.*, Washington, 319–330.

Christol, C. Q., 1981. Inventory of space activities: legal aspects. *Proc. 1st Symp. on Space Activities and their Legal Implications*, Centre for Research of Air and Space Law, McGill University, Montréal, 69–95.

Club de Sahel, 1978. Energy in the Development Strategy of the Sahel; Situation, Perspectives, Recommendations. Paris: Club du Sahel.

CNRS (Centre National de la Recherche Scientifique) 1976. *Télédétection et Environnement Tropical*, Centre d'Etudes de Géographie Tropicale, Travaux et Documents de Géographie Tropicale, No. 25.

Colwell, R. N., 1973. Remote sensing as an aid to the management of earth resources. *Am. Scient.*, **61**, 175–183.

Committee on Remote Sensing for Development, 1977. *Resource Sensing from Space: Prospects for Developing Countries*. Washington: National Research Council-National Academy of Sciences.

Conitz, M. W. and Lowe, D. S., 1977. Remote sensing utilization of developing countries: an appropriate technology. *Proc. 11th int. Symp. Remote Sensing Environ.*, **2**, 1055–1064.

Conitz, M. W., Falconer, A. and Haack, B. N., undated. *Landsat Data for Development*. Nairobi: Regional Remote Sensing Facility.

Dayal, H. H. and Khairzada, B. A., 1976. The first national demographic survey of Afghanistan: the role played by air photos and photo-counting techniques. *ITC Jl*, **1976**-1, 84–97.

Eckholm, E. P., 1976. *Losing Ground*. New York: Norton and Co.

ESCAP, 1984a. Land degradation in Vietnam. *ESCAP Environ. News*, **2**(2), 18–19.

ESCAP, 1984b. Thailand: a losing battle to preserve the nation's green resources. *ESCAP Environ. News*, **2**(4), 19.

ESCAP, 1984c. Factsheet, status of desertification. *ESCAP Environ. News*, **2**(1), 15–16.

Fraysse, G., 1980. European applications, requirements, perspectives. In Fraysse, G. (Ed.), *Remote Sensing Applications in Agriculture and Hydrology*. Rotterdam: A. A. Balkema, pp. 497–502.

Goulet, D., 1977. *The Uncertain Promise: Value Conflicts in Technology Transfer*. Washington: IDOC.

de Haas, W. G. L. and Rijnberg, T. F., (Eds), 1973. Integrated surveys and environmental management in developing countries. *ITC Jl* special issue 1972–3.

Hart, G. F., 1977. A panel discussion on the economics of remote sensing. *Proc. 1st Conf. on the Economics of Remote Sensing Information Systems*, San Jose State University, 356–357.

Hood, V. A., 1982. Satellite remote sensing—ESA's experience with developing countries. In Longdon, N. and Lévy, G. (Eds), *Satellite Remote Sensing for Developing Countries*. Proc. EARSeL-ESA Symp., Igls, Austria, European Space Agency, SP-175, pp. 201–204.

Hossain, A., 1978. Transfer of remote sensing technology for the benefit of developing countries—a case study in Bangladesh. *Proc. 12th int. Symp. Remote Sensing Environ.*, **1**, 213–218.

Howard, J. A., 1977. Satellite remote sensing of agricultural resources for developing countries, present and future: an international perspective. In Peel, R. F., Curtis, L. F. and Barrett E. C. (Eds), *Remote Sensing of the Terrestrial Environment*. London: Butterworths.

Howard, J. A. and Lanly, J. P., 1975. Remote sensing for tropical forest surveys. *Unasylva*, **27**, 32–37.

Jawara, L. K., 1982. ITC and the Third World. *ITC Jl*, **1982**-1, 20–23.

Larin-Alabi, F. B., 1978. Problems of remote sensing in the tropics: an appraisal of the Nigerian situation with regard to forest resources. In Collins, W. G. and van Genderen, J. L., (Eds), *Remote Sensing Applications in Developing Countries*. Birmingham: Remote Sensing Society, pp. 57–61.

Lister, R. G., 1984. The China model, incentive socialism. *N. Z. Jl Geog.*, **76**, 13–16.

Longdon, N. and Lévy, G. (Eds), 1982. Satellite remote sensing for developing countries. *Proc. EARSeL-ESA Symp.*, Igls, Austria, European Space Agency SP-175.

Mabogunje, A. L., 1980. *The Development Process—a Spatial Perspective*. London: Hutchison.

Maitra, R., 1980. *The Mainspring of Economic Development*. London: Croom Helm.

Malingreau, J. P., 1981. Remote sensing and technology transfer in a developing society—or is there such a thing as appropriate advanced technology? *ITC Jl*, **1981**-2, 125–139.

Myers, N, 1980. Kenya's baby boom. *New Scientist*, **87**, 848–850.

NASA, 1976. *Landsat Data Users Handbook*. Greenbelt: Goddard Space Flight Center, National Aeronautics and Space Administration.

NOAA, 1983. *Landsat Data Users Notes*, No. 28, National Oceanic and Atmospheric Administration, US Department of Commerce, Washington.

Ochanda, N. and Epp, H., 1982. Monitoring recent changes in extent of natural forests in Kenya using remote sensing techniques. *Int. Archives of ISPRS*, 24–VII/1, 489–496.

Odingo, R. S., 1983. Natural resources and economic development in developing countries. In Ooi, J. B. (Ed.), *Natural Resources in Tropical Countries*. Singapore: Singapore University Press, pp. 526–541.

Ooi, J. B., 1983. Natural resources in tropical countries: an examination. In Ooi, J. B. (Ed.), *Natural Resources in Tropical Countries*. Singapore: Singapore University Press, pp. 1–69.

Pacheco, R. A., 1980. Applications of remote sensing to agricultural development in tropical countries. In Fraysse, G. (Ed.), *Remote Sensing Application in Agriculture and Hydrology*. Rotterdam: A. A. Balkema, pp. 299–308.

Parry, J. T., 1981. Spaced out—spacecraft remote sensing—costs and benefits. *Proc. 2nd Symp. on Space Activities and their Legal Implications*, Centre for Research of Air and Space Law, McGill University, Montréal, 68–100.

Paul, C. K. and Mascarenhas, A. C., 1981. Remote sensing in development. *Science*, **214**, 139–145.

Persson, R., 1977. *Forest Resources of Africa*. Stockholm: Royal College of Forestry.

Potter, L., 1978. The Amerindians of Guyana and their environment. Paper presented to *10th Conf. Caribbean Historians*, St. Thomas, Virgin Islands, March–April 1978.

Rao, U. R. and Kasturirangan, K., 1985. *Remote sensing for the Commonwealth, Science for Technology for Development: an Expanded Programme of Scientific Cooperation in the Commonwealth*, Commonw. Sci. Coun. Tech. Pub. Ser. No. 140, 311–330.

Reining, P. (Ed.), 1978. *Handbook on Desertification Indicators*. Washington: American Association for the Advancement of Science.

Remote Sensing Working Group, 1978. *Applications of Earth Resources Satellite Data to Development Aid Programmes*, SP-1010 European Space Agency, Paris.

Reutlinger, S. and Selowsky, M., 1976. *Malnutrition and Poverty*, World Bank Occas. Paper 23, Washington.

Rowan, L. C., 1975. Applications of satellites to geologic exploration. *Am. Scient.*, **63**, 393–403.

Schultink, G., 1983. Integrated remote sensing and information management procedures for agricultural production potential assessment and resource policy design in developing countries. *Can. J. Remote Sensing*, **9**, 4–18.

Schumacher, E. F., 1974. *Small is Beautiful*. London: Sphere Books.

Shelton, R. L. and Tilmann, S. E., 1978. Remote sensing, geographical information systems and national land planning: some Central American and Caribbean experiences. *Proc. 12th int. Symp. Remote Sensing Environ.*, **1**, 567–585.

Sillitoe, R. H., 1974. Tectonic segmentation of the Andes: implications for magnetism and metallogeny. *Nature, London*, **250**, 542–545.

Soemerwato, O., 1974. The soil erosion problem in Java. *Proc. 1st Ecology Congr.*, The Hague, 236–242.

SPOT IMAGE, 1984. *Nouvelles de SPOT*, No. 5, Centre National d'Etudes Spatiales, Paris.

Steiner, D., 1972. Remote sensing input to information systems. In Tomlinson, R. F. (Ed.), *Geographical Data Handling*. Ottawa: UNESCO-IGU Commission on Geographical Information Systems, vol. 1, pp. 4–631.

Stocking, M., 1983. Farming and environmental degradation in Zambia: the human dimension. *Appl. Geog.*, **1**, 157–166.

Stowe, R. W., 1977. The United Nations and remote sensing, recent political and legal developments. *Proc. 1st Conf. on the Economics of Remote Sensing Information Systems*, San Jose State University, 12–21.

Tolba, M. K., 1984. Harvest of dust. *ESCAP Environ. News*, **2**, 17–20.

Tomlinson, R. F. (Ed.), 1972. *Geographical Data Handling*, vols. 1 and 2. Ottawa: UNESCO-IGU Commission on Geographical Information Systems.

Tuyahov, A. J., Weinstein, R., Weber, J. *et al.*, 1982. Technology transfer of space applications by the National Aeronautics and Space Administration. In Johannsen, C. J. and Sanders, J. L., (Eds), *Remote Sensing for Resource Management*. Ankeny: Soil Conservation Society of America, pp. 579–588.

United Nations, 1977. *U.N. Conference on Desertification*. New York: Earthscan.

Unwin, T., 1983. Perspectives on 'development'—an introduction. *Geoforum*, **14**-3, 235–241.

Voss, A. de, 1975. *Africa: the Devastated Continent*. The Hague: W. Junk.

Voûte, C., 1982. Satellite remote sensing for developing countries—prospects and constraints. In Longdon, N. and Lévy, G., (Eds), *Satellite Remote Sensing for Developing Countries*. Proc. EARSeL-ESA Symp., Igls, Austria, European Space Agency SP-175, 193–200.

Walker, A. S. and Robinove, C. J., 1981. *Annotated Bibliography of Remote Sensing Methods for Monitoring Desertification*. Geol. Surv. Circular 851, US Department of the Interior, Washington.

Winston, J. S., 1969. Global distribution of cloudiness and radiation as measured from weather satellites. In Rex, D. F. (Ed.), *World Survey of Climatology*, vol. 4. Amsterdam: Elsevier, pp. 247–280.

World Bank, 1979. *International Technology Transfer: Issues and Policy Options*, World Bank Working Paper No. 344, Washington.

World Bank, 1981. *World Bank Atlas: Population, per Capita Product and Growth Rates*. New York: World Bank.

# Index